A Specialist Periodical Report

Electron Spin Resonance

Volume 1

A Review of the Literature Published between January 1971 and May 1972

Senior Reporter

R. O. C. Norman, *Department of Chemistry, University of York*

Reporters

R. F. Adams, University of Leeds

N. M. Atherton, University of Sheffield

P. W. Atkins, University of Oxford

J. E. Bennett, Shell Research Ltd., Chester

T. A. Claxton, University of Leicester

J. F. Gibson, Imperial College, London

B. C. Gilbert, University of York

A. Hudson, University of Sussex

K. D. J. Root, Varian Associates Ltd., Walton-on-Thames, Surrey

C. Thomson, University of St. Andrews

The Chemical Society
Burlington House, London, WIV 0BN

ISBN: 0 85186 751 0

Library of Congress Catalog No. 72-95099

Organic formulae composed by Wright's Symbolset Method

PRINTED IN GREAT BRITAIN BY JOHN WRIGHT AND SONS LTD., AT THE STONEBRIDGE PRESS, BRISTOL

Foreword

This is the first volume on Electron Spin Resonance in the series of *Specialist Periodical Reports*. As such, its aim is in keeping with volumes in the series on other topics: to provide an account of recent progress in the field, with as comprehensive a coverage as possible, oriented toward the specialist research worker.

A two-fold approach has been adopted as a deliberate policy. Some chapters are primarily concerned with the phenomenon of e.s.r. and its underlying theory; examples are those on theoretical aspects of hyperfine splittings and *g*-factors, and on relaxation processes and linewidths. Other chapters, in contrast, are directed more towards applications of the technique in a variety of fields, with respect particularly to results of structural, kinetic, and mechanistic interest; examples are those on radical ions, ion pairs, and dynamic processes, on free radicals in the solid state, and on transition-metal ions.

This volume covers the literature from January 1971 to May 1972. It is intended that Volume 2 will cover the period June 1972 to November 1973, with publication in mid-1974, and that future volumes will be published at 18-month intervals. Volume 2 will differ in presenting material corresponding to Chapters 6, 8, and 9 in reorganized form, and by the inclusion of a section on applications in Biochemistry. Further such changes, either to help the reader or to increase the coverage, will doubtless be made in later volumes, and indeed suggestions would be welcomed.

R. O. C. Norman

January 1973

Contents

1
Theoretical Aspects of Hyperfine Splittings and g-Factors

BY C. THOMSON

1 Introduction

The fundamental theory of e.s.r. spectroscopy has been well reviewed in recent books,[1-3] and will not be dealt with here. This report is concerned, furthermore, with organic free radicals and does not deal with the theory of transition-metal ion e.s.r.,[4, 5] nor with the theory of gas-phase e.s.r.[6] We are concerned primarily with recent advances in the calculation of hyperfine coupling constants (hfcc) of free radicals, both semi-empirically and non-empirically. A limited number of references to experimental papers containing theory which is used in the interpretation of experimental results will also be made. Hyperfine coupling constants are quoted in gauss (G).

The analysis of the e.s.r. spectra of free radicals in solution gives the magnitude of the hyperfine coupling constants in the radical. We shall use the notation a_i^N where this hfcc refers to the coupling with nucleus N, and the subscript i refers to the position of this nucleus in the radical. Since the hyperfine coupling constants are related to the value of the wave-function at the nucleus in question, the study of the molecular electronic structure of free radicals and the interpretation of the experimental e.s.r. parameters are closely related. The absolute signs of the a_i^N cannot be obtained directly from the e.s.r. spectrum, although they may be obtained by observation of linewidth variations in the spectra.[7] The signs are important, however, in theoretical work, as will be seen. Determination of the total hyperfine tensor from single crystal studies also yields the signs of the components, but few radicals are stable enough to be studied in this way.[2, 5]

[1] A. Carrington and A. D. McLachlan, 'Introduction to Magnetic Resonance', Harper and Row, New York, 1967.
[2] P. B. Ayscough, 'Electron Spin Resonance in Chemistry', Methuen, London, 1967.
[3] F. Gerson, 'High Resolution E.S.R. Spectroscopy', John Wiley and Sons, London, 1970.
[4] J. S. Griffith, 'The Theory of Transition Metal Ions', Cambridge University Press, 1961.
[5] 'Spectroscopic Properties of Inorganic and Organometallic Compounds', ed. N. N. Greenwood (Specialist Periodical Reports), The Chemical Society, London, 1970, vol. 3, p. 160; 1971, vol. 4, p. 150.
[6] A. Carrington, D. H. Levy, and T. A. Miller, *Adv. Chem. Phys.*, 1970, **18**, 149.
[7] A. Hudson and G. R. Luckhurst, *Chem. Rev.*, 1969, **69**, 191.

A second parameter which is obtained from an e.s.r. spectrum is the g-factor in the equation for the resonance condition

$$h\nu = g\beta H \tag{1}$$

In equation (1), h is Planck's constant, ν is the resonance frequency, β is the Bohr magneton, and H is the external magnetic field. The paramagnetism of organic free radicals is almost exclusively due to the spin of the unpaired electron, so that the g-value differs only slightly from the value for the free electron $g_e = 2.00232$. However, in careful experimental work,[8] g-values can be measured to $\pm$ 0.000005 and the theoretical interpretation of the small variations in g-values is of some interest. Recent work in this area is reviewed in the next section.

2 The Theory of g-Factors

A general theory of the g-tensors of molecules has been given by Stone,[9] who derived an expression for the principal g-values of a molecule in an orbitally non-degenerate state. The theory could then be applied to hydrocarbon radicals and to semiquinone radical ions,[10, 11] when some simplifications in the equations had been made, with the equation for the g-factor shift being

$$\Delta g = g - g_e = b + \lambda c \tag{2}$$

where λ is the coefficient of the resonance integral β in the energy of the Hückel molecular orbital containing the unpaired electron, and b and c are semi-empirical constants, with values[8] $b = 24.7 \times 10^{-5}$, $c = -19.3 \times 10^{-5}$. In the case of aromatic even-alternant hydrocarbons, the difference between the in-plane g tensor components was equal in magnitude but opposite in sign for the anion and cation radicals. Careful studies by Fassaert and de Boer,[12] in which linewidth variations were used to estimate the g-shifts, gave good qualitative agreement with Stone's theory. However, there are significant quantitative discrepancies, *i.e.* the predicted decrease in $(g_1 - g_2)$ going from anthracene$^{\pm}$ to tetracene$^{\pm}$ is not reflected in the observed values. Also for $C_{10}H_8^{\overline{\cdot}}$, $(g_1 - g_2)$ is predicted to be -9×10^{-4}, whereas it is found to be essentially zero. The deviations may be due to approximations in the Stone theory, or to the assumption of isotropic rotational motion in the linewidth analysis.

Moss and Perry[13] have extended the Stone theory to degenerate radicals, where there is an extra term involving the orbital angular momentum, and in estimating this term, vibronic coupling must also be taken into account.

[8] B. G. Segal, M. Kaplan, and G. K. Fraenkel, *J. Chem. Phys.*, 1965, **43**, 4191.
[9] A. J. Stone, *Proc. Roy. Soc.*, 1963, **A271**, 424.
[10] A. J. Stone, *Mol. Phys.*, 1963, **6**, 509.
[11] A. J. Stone, *Mol. Phys.*, 1964, **7**, 311.
[12] D. J. M. Fassaert and E. de Boer, *Mol. Phys.*, 1971, **21**, 485.
[13] R. E. Moss and A. J. Perry, *Mol. Phys.*, 1971, **22**, 789.

Thermal averaging over the possible vibronic states was carried out, and the computed *g*-value for $C_6H_6^{\overline{\cdot}}$ was in qualitative agreement with the measured value.[3] However, there were a number of approximations made which may affect the numerical results, such as the neglect of possible solvent effects and the use of data from the neutral molecule, but the general conclusions should apply to other degenerate anions of unicyclic polyenes and account for the major difference between the *g*-factors of degenerate and non-degenerate radicals.

The theories developed above have been largely applied, with several approximations, to large aromatic radical ions. Applications to some small radicals were reported some years ago by Glarum,[14] who carried out calculations on $\dot{C}H_3$, $\dot{N}H_2$, and CH_2. However, some multicentre integrals were approximated in this work, and their influence is not negligible. In order to provide a test of the different approximations used in calculations on larger molecules, the theory of the *g*-tensor in one-electron systems has been studied for H_2^+.[15] It was concluded that even for very simple molecules, some of the approximations made by Stone are not quantitatively very good, particularly the neglect of the two-centre terms. This work therefore shows that more accurate calculations on the larger systems with fewer approximations are needed.

The INDO (intermediate neglect of differential overlap) molecular orbital method which is proving very useful in the calculation of approximate hyperfine coupling constants (see Section 4) has recently been applied to the calculation of *g*-tensor components. The applications, within the approximations of Stone's theory, were to the radical $H_2\dot{N}O$,[16] and also to a variety of σ-radicals.[17] This work follows earlier calculations with the CNDO (complete neglect of differential overlap) method.[18] In the INDO study, a variety of excited configurations based on the INDO orbitals was taken into account. Good agreement with experiment was found; for example, in the case of $\dot{N}O_2$, where gas-phase measurements are available $g_{xx} = 1.9908$, $g_{yy} = 2.0024$, and $g_{zz} = 2.0067$ compared with the experimental values of $g_{xx} = 1.9910$, $g_{yy} = 2.0020$, and $g_{zz} = 2.0062$. The agreement with experiment using the INDO method was substantially better in the case of $H_2\dot{N}O$ than with the CNDO/2 calculations.[18] It was pointed out, however, that the results are quite sensitive to the assumed geometry, which should be that which minimizes the energy. The anomalously low *g*-factors for diphenylacetylene and diphenyldiacetylene anions can be explained on the basis of a strong interaction between the phenyl groups and the acetylene *p*-orbitals.[19]

[14] S. H. Glarum, *J. Chem. Phys.*, 1963, **39**, 3141.
[15] Ph. de Montgolfier and J. E. Harriman, *J. Chem. Phys.*, 1971, **55**, 5262.
[16] T. Morikawa, O. Kikuchi, and K. Someno, *Bull. Chem. Soc. Japan*, 1971, **44**, 2256.
[17] T. Morikawa, O. Kikuchi, and K. Someno, *Theor. Chim. Acta*, 1971, **22**, 224.
[18] O. Kikuchi, *Bull. Chem. Soc. Japan*, 1969, **42**, 1187.
[19] I. B. Goldberg and A. J. Bard, *Chem. Phys. Letters*, 1970, **7**, 139.

3 *Ab Initio* Calculations of Hyperfine Splitting Constants

There are two main contributions to the hyperfine interaction in free radicals. The first of these is the anisotropic hyperfine interaction with the Hamiltonian $\mathscr{H}_d$

$$\mathscr{H}_d = g_e \beta_e \sum_{k, N} g_N \beta_N [3(\hat{S}_k r_{kN})(\hat{I}_N . r_{kN}) r_{kN}^{-5} - r_{kN}^{-3}(\hat{S}_k . \hat{I}_N)] \quad (3)$$

where β_e and β_N are the Bohr and nuclear magnetons, g_e and g_N are the electron and nuclear *g*-factors, r_{kN} is the vector which connects electron k and nucleus N, having spin operators $\hat{S}_k$ and $\hat{I}_N$, respectively. This interaction vanishes for radicals in solution in which there is rapid re-orientation of the radicals, and although the elements of the anisotropic coupling tensor may be obtained from single-crystal studies, there have until recently been rather few theoretical studies of this contribution (see later p. 16).

The second contribution is the hyperfine splitting caused by the Fermi contact term with Hamiltonian $\mathscr{H}_c$

$$\mathscr{H}_c = (8\pi/3) g_e \beta_e \sum_{k, N} g_N \beta_N \delta(r_{kN}) \hat{S}_k . \hat{I}_N \quad (4)$$

where the symbols are as defined above and $\delta(r_{kN})$ is the Dirac delta function. The Fermi contact splitting for nucleus N can be calculated as the expectation value of $\mathscr{H}_c$ which leads to the hyperfine coupling constant for nucleus N:

$$a^N = (8\pi/3) g_e g_N \beta_e \beta_N \gamma(r_N) \quad (5)$$

where

$$\gamma(r_N) = \langle \Psi | \sum_k (S_z)^{-1} \hat{S}_{kz} \delta(r_{kN}) | \Psi \rangle \quad (6)$$

In equation (6) Ψ denotes the exact ground-state wavefunction, the other symbols are defined above, and the summation is over all the electrons. $\gamma(r_N)$ is the spin density at the nucleus, or the net density of electron spin at nucleus N, *i.e.* the number of electrons (Bohr)$^{-3}$ with spin α minus the number of electrons (Bohr)$^{-3}$ with spin β. For a strict comparison with experimental data the computed splittings should be averaged over the thermally populated vibronic states.[20] Most computations, however, refer to the evaluation of a^N using $\Psi = \Psi_{el}$ where Ψ_{el} is the electronic part of the wavefunction in the Born–Oppenheimer approximation.

Semi-empirical calculations of a^N for various magnetic nuclei have been widely used in the past, and recent work is discussed in Section 4. However, it is clearly desirable to calculate hyperfine coupling constants non-empirically in order to assess the quality of molecular wavefunctions for open-shell species, and to test the validity of the various approximate calculations of a^N. Until relatively recently such calculations were not practicable without many approximations, but with the availability of more

[20] S. Y. Chang, E. R. Davidson, and G. Vincow, *J. Chem. Phys.*, 1970, **52**, 5596.

powerful computers there have been many recent studies of hyperfine splittings in small molecules. Before discussing these calculations it is appropriate to consider the calculations for atoms, since it is on atoms that the various theoretical methods are usually tested. In view of the large literature on the subject,[21] we have confined our attention to those results which bear directly on the molecular calculations.

For atoms in other than ^2S states, the restricted Hartree–Fock method (RHF), with a wavefunction Ψ_0 [(equation (7)] which has $k - 1$ doubly occupied orbitals and where the odd electron occupies orbital ϕ_k, gives

$$\Psi_0 = |\,\phi_0\,\bar{\phi}_0 \ldots \phi_{k-1}\,\bar{\phi}_{k-1}\,\phi_k\,| \tag{7}$$

zero spin densities at the nucleus, and it is necessary to include excited configurations $\Phi^{ij\ldots}_{ab\ldots}$ of the same symmetry, in which electrons have been excited from spin orbitals $a, b, \ldots$ to spin orbitals $i, j, \ldots$. Then the complete configuration interaction (CI) wavefunction can be written as

$$\Psi = \Psi_0 + \sum C^i_a\,\Phi^i_a + \sum C^{ij}_{ab}\,\Phi^{ij}_{ab} + \ldots \tag{8}$$

The summations run over all possible values of $abc \ldots$, $ijk \ldots$, and configurations may be defined as singly excited (Φ^i_a), doubly excited (Φ^{ij}_{ab}) *etc.* with respect to Ψ_0. It is easy to construct excited configurations which are eigenfunctions of $\hat{S}^2$ and $\hat{S}_z$, but it is not obvious which configurations are the most important in the CI expansion, and even with extensive CI the resulting a^N are not always in good agreement with experiment.[22] To allow for spin polarization of the inner shells by the unpaired electron, one can allow different spatial orbitals ϕ_i, ϕ'_i for different spins in a single determinant. In this case, Ψ_{UHF} is called an unrestricted Hartree–Fock (UHF) function but is no longer an eigenfunction of $\hat{S}^2$. This deficiency

$$\Psi_{UHF} = |\,\phi_0\,\bar{\phi}'_0\,\phi_1\,\bar{\phi}'_1 \ldots \phi_{k-1}\,\bar{\phi}'_{k-1}\,\phi_k\,| \tag{9}$$

can be remedied using a projection operator,[22] but the projected UHF wavefunction (PUHF) is no longer a single determinant. Furthermore, the optimal UHF orbitals are not optimal for the PUHF function. The extended Hartree–Fock (EHF) method optimizes the orbitals after projection[23] but has only recently become feasible in the form known as the GF method,[24, 25] although other authors have used essentially equivalent methods.[26–29] Other sophisticated methods have been used,

[21] P. J. Hay and W. A. Goddard, *Chem. Phys. Letters*, 1971, **9**, 356, gives many references to earlier work.

[22] See, *e.g.* the discussion in A. J. Freeman and R. B. Frankel, 'Hyperfine Interactions', Academic Press, New York, 1967.

[23] P. O. Löwdin, *Phys. Rev.*, 1955, **97**, 1509.

[24] W. A. Goddard, *Phys. Rev.*, 1967, **181**, 157; 1969, **182**, 48, 196.

[25] W. A. Goddard, *J. Chem. Phys.*, 1968, **48**, 450; 1968, **48**, 5337.

[26] U. Kaldor, *J. Chem. Phys.*, 1968, **48**, 835; 1968, **49**, 6.

[27] U. Kaldor, *Phys. Rev.*, 1968, **176**, 19.

[28] S. Lunnell, *Phys. Rev.*, 1968, **173**, 85; *Phys. Rev.* (*A*), 1970, **1**, 360.

[29] S. Larsson, *Phys. Rev.*, 1968, **49**, 169.

such as the Bethe–Goldstone method,[30] and the method of Meyer,[31] to which we refer later.

For molecular calculations, it is not feasible to obtain numerical approximations to the orbitals ϕ_i, and these are usually expanded in a basis set of M basis functions

$$\phi_i = \sum_{j=1}^{M} C_{ij}\,\chi_j \tag{10}$$

and the coefficients C_{ij} obtained variationally. Since the Slater orbitals (STO) have the correct asymptotic behaviour at the nuclei, until recently these have been the most widely used basis functions, and calculations for atoms using these basis sets will now be discussed.

One of the problems associated with the theoretical calculation of a^N is that it depends on the electron spin density at a particular point and is therefore very sensitive to the wavefunction. However, the wavefunctions are obtained *via* the variation method, and depend on evaluation of the integrals over all space, and wavefunctions which give very good energies may give very poor values of the spin density at the nucleus.[22] Slater basis functions have been used in calculations with the methods detailed above, with basis sets ranging from minimal STO to basis sets of close to Hartree–Fock quality. The recent configuration interaction calculations of Schaefer and co-workers,[32] using first-order wavefunctions, give results for the spin densities in error by *ca.* 25%, and this is typical of the accuracy attainable with CI calculations, although they account for a large fraction of the correlation energy.

On the other hand, the UHF method has probably been the most widely used,[33] and comparisons with numerical UHF calculations have been carried out by Bagus, Liu, and Schaefer.[34] These authors found that there were very large errors even with numerical UHF wavefunctions, although these were usually within about 20% of the results obtained for analytic expansions to UHF orbitals. Their conclusions were thus rather pessimistic regarding the reliability of calculated UHF spin densities. In an important paper, Meyer [31] showed that the errors in the UHF procedure are due to a partial and incorrect inclusion of pair correlation effects. He presented a version of the extended Hartree–Fock (EHF) method [23] which uses the natural orbitals from UHF calculations, but which used a Gaussian function basis (GTO). We return to this method later. Projected (PUHF) wavefunctions sometimes give better agreement with experimental than UHF,[22, 31] but may give worse agreement. One problem, however, is the difficulty of carrying out the projected calculations without further approximations.

[30] R. K. Nesbet, *Phys. Rev.* (*A*), 1970, **2**, 661.
[31] W. Meyer, *J. Chem. Phys.*, 1969, **51**, 5149.
[32] H. F. Schaefer, R. A. Klemm, and F. E. Harris, *Phys. Rev.*, 1968, **176**, 49; 1969, **181**, 137.
[33] For a review, see C. M. Moser in ref. 22.
[34] P. S. Bagus, B. Liu, and H. F. Schaefer, *Phys. Rev.* (*A*), 1970, **2**, 555.

EHF calculations by Kaldor,[26, 27] and the similar calculations by the GF method of Goddard[24, 25] have also given quite good agreement in some cases, but again the results are very dependent on the basis set size. One interesting conclusion from the latter calculations is that the core polarization contribution is larger than the experimental value, and therefore correlation effects significantly decrease the electron spin density at the nucleus. However, in the usual UHF method the orbitals are allowed to be different but are still required to be symmetry functions. Goddard *et al.*[21] have recently shown that relaxation of this constraint improves the anisotropic contributions to hyperfine interaction but makes little difference to the spin densities and therefore the isotropic contribution.

A similar method to the EHF, the spin-optimized (SO–SCF) method has been developed by Kaldor and Harris,[35, 36] in which optimum spatial orbitals and spin functions are determined. This method gives much improved agreement with experiment. An interesting application to the carbon atom ^{13}C coupling constant in the sp^3 (^5S) state has been used to examine the assumptions made in semi-empirical methods such as INDO.[37] Extension of this work to molecules, however, seems to be rather difficult.

Calculations using many-body theory have also been widely used, also with good results.[38] It therefore seems clear that in order to calculate hyperfine splittings accurately for atoms, one must use quite elaborate methods with large basis sets in order to obtain good agreement with experiment, but of the methods used only a few are readily extended to molecules without considerable computational difficulties. No doubt, however, the calculations will be further extended to more complicated atoms and molecules with higher accuracy as faster computers become available.

Turning now to hyperfine splittings in molecules, it might seem that the prospect for fairly accurate calculations was rather poor. However, in molecules the wavefunction even near the nucleus is distorted compared with the atom case and it might be hoped that with large basis sets including polarization functions a reasonable description of the spin density distribution could be obtained.

One idea which will probably prove useful is to modify a good quality atomic basis set to give good spin densities for the atoms, and to use such a basis set in molecular calculations. Konishi and Morukuma[39] have calculated the spin densities at the nuclei of first-row atoms using minimal, double zeta (DZ) and Hartree–Fock (HF) Slater basis sets, performing a CI calculation with all singly and doubly excited configurations. The spin density contributions near the nuclei were examined and it was shown that addition of a third 2*s* orbital to the minimal and DZ basis sets, with

[35] U. Kaldor and F. E. Harris, *Phys. Rev.*, 1969, **183**, 1.
[36] U. Kaldor, *Phys. Rev.* (*A*), 1970, **1**, 1586; 1970, **2**, 1267.
[37] U. Kaldor, *Chem. Phys. Letters*, 1972, **14**, 124.
[38] R. K. Nesbet, *Phys. Rev.* (*A*), 1970, **2**, 1208.
[39] H. Konishi and K. Morukuma, *Chem. Phys. Letters*, 1971, **12**, 408.

a large exponent, effectively allows the in–out correlation which builds up spin density at the nucleus. It should be emphasized that the idea is to determine a basis set which will simulate the RHF results, whilst being small enough to use in *molecular* hyperfine splitting constant calculations, remembering that the *atom* RHF results may be quite far from experiment.

Meyer has also suggested a similar approach using GTO.[31] Since the basis set should be able to be flexible enough to describe the spin polarization of the core region, and the wavefunction itself near the nucleus, a GTO basis should be suitable since it can be particularly flexible in the valence region, but this flexibility has little influence near the nucleus. Therefore in order to describe the nuclear region well, a group of Gaussians in the inner regions are used with fixed coefficients chosen to represent the correct wavefunction at the nucleus for a one-electron atom. The loss of variational freedom is only slight. Results using this approach are very good for first-row atoms. However, Meyer[31] also studied the radicals $\dot{O}H$, $\dot{N}H_2$, and $\dot{C}H_3$ using a small basis set, owing to computer limitations. In view of the latter, the agreement with experiment was reasonable. Results with large basis sets should be of interest.

The importance of π-electron radicals in e.s.r. spectroscopy led[40, 41] to the McConnell equation [equation (11)] in which experimental proton hyperfine coupling a_k^H constants are proportional to π-electron spin

$$a_k^H = Q\rho_{kk}^{\pi} \qquad (11)$$

densities ρ_{kk}^{π}. The derivation of this equation rested on several approximations and semi-empirical arguments and it is of particular importance that a thorough non-empirical study be carried out of the theory of proton hyperfine splittings in π-electron radicals. An important series of papers by Chang, Davidson, and Vincow,[42–45] summarized by Vincow,[46] has examined this question. The calculations were carried out with the spin-restricted Hartree–Fock method plus configuration interaction (RHF–CI). The general formulation of the problem was first investigated and applications to date have been to the specific case of π-electron radicals, using $\dot{C}H$ and $\dot{C}H_3$ as prototype π-radicals. There have been a number of previous approximate calculations[41] and Vincow *et al.* have considered a number of extensions of these.

A general expression for the hyperfine splitting constants can be obtained using perturbation theory and the importance of both first- and second-order contributions has been assessed. The dominant term in the wave-

[40] H. M. McConnell and D. B. Chesnut, *J. Chem. Phys.*, 1958, **28**, 107.
[41] For several reviews, see 'Radical Ions', ed. E. T. Kaiser and G. Vincow, Interscience, New York, 1968.
[42] S. Y. Chang, E. R. Davidson, and G. Vincow, *J. Chem. Phys.*, 1968, **49**, 529.
[43] S. Y. Chang, E. R. Davidson, and G. Vincow, *J. Chem. Phys.*, 1970, **52**, 1740.
[44] S. Y. Chang, E. R. Davidson, and G. Vincow, *J. Chem. Phys.*, 1970, **52**, 5596.
[45] S. M. Poling, E. R. Davidson, and G. Vincow, *J. Chem. Phys.*, 1971, **54**, 3005.
[46] G. Vincow, *J. Phys. Chem.*, 1971, **75**, 3400.

function is the RHF determinant

$$\psi^{(0)} = | \phi_1(1)\, \bar{\phi}_1(2) \ldots \phi_{k-1}(2k - 3)\, \bar{\phi}_{k-1}(2k - 2)\, \phi_k(2k - 1) | \quad (12)$$

where $\{\phi_j\}_{j=1}^{k-1}$ denote the doubly occupied molecular orbitals, ϕ_k is the singly occupied orbital and $\{g_m\}_{m-k+1}^{\infty}$ are a set of functions which taken with the $\{\phi_i\}_{i=1}^{k}$ form a complete set. In practice the g_m are taken to be the SCF virtual orbitals, but this need not necessarily be the case. In first order, only configurations arising from single replacements $\phi_i \rightarrow g_m$ contribute to the hyperfine splitting: in second order, there are, of course, many other possible excitations. To illustrate the approach, we consider the first-order case only.

The excitation $\phi_i \rightarrow g_m$ gives rise to two doublets, and the conventional choices of spin functions for the $\phi_j\, g_m\, \phi_k$ part of the wavefunction are

$$2^{-\frac{1}{2}}(\alpha\beta\alpha - \beta\alpha\alpha) \quad (13)$$

$$6^{-\frac{1}{2}}(\alpha\beta\alpha + \beta\alpha\alpha - 2\alpha\alpha\beta) \quad (14)$$

However, other possible spin doublets were also investigated, but the calculations are insensitive to the choice provided they are taken to second order and the usual doublets are used in the following discussion. The calculations, to first order, are applicable to any free radical, and it can be shown that only the second configuration with the spin function of equation (14) contributes to the hyperfine splitting in first order. For a planar π-radical, it turns out that

$$\gamma(r_{\mathrm{N}}) = \sum_{m,j} \frac{2[g_m\, \phi_k \,|\, \phi_k\, \phi_j]}{\Delta E_{2jm}}\, \phi_j(r_{\mathrm{N}})\, g_m(r_{\mathrm{N}}) \quad (15)$$

where $[ij/kl]$ represents a conventional electron repulsion (exchange) integral and ΔE_{2jm} is the energy difference between the ground configuration and the excited configuration, *i.e.*

$$\Delta E_{2jm} = \langle \psi_{2jm} \,|\, \mathscr{H} \,|\, \psi_{2jm} \rangle - \langle \psi^{(0)} \,|\, \mathscr{H} \,|\, \psi^{(0)} \rangle \quad (16)$$

If we expand the singly occupied MO in an atomic orbital basis $\{\chi_i\}$

$$\phi_k = \sum_i C_{ik}\, \chi_i \quad (17)$$

we find that the zeroth-order spin density function is

$$\gamma^{(0)}(r) = \phi_k^2(r) = \sum_{i,i'} \gamma_{i,i'}^{(0)}\, \chi_i(r)\, \chi_{i'}^{*}(r) \quad (18)$$

where

$$\gamma_{i,i'}^{(0)} = C_{ik}\, C_{i'k} \quad (19)$$

are the elements of the spin density coefficient matrix. Substitution into equation (15) gives the general expression for a^{N}:

$$\begin{aligned} a^{\mathrm{N}} &= \sum_{i,i'} \left(\frac{8\pi}{3}\right) g_{\mathrm{e}}\, \beta_{\mathrm{e}}\, g_{\mathrm{N}}\, \beta_{\mathrm{N}} \sum_{m,j} \left\{ \frac{2[g_m\, \chi_{i'} \,|\, \chi_i\, \phi_j]}{\Delta E_{2jm}}\, \phi_j(r_{\mathrm{N}})\, g_m(r_{\mathrm{N}}) \right\} \gamma_{ii'}^{(0)} \\ &= \sum_{i,i'} G_{i'i}\, \gamma_{ii'}^{(0)} = \mathrm{tr}(G^{\mathrm{N}}\, \gamma^{(0)}) \quad (20) \end{aligned}$$

The McConnell equation can be obtained by simplifying equation (20). If the π-orbitals are described using a minimal basis set, and a particular proton p is bonded to the carbon atom with $2p\pi$ basis orbital χ_p, then we can neglect all terms in the double sum except those for which $i = i' = p$ and

$$a^{\mathrm{H}} = G^{\mathrm{H}}_{pp}\,\gamma^{(0)}_{pp} \tag{21}$$

If G^{H}_{pp} is an approximate constant for hydrocarbon radicals for the various types of position and is given a semi-empirical value Q, then we obtain the McConnell equation

$$a^{\mathrm{H}}_{p} = Q\gamma^{(0)}_{pp} = Qc^{2}_{pp} = Q\rho^{\pi}_{pp} \tag{22}$$

This equation has been used to correlate a vast amount of data,[41] but the extent of its validity depends on (*i*) the validity of using a minimum basis set for the π-electrons, (*ii*) the size of the errors involved in neglecting the other terms $G_{kk'}\,\gamma^{(0)}_{k'k}$ $(k \neq k')$ and assuming the constancy of G^{H}_{pp}, and (*iii*) the error in calculating only to first order. For nuclei such as ^{13}C, ^{14}N, and ^{19}F an expression for their hyperfine couplings can be obtained from equation (20), *i.e.* for ^{13}C

$$a^{\mathrm{C}}_{p} = G^{\mathrm{C}}_{pp}\,\gamma^{(0)}_{pp} + \sum_{i} G^{\mathrm{C}}_{ii}\gamma^{(0)}_{ii} \tag{23}$$

where i denotes the neighbouring carbon atoms. This form is equivalent to the Karplus–Fraenkel [47] equation and similar comments on the quantitative validity of this equation apply.

Extension of the calculations to second order introduces an extra six terms in the expression for the hyperfine splitting, and was first carried out by Malrieu [48] (see ref. 46 for the detailed expressions).

The experimental evidence strongly confirms the leading term in the first-order contribution to $\gamma(r_{\mathrm{H}})$ to be $G^{\mathrm{H}}_{pp}\,\gamma^{(0)}_{pp}$, but there is experimental evidence that for most even alternant radical ions, cation splittings are *ca.* 10—20% larger than in the corresponding ring positions in the anions.[41] Semi-empirical π-electron theories predict that $\gamma^{(0)}_{pp}$ should be the same in cations and anions and therefore the effective Q must differ. There have been two extensions of the McConnell relationship proposed to account for this. In the Colpa–Bolton equation,[49] Q is replaced by

$$a^{\mathrm{H}}_{p} = (Q_0 + K\varepsilon^{\pi}_{p})\rho^{\pi}_{pp} \tag{24}$$

where the new proportionality factor $(Q_0 + K\varepsilon^{\pi}_{p})$ depends on the atom p, and on the excess π charge density ε_p on the atom: $\varepsilon_p = 1 - q^{\pi}_{p}$ where q_p is the total π-electron charge density on atom p.

A second two-parameter equation, due to Giacometti, Nordio, and Pavan,[50] corresponds to inclusion of nearest-neighbour 'bond spin density'

[47] M. Karplus and G. K. Fraenkel, *J. Chem. Phys.*, 1961, **35**, 1312.
[48] J.-P. Malrieu, *J. Chem. Phys.*, 1967, **46**, 1654.
[49] J. P. Colpa and J. R. Bolton, *Mol. Phys.*, 1963, **6**, 273; J. R. Bolton, *J. Chem. Phys.*, 1965, **43**, 309.
[50] G. Giacometti, P. L. Nordio, and M. V. Pavan, *Theor. Chim. Acta*, 1963, **1**, 404.

terms

$$a_p^{\mathrm{H}} = Q_0\, \rho_{pp}^{\pi} + Q_1(\rho_{p,p+1}^{\pi} + \rho_{p,p-1}^{\pi}) \qquad (25)$$

Both these equations have been critically discussed by Moss and Fraenkel[51] who, using Hückel molecular orbital theory, concluded that the GNP relation reduces to a form of the CB equation.

The quantitative validity of the above extensions to the McConnell equation has been assessed by Chang *et al*,[42] using calculations on the $\dot{C}H$ fragment, but unlike the earlier workers, they explicitly included overlap and a properly orthogonalized 2*s* orbital on carbon, and also considered the effect of the orbital exponents on ΔE_{2jm}. These calculations show that the difference in proton splittings between anions and cations is not principally due to the excess charge effect on the exponents of the 2*s* and 2*p* orbitals, since the computed value of $Q_0/K = 27$, an order of magnitude less than calculated by Bolton.[49] A similar conclusion was reached in some semi-empirical work by Melchior.[52] Calculations on the importance of the nearest-neighbour effect, which evaluated the integrals accurately, gave similar conclusions: namely, that neither of these possible mechanisms is responsible for the leading correction to the McConnell equation. Clearly more quantitative work is needed.

A quantitative evaluation of the second-order contributions was made for $\dot{C}H_3$,[43] and their contribution was estimated to be *ca.* 10%. Second-order terms are expected to be more important for ^{13}C, ^{19}F, *etc.*, splittings.

The assessment of the above relations was carried out using variationally calculated wavefunctions for $\dot{C}H$ and $\dot{C}H_3$. CI was performed using both single and single plus double excitations. A variety of basis sets was used, ranging from minimum basis to the best near Hartree–Fock basis for $\dot{C}H$ obtained by Cade and Huo.[53] Since the semi-empirical calculations are carried out using a minimum basis, then it is of importance to see how well the *ab initio* calculations perform with this basis. As expected, the results are very sensitive to the 1*s* orbital exponent on hydrogen, and the authors suggest the use of a hybrid basis set with atom optimized exponents on carbon but with $\zeta_{\mathrm{H}} = 1.0$, which gives good agreement for the proton-splittings. The use of a modified basis set has been mentioned above in the work of Morukuma[39] and Meyer.[31] The results of the calculated spin densities were in good agreement with the INDO calculations,[54] and thus support the neglect of the 1*s* shells in the latter. It should be noted that the spin-projected UHF (INDO)[55] results are in much poorer agreement with the *ab initio* calculations and therefore despite the contamination with other spin states in the INDO[55] method, the results are more reliable.

[51] R. E. Moss and G. K. Fraenkel, *J. Chem. Phys.*, 1969, **50**, 252.
[52] M. T. Melchior, *J. Chem. Phys.*, 1969, **50**, 511.
[53] P. E. Cade and W. Huo, *J. Chem. Phys.*, 1967, **47**, 614.
[54] J. A. Pople, D. L. Beveridge, and P. A. Dobosh, *J. Amer. Chem. Soc.*, 1968, **90**, 4201.
[55] D. L. Beveridge and P. A. Dobosh, *J. Chem. Phys.*, 1968, **48**, 5532.

Extended basis set calculations still do not give good agreement with experiment for $\dot{C}H_3$ if the atomic basis sets are used, but by adjustment of the H exponents (hybrid basis set) it is possible to improve the coupling constant considerably with very little change in energy. A larger basis set is clearly needed here. In the case of $\dot{C}H$, a similar hybrid basis set gives essentially the same results as the much larger HF basis set.[53]

^{13}C Splittings, however, are rather poorly calculated with the minimal basis, since the valence shell contributions are overestimated whereas the inner shell polarization terms are underestimated. Even with a DZ basis, there is a lot of cancellation and the conclusion is that quite extended basis sets are needed to calculate ^{13}C splittings reliably.

Extensions of these calculations to the pyramidal form of $\dot{C}H_3$ [44] and appropriate vibrational averaging has given good agreement with the observed temperature dependence of the hyperfine coupling constants. The temperature dependence of the ^{13}C splitting was also in reasonable agreement with experiment. This series of papers has thus highlighted some of the difficulties of *ab initio* calculations, but has also shown that semi-empirical methods such as INDO can be put on a firmer basis, particularly for proton hfcc.

The use of Gaussian basis functions in molecular calculations is well known, but it is only recently that they have been used in hyperfine coupling constant calculations. Since Gaussians have the wrong asymptotic behaviour at the nucleus, a rather large basis set has to be used. Planar π-radicals have zero spin density at the in-plane nuclei in the RHF open-shell approximation, but σ-radicals do not and some of the first applications using Gaussian bases were to such radicals, but this earlier work was restricted to calculations with rather small basis sets.[56, 57] Millie *et al.*[58] investigated the vinyl radical with a (10*s*6*p*) basis set and standard bond lengths and angles for the $CH_2=$ group but found that the computed proton coupling constants were too small and in the wrong order. However, the ^{13}C coupling constants were quite reasonable, but it should be noted that only the $\widehat{HCC}$ angle was varied in the calculations. The same authors have gone beyond the RHF approximation and also investigated spin polarization in $\dot{C}H_3$ and in vinyl using a first-order perturbation treatment.[59] In both cases a (10, 6, 2/5, 1) basis was used which was contracted [6, 4, 2/3, 1] for $\dot{C}H_3$, and [53/3] for vinyl, where polarization functions were not included. The results (Table 1), where two different SCF–MO's were used, show that the spin polarization contribution is large. Agreement with experiment varies from position to position. For methyl the results are too small, in agreement with the calculations of Chang *et al.*,[43] probably partly because the molecule-optimized H-exponents were not used and also because the calculations were carried

[56] K. Morukuma, L. Pederson, and M. Karplus, *J. Chem. Phys.*, 1968, **48**, 4801.
[57] D. B. Cook, A. Hinchcliffe, and P. Palmieri, *Chem. Phys. Letters*, 1969, **3**, 223.
[58] P. Millie and G. Berthier, *Internat. J. Quantum Chem.*, 1968, **2S**, 67.

Table 1 Ab initio *hyperfine splittings in vinyl* ($C^1\widehat{C^2}H = 138°$) *and methyl radicals* [59]

(*a*) *Vinyl radical*

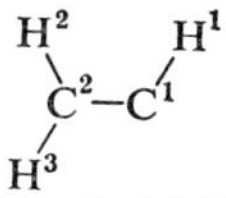

	Zeroth order	Canonical MO's First order	Canonical MO's Total	Equivalent MO's First order	Equivalent MO's Total	*Exp.* (ref. 60)
a^{H^1}	21.75	−9.25	12.50	−12.38	9.37	16.0
a^{H^2}	21.50	11.45	32.95	15.59	37.09	68.0
a^{H^3}	13.31	7.13	20.44	9.43	22.74	34.0
a^{C^1}	109.07	7.09	116.16	−11.81	97.26	107.5
a^{C^2}	10.24	−8.39	1.85	−1.86	8.38	8.5

(*b*) *Methyl radical*

CH_3

	Zeroth order	Canonical First order	Equivalent First order	*Exp.* (*ref.* 60)
a^H	0	−20.61	−13.01	−24.2
a^C	0	10.16	−4.97	+28.3

out to first order only. The ^{13}C coupling constants are less influenced by spin polarization effects. The results are in better agreement with experiment than *ab initio* UHF calculations which are considered below. The same group [61] have also examined the interplay of direct and indirect contributions to the proton hyperfine coupling constants in RHF–CI calculations on CH_3CH_2, taking the geometry from C_2H_6 and the paramagnetic centre to be planar. A slightly smaller (9, 5/4) basis set was used and a first-order calculation of the spin polarization contribution was also carried out. The results show that delocalization and spin polarization contribute about the same amount to the β-proton splittings. The computed values of the splitting constants were $a_{\beta H} = 16.7$ ($a_{\beta H}^{exp} = 26.87$ G), $a_{\alpha H} = -33.8$ G ($a_{\alpha H}^{exp} = 22.38$ G), in error by about the same amount as in the $\dot{C}H_3$ calculations. One conclusion of interest is that *ca.* 90% of the indirect coupling arises from the $\sigma_{C-H} \rightarrow \sigma_{C-H}^*$ excitation. Furthermore, it is shown that the $\cos^2 \theta$ dependence of the β-proton splittings is independent of the actual nature of the coupling.

Of the known σ-radicals, the simplest to exhibit internal rotation is the acetyl radical $CH_3\dot{C}O$. Calculations with the RHF method and also by the UHF method using a (10, 6/5) → [42/2] basis set have been carried out by Veillard and Rees,[62] without geometry optimization. Calculated

[59] P. Millie, B. Levy, and G. Berthier, *Internat. J. Quantum Chem.*, 1972, **6**, 155.
[60] R. W. Fessenden, *J. Phys. Chem.*, 1967, **71**, 74.
[61] Y. Ellinger, A. Rassat, R. Subra, G. Berthier, and P. Millie, *Chem. Phys. Letters*, 1971, **11**, 362.
[62] H. Veillard and A. Rees, *Chem. Phys. Letters*, 1971, **8**, 267.

coupling constants for the UHF calculation after spin annihilation are in satisfactory agreement with experiment, and further indicate that Gaussian basis set calculations can give reasonable values of the coupling constants if a large enough basis set is taken and the UHF method is used. In an application of the RHF method to the hyperfine couplings in NO_2,[63] good agreement with the nitrogen coupling constant was obtained if *d*-orbitals (polarization functions) were added to a (9, 5) basis set, but the ^{17}O hfcc was consistently low. These authors conclude that electron correlation must be included for high accuracy, but useful qualitatively correct values of the parameters can be obtained with a properly chosen basis set even within the RHF method.

Since spin polarization is included in a single configuration UHF wavefunction it has proved a popular method for hfcc calculations, and Claxton and co-workers[64-69] have reported the results of calculations on a variety of small molecules. In these calculations, a minimum basis set of SCF atomic orbitals, expanded as a linear combination of Gaussians, was used, and the wavefunction was subject to spin annihilation to remove the quartet state which is the most important contaminant of the UHF wavefunction. Calculations on $\dot{N}H_2$ give good agreement with experiment after the angle and hydrogen atom exponent were optimized.[65] Similar calculations[67] on $\overset{+}{NH}{}^{\bullet}$ and $\overset{-}{NH}{}^{\bullet}$ give results which indicate that the trapped radical observed experimentally in γ-irradiated $(NH_4)_2HPO_4$ crystals is $\overset{+}{NH}{}^{\bullet}$. Calculations on the planar $\overset{+}{NH_3}{}^{\bullet}$ radical indicate that out-of-plane vibrational modes contribute significantly to the heavy-atom coupling, but with a large basis set the results agree well with experiment.[68]

In an interesting discussion[69] of the radicals $\overset{-}{HBO}{}^{\bullet}$, $H\dot{C}O$, $\overset{-}{HCN}{}^{\bullet}$, and $H_2C\dot{N}$ it is found that the unpaired electron is extensively delocalized and the first three radicals show a larger proton coupling constant than in $H_2C\dot{N}$ where hyperconjugation is believed to be important. The results, in good agreement with experiment, show that hyperconjugation in $H_2C\dot{N}$ and delocalization in $\overset{-}{HCN}{}^{\bullet}$ have an identical origin. The results are sensitive, however, to geometrical changes, a fact noted previously in INDO calculations[70] (see below).

An investigation[71] of the isoelectronic series $\overset{-}{BH_3}{}^{\bullet}$, $\dot{C}H_3$, and $\overset{+}{NH_3}{}^{\bullet}$ resulted in rather poor agreement with experiment in the case of $\overset{-}{BH_3}{}^{\bullet}$

[63] H. F. Schaefer and S. Rothenberg, *J. Chem. Phys.*, 1971, **54**, 1423.
[64] T. A. Claxton, D. McWilliams, and N. A. Smith, *Chem. Phys. Letters*, 1970, **4**, 505.
[65] T. A. Claxton and D. McWilliams, *Trans. Faraday Soc.*, 1970, **66**, 513.
[66] T. A. Claxton, *Trans. Faraday Soc.*, 1970, **66**, 1537.
[67] T. A. Claxton, *Trans. Faraday Soc.*, 1970, **66**, 1.
[68] T. A. Claxton and N. A. Smith, *Trans. Faraday Soc.*, 1970, **66**, 1825.
[69] T. A. Claxton, *Trans. Faraday Soc.*, 1971, **67**, 897.
[70] C. Thomson, *Theor. Chim. Acta*, 1970, **17**, 320.
[71] T. A. Claxton, M. J. Godfrey, and N. A. Smith, *J. C. S., Faraday II*, 1972, **68**, 181.

compared with the other two radicals. Out-of-plane vibrations complicate the calculation, but one conclusion is that environmental effects are probably important, particularly for $BH_3^{\bullet-}$, which is bent in isolation, but could be planar in a positively charged environment.

Earlier calculations on the $\dot{C}N$ radical gave ^{14}N coupling constants in good agreement with experiment, but the ^{13}C splitting constant was much larger than observed (339.2 G *versus* 210 G). In a recent paper,[72] the rather large amount of spin contamination was reduced and the radical geometry was optimized, leading to a much closer agreement with experiment. Calculations on vinyl[73] with this method are in better agreement with experiment than RHF calculations.[58]

Few calculations on the same radical have examined the effect of using different basis sets and different methods. In a recent note, however, Bendazzoli *et al.*[74] have investigated $\dot{N}H_2$, $\dot{O}H$, and $\dot{S}H$ using PUHF and RHF–CI wavefunctions with (*i*) a minimal basis set of atomic orbitals, (*ii*) a DZ basis, and (*iii*) a contracted GTO basis set. In cases (*i*) and (*ii*) each atomic orbital was expressed as a linear combination of four Gaussians. It was concluded that proton couplings are easier to calculate than those for other nuclei, for which very large bases are needed. In addition, the choice of the wavefunction form is related to the nature of the basis. With Slater orbitals, the UHF method seems to provide a better description of spin polarization, but the RHF–CI method seems to possess some advantages with Gaussians. In a study of one of the simplest anion radicals, Hinchcliffe[75] has carried out calculations on butadiene and its ions using a Gaussian basis set.

To summarize, these calculations show that quite good agreement with experiment may be obtained from *ab initio* calculations of isotropic hfcc providing large basis sets are used, but the calculations are very sensitive to the geometry, to the basis set used, and particularly to the orbital exponents, but can provide very useful information on the mechanisms involved and place the semi-empirical methods which must be used for bigger molecules on somewhat surer ground.

There have been numerous experimental measurements on oriented free radicals, from which the principal values of the dipolar hyperfine interaction tensor may be obtained.[2,5] The general theory was formulated by McConnell and Strathdee,[76] and with the interest in *ab initio* calculations of isotropic hyperfine interactions it is natural that a quantitative assessment of the earlier models be carried out non-empirically. Semi-empirical investigations are described in Section 4.

The principal values of the dipolar hyperfine tensor for nucleus N are

[72] T. A. Claxton, M. J. Godfrey, and B. L. J. Weiner, *J. C. S. Faraday II*, 1972, **68**, 566.
[73] T. A. Claxton, *Internat. J. Quantum Chem.*, 1970, **4**, 337.
[74] G. L. Bendazzoli, F. Bernardi, and P. Palmieri, *Mol. Phys.*, 1972, **23**, 193.
[75] A. Hinchcliffe, *J. Mol. Structure*, 1971, **10**, 379.
[76] H. M. McConnell and J. Strathdee, *Mol. Phys.*, 1959, **2**, 129.

given by

$$T^{ii} = \sum_{\lambda,\mu} T^{ii}_{\mu\lambda} = -g_e \beta_e g_N \beta_N \sum_{\lambda,\mu} \gamma_{\lambda\mu} \langle \phi_\mu | (r')^{-3}[1 - 3ii(r')^{-2}] | \phi_\lambda \rangle \quad (26)$$

where $i = x, y, z$; $\gamma_{\lambda\mu}$ are the elements of the spin density coefficient matrix in the AO representation $\{\phi_\mu\}$, and r' is the nuclear–electron distance. Calculations by Vincow *et al.*[77] on $\dot{C}H_3$, using the method described above, give good agreement for T^{yy}, where the z axis is along the CH_1 bond and the y axis is perpendicular to the molecular plane. It should be noted that the three-centre integrals in equation (26) were evaluated approximately, using the Mulliken approximation. The dominant contribution comes from the $2p\pi$ atomic orbital, *i.e.* $T^{yy}_{2py,\,2py} = +53.8$ for the Slater orbital basis and + 39.26 for the optimized basis set, with the next largest term being associated with spin density in the $2p\sigma$ orbital. The experimental results give $T^{yy} = 45 \pm 0.5$ G, whereas the calculated total contributions are $T^{yy} = 51.4$ (Slater) and 36.8 (optimized). Values for the proton interaction are small in the case of T^{yy} and of the wrong sign, but it is expected that this contribution will be difficult to calculate since a large number of terms make significant contributions. The values of T^{xx} and T^{zz}, on the other hand, are quite reasonable.

Claxton[78] has carried out similar calculations using Gaussian basis functions and the UHF method. $\overset{+}{NaH}{}^{\bullet}$, $N_2^{\bar{\bullet}}$, $\dot{N}O$, $\dot{C}N$, $H\dot{C}O$, $HCN^{\bar{\bullet}}$, $HBO^{\bar{\bullet}}$, $H_2C\dot{N}$, and vinyl were all considered. The agreement with experiment was as good as in the case of the isotropic coupling constants and these calculations thus show that the observable hyperfine parameters can be calculated using *ab initio* methods; although quantitative accuracy is not easy to attain, the calculations carried out to date give grounds for optimism concerning future calculations, despite the difficulties found for atoms.

4 Semi-empirical Calculation of Hyperfine Coupling Constants

All-valence-electron Calculations.—Despite the successes of the *ab initio* calculations referred to above, they are necessarily restricted to small molecules and semi-empirical methods will continue to be important for calculations on larger radicals.

Of the various approximate methods, those closest to *ab initio* are the all-valence-electron approximate SCF–MO methods, of which the INDO[54, 55] method has been developed particularly to calculate hyperfine coupling constants. This method computes a spin unrestricted SCF single-determinant wavefunction using a minimum STO basis of valence shell orbitals. Approximations involving neglect of all repulsion integrals except Coulomb and some one-centre exchange integrals and semi-empirical evaluation of other integrals make solution of the Fock equations feasible

[77] G. Vincow, S. Y. Chang, and E. R. Davidson, *J. Chem. Phys.*, 1971, **54**, 4121.
[78] T. A. Claxton and N. A. Smith, *Trans. Faraday Soc.*, 1971, **67**, 1859.

for quite large molecules. The hyperfine splitting is given by

$$a^{\mathrm{N}} = [(8\pi/3)\, g_{\mathrm{e}}\, g_{\mathrm{N}}\, \beta_{\mathrm{e}}\, \beta_{\mathrm{N}}\, S^2(r_{\mathrm{N}})]\, \gamma_{S_{\mathrm{N}}S_{\mathrm{N}}} \quad (27)$$

where S_{N} is the *s*-type valence AO on nucleus N. All other terms in the AO expansion of $\gamma(r_{\mathrm{N}})$ are neglected.

The quantity in square brackets is assumed constant for all radicals for hyperfine splittings from a particular nucleus N and the appropriate value for $S_{\mathrm{N}}^2(r_{\mathrm{N}})$ is obtained from a least-squares fit of a plot of a^{N} *vs.* calculated *s*-orbital spin densities $\gamma_{S_{\mathrm{N}}S_{\mathrm{N}}}$. The values of γ_{HH} from INDO are in good agreement with the *ab initio* results for $\dot{C}H_3$ cited above,[46] and in fact spin-annihilation gives much poorer values of γ_{HH}. A similar conclusion as to the effect of spin contamination was reached by Beveridge and Dobosh.[55] However, for ^{13}C splittings, since the inner shell polarization is important, Vincow [46] questions the accuracy of INDO-computed ^{13}C and other hyperfine coupling constants. The calculated value of $\gamma_{2s_{\mathrm{C}},\,2s_{\mathrm{C}}}$ for $\dot{C}H_3$, 0.0542, is quite different from the *ab initio* values of 0.1637 (STO) or 0.1105 (best minimal basis set and single excitations). However, despite possible shortcomings, this method has been used to study a variety of problems in e.s.r.

Thomson [70] has calculated the splittings in $HCN^{\overline{\cdot}}$, $HBO^{\overline{\cdot}}$, and $H\dot{C}O$. Good agreement with experiment is found for the proton coupling constants; in particular, the calculations give a low value for a^{H} in $HBO^{\overline{\cdot}}$ as observed; a^{H} in the formyl radical, however, is lower than observed. The ^{13}C couplings are in poor agreement for $HCN^{\overline{\cdot}}$ although the small value of a^{N} is predicted. It is clear that careful optimization of the geometry is necessary, and the assumption of 'standard' bond lengths and angles is a poor one for free radicals. Pederson [79] carried out a similar study with similar conclusions, although his calculated bond lengths were somewhat longer than those given in ref. 70. He also studied the σ-radical $F\dot{C}O$, where the computed angle is some 10° larger than e.s.r. data indicates.

Owens and Vogel [80] have used INDO in an attempt to determine the equilibrium geometries and coupling constants in $N_2^{\overline{\cdot}}$ and $N_4^{\overline{\cdot}}$. A bond length in the former of 1.20 Å gives a calculated a^{N} in excellent agreement with experiment. $N_4^{\overline{\cdot}}$ was found to be rectangular rather than linear, with equal coupling to the four nitrogen nuclei. Calculations on $\dot{B}F_2$ also gave good agreement with experiment.[81]

The original formulation of the method was restricted to first-row atoms (Li to F) but Benson and Hudson [82] have extended it to the second-row atoms without, however, inclusion of *d*-orbitals. The parametrization involved similar considerations for both first- and second-row atoms and resulted in new values for the scaling factor in equation (27). Results with and without spin-annihilation were essentially the same, and calculations on a variety of triatomic radicals containing Si and Cl gave good

79 L. J. Pederson, *J. Mol. Structure*, 1970, **5**, 21.
80 F. J. Owens and V. L. Vogel, *J. Chem. Phys.*, 1972, **56**, 2487.
81 M. Zaucer and A. Azman, *Croat. Chem. Acta*, 1971, **43**, 139.
82 H. G. Benson and A. Hudson, *Theor. Chim. Acta*, 1971, **23**, 259.

agreement with experiment. Equilibrium geometries were calculated in this work in order to compute the hyperfine couplings. The large ^{35}Cl coupling in 2-chloroethyl is in good agreement with experiment if the radical is assumed to be locked in a conformation with the chlorine eclipsing the odd electron orbital. $\dot{S}iH_3$ and $\dot{C}Cl_3$ are both predicted to be pyramidal, as in $\dot{S}iCl_3$.

Numerous previous calculations of the spin distribution in benzyl,[83–85] usually π-electron calculations, have failed to give good agreement with experiment, and usually predict $\rho_p < \rho_o$, in reverse order to experiment. An INDO calculation in the original applications of the method also failed to predict this ordering.[54] Explanations of these deviations in terms of deviations of the ring geometry from that of a regular hexagon have been made and Beveridge and Guth,[86] and Benson and Hudson,[87] have independently carried out INDO calculations with the ring non-hexagonal.

Complete geometry optimization was not feasible, but the calculated geometry, with r_{23}, r_{56}, and r_{17} reduced and the others extended, gives the spin densities in the right order and a much better agreement with experiment, with the calculations of Beveridge *et al.* rather closer than those of Benson *et al.* (Figure 1). The bond length changes are similar but not

Figure 1 *Hyperfine coupling constants and calculated bond lengths in the benzyl radical*

Benson and Hudson (ref. 87) | Beveridge and Guth (ref. 86)

(*a*) *Bond lengths* (Å)

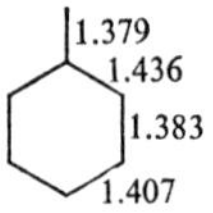

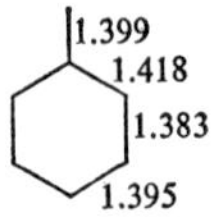

(*b*) *Hyperfine coupling constants*

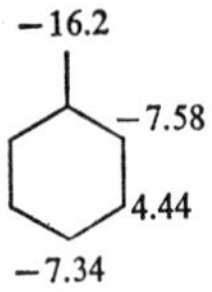

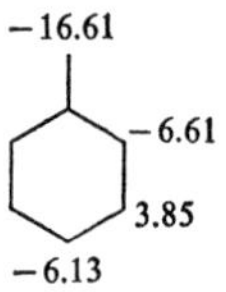

Experimental hyperfine couplings

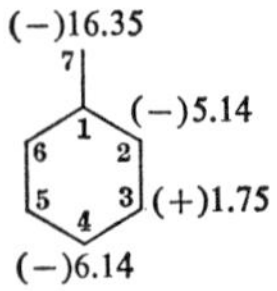

[83] A. Carrington and I. C. P. Smith, *Mol. Phys.*, 1965, **9**, 137.
[84] V. A. Kuprievich, Y. A. Kruglyak, and E. V. Mazdor, *Internat. J. Quantum Chem.*, 1970, **4**, 73.
[85] A. T. Amos and B. L. Burrows, *J. Chem. Phys.*, 1970, **52**, 3072.
[86] D. L. Beveridge and E. Guth, *J. Chem. Phys.*, 1971, **55**, 458.
[87] H. G. Benson and A. Hudson, *Mol. Phys.*, 1971, **20**, 185.

equal in the two sets of calculations. Thus there is now strong evidence that careful geometry optimization is necessary for the highest accuracy with this method.

A similar investigation of the fluorobenzyl and anilino radicals [88] gave good results, when the geometry was optimized, for the values of a^{F} and a_i^{H}. These calculations also confirm that the dominant contribution to a^{F} comes from Q_{FF}^{F} (see later).

The isoelectronic azabenzyl radicals have also been investigated by the INDO method. However, regular hexagonal geometry was assumed and, as might be expected from the above discussion, agreement with experiment was not very good.[89] Limited geometry variation was carried out for α-azabenzyl, but the spin distribution is only slightly improved, the calculated spin densities being similar to those in benzyl. With the exception of Benson and Hudson, most investigators using INDO have used the original scaling factor of Pople *et al.*[54] Hirst,[90] in a study of nitrogen radicals, has redetermined this factor for ^{14}N and obtained 332 G compared with Pople's value of 379 G. An improved correlation was obtained, particularly because the total number of radicals studied was higher. A variety of pyridine anions was studied [91] in the work which led to this improved correlation, but the calculated hfcc were uniformly too high. It should be noted in passing that an INDO calculation on pyridine gives atomic populations significantly different from Clementi's *ab initio* results.[92] However, it is now well established that atomic populations are very difficult to calculate accurately with small basis sets. In later work by Hirst [93] a variety of radical ions derived from derivatives of styrene was studied, but the agreement with experiment was only fair with this series of radicals. However, no geometry optimization was attempted.

There have also been applications of the INDO method to a variety of radicals derived from acetylenes and nitriles.[94] Among the radicals considered were $\dot{C}H_2CH$, $\dot{C}H_2C{\equiv}CH$, $\dot{C}H_2C{=}CCH_3$ and $\dot{C}H_2C{\equiv}CCH_2Cl$. The calculated coupling constants were in reasonable agreement with experiment.

Some rather larger ring systems have also been studied. The cyclo-octatrienyl free radical,[95] produced by γ-irradiation of cyclo-octatetraene in an adamantane matrix, was shown by INDO calculations to be planar with the two methylene groups locked in a fixed orientation with respect to the rest of the molecule. Although the aziridino-radical (1), isoelectronic with cyclopropyl, was expected to be a σ-radical, the e.s.r. evidence together

[88] H. G. Benson, A. Hudson, and J. W. E. Lewis, *Mol. Phys.*, 1971, **21**, 935.
[89] R. V. Lloyd and D. E. Wood, *Mol. Phys.*, 1971, **20**, 735.
[90] D. M. Hirst, *Theor. Chim. Acta*, 1971, **20**, 409.
[91] D. M. Hirst, *Theor. Chim. Acta*, 1971, **20**, 292.
[92] E. Clementi, *J. Chem. Phys.*, 1967, **46**, 4731.
[93] D. M. Hirst, *Theor. Chim. Acta*, 1972, **23**, 378.
[94] H. G. Benson, A. J. Bowles, A. Hudson, and R. A. Jackson, *Mol. Phys.*, 1971, **20**, 713.
[95] A. R. McIntosh, D. R. Gee, and J. K. S. Wan, *Spectroscopy Letters*, 1971, **4**, 217.

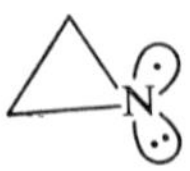

(1)

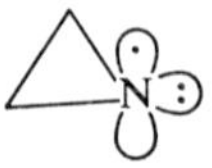

(2)

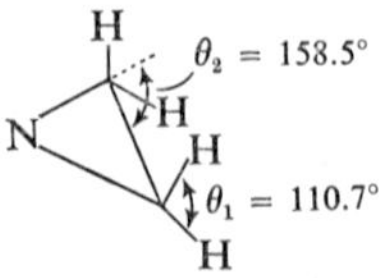

(3)

with a complete INDO energy minimization established that it is a π-radical (2).[96] The four-membered-ring azetidino-radical is very similar. The calculated geometry is illustrated in (3) for aziridino.

Long-range couplings in e.s.r. are well established, particularly in rigid bicyclic semidiones,[97] and Underwood and Vogel[98] have used the INDO method to calculate ^{1}H, ^{13}C, and ^{17}O coupling constants with good results, although geometry optimization was not carried out. γ-Proton splittings result from several mechanisms, but α-carbon hfcc values are due primarily to spin polarization and the β-carbon hfcc to carbon–carbon hyperconjugation. These authors point out the need for a redetermination of the ^{13}C scaling factor.

The fluoromalonyl radical has been studied.[99] In this study, however, spin annihilation was found to make a significant difference to the results. There have also been applications of the INDO method to complexes,[100] and to ^{13}C couplings in radicals derived from carboxylic acids.[101]

An earlier INDO investigation of $\dot{C}H_3$,[102] in which the coupling constant was investigated as a function of the out-of-plane bending vibration, has been shown to contain an error in the scale factor of the vibrational wavefunction. Corrections to this calculation have appeared,[103] and the results compared with a valence bond calculation. A critical comparison does not lead to a clear conclusion as to which method is most suitable for this type of calculation.

The stable geometries of a large number of ketyls, semidiones, and semiquinones have also been studied by INDO.[104] In the case of the semidiones and semiquinones the carbonyl carbons are planar, but the ketyls deviate from planarity at this carbon by as much as 32°. Since formaldehyde ketyl is non-planar, an earlier explanation that the non-planarity in hexamethylacetone ketyl is due to steric effects may be incorrect.[104] It is suggested that the non-planarity is due to the presence of the odd electron in the antibonding orbital of the carbonyl group.

[96] W. C. Danen and T. T. Kensler, *Tetrahedron Letters*, 1971, 2247.
[97] G. R. Underwood and R. S. Givens, *J. Amer. Chem. Soc.*, 1968, **90**, 3713.
[98] G. R. Underwood and V. L. Vogel, *J. Amer. Chem. Soc.*, 1971, **93**, 1058.
[99] F. C. Herring, W. C. Lin, and M. R. Mustafa, *J. Magn. Resonance*, 1970, **2**, 9.
[100] M. J. Scarlett, A. T. Casey, and R. A. Craig, *Austral. J. Chem.*, 1970, **23**, 1333; 1971, **24**, 31.
[101] G. P. Laroff and R. K. Fessenden, *J. Chem. Phys.*, 1971, **55**, 5000.
[102] D. L. Beveridge and K. Miller, *Mol. Phys.*, 1968, **14**, 401.
[103] D. M. Schrader and K. Morukuma, *Mol. Phys.*, 1971, **21**, 1033.
[104] G. R. Underwood, *Mol. Phys.*, 1971, **22**, 729.

INDO calculations[105] on two new α-substituted vinyl radicals (4) and (5) show that these are π-radicals, whereas α-methylvinyl is a σ-radical. Another radical whose structure has been confirmed by INDO is (6), which is produced during photolysis of phenyliodoacetylene.[106] The

$H_2C{=}\dot{C}{-}CO_2H$ (4) $H_2C{=}\dot{C}{-}CO_2Me$ (5) $Ph{-}C{\equiv}C\cdot$ (6)

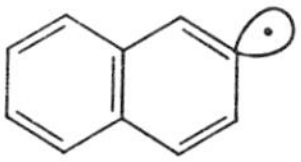

(7)

coupling constants observed are not consistent with an expected σ-radical. However, Bennett and Howard[107] showed that α-styryl is a π-radical and INDO calculations on phenylethynyl (6) indicate that this radical is also a π-radical. The 2-naphthyl radical (7) which has recently been prepared is a σ-radical, and its structure has also been studied using INDO.[108] In this case also, the calculated coupling constants were strongly dependent on the geometry chosen.

The structure of dialkylamino free radicals has been studied both experimentally and by means of INDO calculations[109] on the dimethyl-amino-radical. The optimum geometrical parameters are given in (8), and the CNC angle is less than 120°. The calculated ^{14}N coupling constant is very sensitive to the bond angle in this radical.

Among other applications of the method are the calculations on the nitroxides $H_2\dot{N}O$ and $(CH_3)_2\dot{N}O$ by Abronin *et al.*,[110] in which both isotropic and anisotropic components were calculated, a study of σ and π delocalization into the phenyl rings in methoxycarbonylcyanobenzyl radicals, using cyanobenzyl (9) as a model,[111] and studies of the geometry

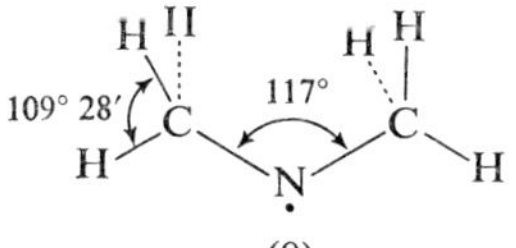

(8)

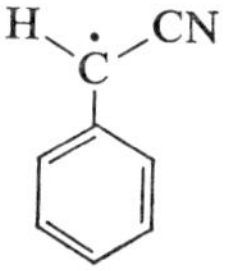

(9)

[105] L. Bonazzola, S. Fenistein, and R. Marx, *Mol. Phys.*, 1971, **22**, 689.
[106] J. S. Coleman, A. Hudson, K. D. J. Root, and D. R. M. Walton, *Chem. Phys. Letters*, 1971, **11**, 300.
[107] J. E. Bennett and J. A. Howard, *Chem. Phys. Letters*, 1971, **9**, 460.
[108] R. V. Lloyd and D. E. Wood, *J. Chem. Phys.*, 1972, **56**, 916.
[109] D. W. Pratt, J. J. Dillon, R. V. Lloyd, and D. E. Wood, *J. Phys. Chem.*, 1971, **75**, 3486.
[110] I. A. Abronin, N. A. Sysoeva, G. M. Zhidomirov, A. L. Buchachenko, and N. D. Churylkin, *Zhur. strukt. Khim.*, 1971, **12**, 985.
[111] W. J. van den Hoek, B. A. C. Rousseeuw, J. Smidt, W. G. B. Huysmans, and W. J. Mijs, *Chem. Phys. Letters*, 1972, **13**, 429.

of *O*-substituted alkyl radicals, particularly $\dot{C}H_2OH$.[112] Work on long-range couplings in bicyclic nitroxides has also appeared and given good agreement with experiment for specific ring conformations.[113]

Anisotropic hyperfine splittings have also been studied using the INDO method, usually with good agreement with experiment. A preliminary account of this work[114] was followed by a detailed report and applications to $\dot{C}H$, $\dot{C}H_3$, and the radical $\dot{C}H(CO_2H)_2$, derived from malonic acid.[115] In the INDO method, since all multicentre electron repulsion integrals involving overlap distributions $\phi_\mu^* \phi_{\nu,\mu}$ on A, ν on B are neglected, the analogous dipole–dipole interaction integrals are also assumed to be negligible. This leaves only two-centre integrals involving charge distribution $\phi_\mu^* \phi_\nu$ with ϕ_μ and ϕ_ν centred on a single atom A interacting with nucleus N, and one-centre integrals with ϕ_μ and ϕ_ν on atom A interacting with nucleus A. Two different ways of calculating these integrals were tried, and the method in which a dipole approximation to these was used seemed to be the most appropriate. Results were in good agreement with experiment: for example, for the $\dot{C}H(CO_2H)_2$ radical, Table 2 gives the calculated and observed coupling constants. The *y*-axis is parallel to the $2p\pi$ orbital and the *z*-axis directed along the C—H bond.

Table 2 *Calculated and observed hfcc for* $H\dot{C}(CO_2H)_2$ (*in* G)

		Calculated		
		Analytical evaluation	*Dipole approximation*	*Observed*
^1H	A^H	−18.2		−21.1
	T^{xx}	−12.8	−13.6	−11.4
	T^{yy}	−2.8	0.6	0.3
	T^{zz}	15.7	12.8	11.1
^{13}C	A^C	37.8		33.2
	T^{xx}	−21.4	−21.4	−25
	T^{yy}	43.6	43.6	42.8
	T^{zz}	−22.1	−22.1	−17.8

Zhidomirov *et al.*[116] have pointed out the difficulties of obtaining a quantitative fit of experimental and theoretical values of isotropic and anisotropic hfcc in $\dot{C}H_3$. They have carried out INDO calculations on $\dot{C}H_3$ and $\dot{N}H_2$ but found the problem remains. The authors considered possible reasons for this and, in particular, the problem of representing the π-orbital in terms of a single STO, and showed that if two STO's are used better agreement is obtained.

[112] R. O. C. Norman, B. C. Gilbert, and A. J. Dobbs, *J. Chem. Soc.* (*A*), 1971, 124.
[113] A. Rassat and J. Ronzaud, *J. Amer. Chem. Soc.*, 1971, **93**, 5041.
[114] D. L. Beveridge and J. W. McIver, jun., *Chem. Phys. Letters*, 1969, **4**, 126.
[115] D. L. Beveridge and J. W. McIver, jun., *J. Chem. Phys.*, 1971, **54**, 4681.
[116] G. M. Zhidomirov, N. D. Churylkin, and O. L. Brotikovskii, *Chem. Phys. Letters*, 1971, **10**, 341.

Barfield[117] has made a detailed comparison of INDO and a semi-empirical VB formulation of ^{13}C and ^{1}H anisotropic coupling constants for $\dot{C}H_3$ and $CH_3\dot{C}H_2$. Comparison with experimental data shows the necessity of including contributions from the σ-framework. However, discrepancies between this work and that of Vincow *et al.*[77] lead to errors in the ^{13}C tensor. Furthermore, off-diagonal spin-density matrix elements and inner-shell contributions were neglected, an approximation which was not justified in the light of Vincow's calculations.[77]

Morikawa *et al.*[118] have also used the INDO method to calculate the anisotropic hfcc tensor components for a variety of σ-radicals. The radicals studied included $H\dot{C}O$, $F\dot{C}O$, $\dot{N}O_2$, $CO_2^{\bar{\cdot}}$, $\dot{C}N$, $C_6H_5^{\cdot}$, and $H_2C{=}\dot{C}H$. The calculated values were in reasonable agreement with the available experimental data, particularly for nuclei one or more bonds away from the radical centre. However, there were some significant discrepancies, *e.g.* ^{19}F in $F\dot{C}O$ and ^{1}H in $H\dot{C}O$.

The temperature dependence of the hfcc in bis(trifluoromethyl) nitroxide has been investigated by INDO, with full geometry optimization.[119] The minimum energy structure is non-planar at the nitrogen with a low barrier to inversion. The increased vibrational amplitude away from planarity with increasing temperature gave hfcc changes in the right direction. Similar investigations of temperature dependence have been carried out for a variety of alkyl and substituted-alkyl radicals.[120] This work yields also the rotational barrier in these radicals.

Although the INDO method is to be preferred over other all valence electron methods for calculating hfcc, there have been some CNDO/2 calculations of the spin distribution and the optimized geometries of a variety of free radicals containing second-row atoms. A number of phosphorus-containing radicals were studied by Kilcast and Thomson,[121] together with their first-row counterparts. Problems were experienced in convergence, particularly for two second-row atoms in the radical. From the data on $\dot{P}H_2$ it was estimated that $Q^{P}_{PP} = 80$ G. Some other phosphorus σ–π spin polarization parameters were also estimated, *i.e.* $Q^{H}_{PH} = 18$ G. Further work is needed on second-row molecules, particularly more careful parametrization for charged species. The CNDO/2 method has also been used to investigate the structure of the radical (10), obtained from γ-irradiated mesitylene.[122] A variant of the CNDO method, in which spin polarization is included, has been formulated by Zhidomirov and Chuvylkin[123] and tested on a variety of radicals. The results are in good qualitative accordance with the original INDO method and the method should be more economical.

117 M. Barfield, *J. Chem. Phys.*, 1970, **53**, 3836.
118 T. Morikawa, O. Kikuchi, and K. Someno, *Theor. Chim. Acta*, 1971, **21**, 768.
119 G. R. Underwood and V. L. Vogel, *Mol. Phys.*, 1971, **19**, 621.
120 P. J. Krusic, P. Meakin, and J. P. Jesson, *J. Phys. Chem.*, 1971, **75**, 3438.
121 D. Kilcast and C. Thomson, *J. C. S. Faraday II*, 1972, **68**, 435.
122 R. Fautechi and G. A. Helké, *Mol. Phys.*, 1971, **22**, 737.
123 G. M. Zhidomirov and N. D. Chuvylkin, *Chem. Phys. Letters*, 1972, **14**, 52.

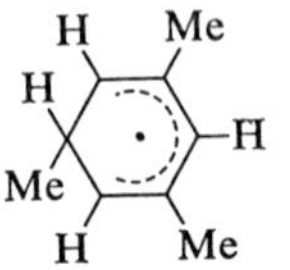
(10)

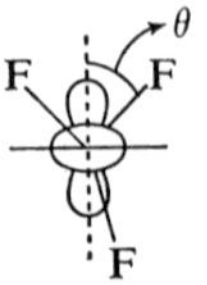
(11)

β-Proton hyperfine splittings, and the mechanism responsible for these, have been extensively studied,[124] but in the case of β-fluorine couplings the detailed mechanism has been in some doubt. It has been suggested that the angular dependence follows an equation of the form

$$a_\beta^F = (B_0 + B_x \cos^2 \theta)\, \rho_x^\pi \tag{28}$$

where ρ_x^π is the π-electron spin density on atom x adjacent to the carbon atom to which the β-fluorine is attached, and θ is the dihedral angle between the C—F bond and the π-system (11). The two major mechanisms proposed to account for the hyperfine interaction are hyperconjugation, which is a direct through-space interaction with a maximum value when $\theta = 0°$, and secondly, a 1–3 conjugation between the unpaired electron's p_z orbital on atom 1 and that lone pair on the fluorine which is perpendicular to the plane formed by 1, C_2, and F. The latter would have a maximum at $\theta = 90°$.

Two recent INDO investigations have studied β-fluorine hyperfine splittings, although there have been several earlier papers on the subject.[125–128] In the first paper, Underwood *et al.*[129] studied a series of t-butylperfluoroalkyl nitroxides. As the value of θ increases, the fluorine hfcc decreases. INDO calculations on $(CF_3)_2\dot{N}O$ in which the geometry was optimized (see above) show that for $\theta = 90°$ there is a negative spin density at the fluorine and therefore the 1–3 conjugation mechanism cannot be responsible for the spin delocalization. The calculated values of $B_0 = -6.2$ G and $B_1 = 114.9$ G were obtained from the INDO densities and the experimental coupling constants.

Knolle and Bolton, on the other hand, carried out an investigation of the ketyl anions of hexafluoroacetone, perfluorodiethyl ketone, and perfluorocyclobutanone in conjunction with INDO calculations.[130] Of note in this work is that there is virtually no temperature dependence of the splittings for some of the radicals, but there is for others. The calculations give a good correlation with experiment and predict a $\cos^2 \theta$ dependence of a_β^F, but show that in general the splittings will not be proportional to the π-electron density ρ_x^π. Full agreement demands the use of the INDO method and the primary mechanism of the spin delocalization is found to

[124] For a review, see D. H. Geske, *Progr. Phys. Org. Chem.*, 1967, **4**, 125.
[125] E. T. Strom and A. Bluhm, *Chem. Comm.*, 1966, 115.
[126] E. T. Strom, G. A. Russell, and R. D. Stephens, *J. Phys. Chem.*, 1965, **69**, 2131.
[127] P. J. Scheidler and J. R. Bolton, *J. Amer. Chem. Soc.*, 1966, **88**, 371.
[128] K. Morukuma, *J. Amer. Chem. Soc.*, 1969, **91**, 5412.
[129] G. R. Underwood, V. L. Vogel, and I. Krefting, *J. Amer. Chem. Soc.*, 1970, **92**, 5019.
[130] W. R. Knolle and J. R. Bolton, *J. Amer. Chem. Soc.*, 1971, **93**, 3337.

be overlap of the fluorine *p* orbitals with the $2p_z^{\pi}$ orbitals. The overlap leads to a $\cos^2\theta$ dependence but the proportionality constant is expected to vary from molecule to molecule. These authors considered, however, that the constant B_0 was zero. There are experimental values of $B_0 = 9$ G, $B_1 = 38$ G from single-crystal studies,[131] but these are expected to depend on the molecule and are quite different to the values for the nitroxide cited above. Different values were also found for the ketyl radicals.

Q **Factors and π-Electron Spin Density Calculations.**—Despite the availability of the INDO method, there are still many correlations of hyperfine splitting constants with calculated π-electron spin densities, by use of the McConnell [40] or Karplus–Fraenkel [47] equations and semi-empirical *Q* factors. Explicit calculations of the latter have been possible for $Q_{\mathrm{CH}}^{\mathrm{H}}$ (see Section 3) and there is still a need for approximate calculations which account for the main factors determining the *Q* parameters. The experimental data in conjunction with semi-empirical values of the *Q* factors have been used to obtain experimental spin densities which can be compared with purely theoretical values.[41]

The McConnell–Strathdee treatment of anisotropic hyperfine interactions has been mentioned above. Some previous calculations on perinaphthenyl [132] using this theory were not in good agreement with experiment. Further calculations using this model have been carried out for $\dot{\mathrm{C}}\mathrm{H}$, allyl, and perinaphthenyl but using STO's with different exponents and taking σ-spin density into account.[133] However, the values of $\langle r^{-3}\rangle$ which govern the hyperfine tensor were then too small, although the σ-contributions were shown to be substantial. It is clear from this work that calculations of these quantities are not easy, a conclusion in accord with the *ab initio* calculations.

In semi-empirical calculations the G^{N} matrix of equation (20) is approximated in various ways, leading to, for example, the McConnell equation and the GNP relation. Waselewski *et al.*[134] have further examined the consequences of non-orthogonality of the basis functions in such calculations, and conclude that proper inclusion of this is necessary to obtain reliable description of the e.s.r. spectra.

There have been other investigations of the $Q_{\mathrm{CH}}^{\mathrm{H}}$ parameter,[135] and of the spin polarization contribution to the spin distribution in a series of aliphatic $\dot{\mathrm{C}}\mathrm{H}_2(\mathrm{CH}_2)_n\mathrm{CH}_3$ chains.[136] Experimental investigations have led to a value of $Q_{\mathrm{OH}}^{\mathrm{H}}$ for systems containing free and intramolecular-bonded OH groups.[137] The radical fragment $\mathrm{C{-}\dot{C}H{-}NH_2}$, which occurs in many

131 M. Iwasaki, *Mol. Phys.*, 1971, **20**, 503.
132 H. R. Falle and G. R. Luckhurst, *Mol. Phys.*, 1966, **11**, 299.
133 H. R. Falle and M. A. Whitehead, *Canad. J. Chem.*, 1972, **50**, 139.
134 J. Wasilewski and W. Wožnicki, *Theor. Chim. Acta*, 1972, **23**, 360.
135 I. Mayer, *Magyar Kém. Folyóirat*, 1970, **76**, 607.
136 I. A. Abronin and G. M. Zhidomirov, *Zhur. strukt. Khim.*, 1970, **11**, 1119.
137 A. E. Lutskii and I. S. Romodanov, *Spktrosk. At. Mol.*, 1969, 366.

amino-acids, has been considered by Poupko and Silver.[138] The method of Melchior[52] was used, in which the set of interatomic σ-bonds was orthonormalized. The Slater orbital exponents were taken from atomic

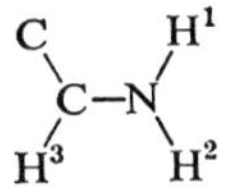

(12)

SCF calculations. If we label the fragment as in formula (12) the calculations yield the following values:

$$Q^{H^1}_{NN} = -24.3\text{ G},\quad Q^{H^2}_{NN} = -24.5\text{ G},\quad Q^{H^1}_{\text{cross (CN)}} = -3.2\text{ G},$$
$$Q^{H^3}_{CC} = -26.1\text{ G},\quad Q^{H^3}_{NN} = -0.4\text{ G},\quad Q^{N}_{NN} = 19.4\text{ G},$$
$$Q^{N}_{\text{cross (CN)}} = 1.9\text{ G},\quad Q^{N}_{CC} = -5.0\text{ G},\quad Q^{C}_{CC} = 39.5\text{ G},$$
$$Q^{C}_{\text{cross (CN)}} = 3.0\text{ G},\quad \text{and}\quad Q^{C}_{NN} = -11.5\text{ G}.$$

The main error was believed to be due to the difficulty of estimating the σ(C—H) $\rightarrow$ σ^*(C—H) and σ(C—H) $\rightarrow$ σ^*(C—H′) excitation energies for the singlet–triplet excitation. Calculations on CH_3 and NH_3^+ were used to estimate these quantities. Calculations for the radical-cation $N_2H_4^{\overset{+}{\cdot}}$ were also carried out, but the results seem to indicate that this ion is non-planar. Applications of these Q values in the analysis of the e.s.r. spectra from other amino-acids should be useful.

Calculations of the various nitrogen Q factors in the Karplus–Fraenkel equation require the use of the relation $a^{H} = Q^{H}_{CH}\,\rho^{\pi}_{C}$ and the value of this equation has been investigated for the symmetrical diazanaphthalenes.[139] Spin densities were calculated by either the Hückel or PPP–SCF methods and it was concluded that the procedures to calculate ρ_N by using the McConnell equation and the normalization condition on the spin densities may lead to serious errors.

There is still some controversy over the use of a simple McConnell-type equation for ^{14}N splittings in aromatics. Zeiss and Whitehead[140] have carried out an investigation of 13 anions and cations of heteroatomic conjugated molecules using π-electron densities calculated by a variable β ZDO–PPP method, using RHF–CI or UHF-type wavefunctions. The former method gives better agreement and a McConnell-type equation for proton splittings is only slightly inferior to a GNP-type relation. The Karplus–Fraenkel equation was rather superior to a simple McConnell

138 R. Poupko and B. L. Silver, *J. Amer. Chem. Soc.*, 1971, **93**, 575.
139 D. M. W. Van den Han, J. J. den Sart, and D. Van der Meer, *Mol. Phys.*, 1971, **21**, 989.
140 G. D. Zeiss and M. A. Whitehead, *J. C. S. Faraday II*, 1972, **68**, 526.

equation for ^{14}N splittings. Other investigations of the parameters for hydrazine-type cation-radicals have also appeared.[141, 142]

As mentioned above, for β-hyperfine splittings equation (28) is expected to apply. There have been a number of investigations of this question apart from the INDO results referred to above. A critical evaluation of the relation

$$a_{\mathrm{H}}^{\beta} = B\rho_{\mathrm{C}} \cos^2 \theta \qquad (29)$$

has been made in the light of the data for hydrocarbon radicals and an approximate treatment of hyperconjugation.[143] It is found that if the variation of B with the energy of the occupied MO is taken into account, good correlations are obtained. Explicit equations are given for B as a function of the geometrical parameters; this treatment neglects B_0.[143] However, an estimate of the latter in the isopropyl radical (in fact the ratio $B_0/B_x = +0.0985$) shows that B_0 is not negligible.[144] Further evidence that this is the case comes from an analysis of cyclopropyl β-hyperfine structure in radicals for which the cyclopropyl group is bonded to a digonal carbon.[145] In this case it was also shown that B_2 for cyclopropyl is greater than B_2 for CH_3 groups.

β-Hyperfine interactions with many other nuclei have been studied experimentally[146] and the results used to obtain information on barriers to rotation. There were several unusual observations, *e.g.* β-alkylmercapto-ethyls have very low β coupling constants, which was interpreted in terms of conformational effects. Few of the numerous studies, however, were concerned with the theory of β-splittings.

There is some controversy regarding the mechanism of such interactions. Kawamura and Kochi[147] conclude from a study of alkyl radicals with β-Si, Ge, or Sn substituents that *p–d* homoconjugation is responsible, whereas Lyons and Symons[148] believe that this mechanism is ruled out in favour of hyperconjugation. Damerau[149] has presented evidence for the angular dependence of β-phosphorus hyperfine splittings in some phosphorus radicals derived from phosphonic acid esters. Similar conclusions were reached in a study of some nitroaromatic phosphorus anions.[150] Phosphorus σ–π spin polarization parameters have been estimated by Thomson and Kilcast.[151]

β-Chlorine hyperfine splittings in chloroalkyl radicals have shown that these radicals are frozen in unique conformations. β-Chloroethyl is

141 S. F. Nelson and P. J. Hintz, *J. Amer. Chem. Soc.*, 1971, **93**, 7105.
142 P. Smith, R. D. Stevens, and R. A. Kaba, *J. Phys. Chem.*, 1971, **75**, 2048.
143 M. Brustolon, C. Corvaja, and G. Giacometti, *Theor. Chim. Acta*, 1971, **22**, 90.
144 N. L. Bauld, C. E. Hudson, and J. S. Hyde, *J. Chem. Phys.*, 1971, **54**, 1834.
145 C. E. Hudson and N. L. Bauld, *J. Amer. Chem. Soc.*, 1972, **94**, 1158.
146 P. J. Krusic and J. K. Kochi, *J. Amer. Chem. Soc.*, 1971, **93**, 846.
147 T. Kawamura and J. K. Kochi, *J. Amer. Chem. Soc.*, 1972, **94**, 648.
148 A. R. Lyons and M. C. R. Symons, *Chem. Comm.*, 1971, 1068.
149 W. Damerau, G. Lassman, and K. Lohns, *J. Magn. Resonance*, 1971, **5**, 408.
150 W. M. Kulick, jun., *J. Amer. Chem. Soc.*, 1972, **94**, 29.
151 C. Thomson and D. Kilcast, *Chem. Comm.*, 1971, 214.

particularly interesting since it has a *g*-factor of 2.00214, which is less than the free-spin value; *p–p* homoconjugation was put forward as an explanation, and calculations substantiate this.[152]

Further study of the *Q* parameters of C—F bonds has come from single-crystal e.s.r. spectra of 1- and 2-fluoronaphthalene[153] and various fluorinated biphenyls[154] excited into the lowest triplet state. These papers contain a detailed discussion of the contributions to the fluorine hyperfine tensor, and two recent reviews by Iwasaki[155] and by Hudson and Root[156] have surveyed the available work on fluorine hyperfine splittings.

Most approaches to *Q* parameter calculations have used either VB or MO methods. The group function[157–159] approach allows the calculation of the wavefunctions for the π-system and a methyl group, before coupling occurs, in the best possible way. A discussion of CH_3 couplings by use of this approach was given by Van der Hart and compared with more conventional hyperconjugation and pseudo-heteroatom approaches.[160]

Turning now to π-electron calculations of spin densities, there are still many calculations using well-established methods and some with rather newer methods. Only a few of the many calculations will be mentioned here. Trinajstic[161] has used the half-electron SCF–MO method of Dewar[162, 163] (with variable β) to calculate the hyperfine coupling constants in benzyl. The results are in good agreement with experiment and rather better than the INDO calculations. The tropone radical-anion[164] and various other anions were also considered,[165] with results in excellent agreement with experiment.

The McLachlan[166] method for π-electron spin density calculations has proved successful in the past, and various extensions to it have been considered recently. The original method needs modification in its application to radical-cations, as shown by Honeybourne[167, 168] who has derived new expressions and parameter values. Bickerton and Moss[169] have also

[152] T. Kawamura, D. J. Edge, and J. K. Kochi, *J. Amer. Chem. Soc.*, 1972, **94**, 1752.
[153] J. Mispelter, J.-Ph. Grivet, and J.-M. Lhoste, *Mol. Phys.*, 1971, **21**, 999.
[154] J. Mispelter, J.-Ph. Grivet, and J.-M. Lhoste, *Mol. Phys.*, 1971, **21**, 1015.
[155] M. Iwasaki, *Fluorine Chem. Rev.*, 1971, **5**, 1.
[156] A. Hudson and K. D. J. Root, *Adv. Magn. Resonance*, 1971, **5**, 1.
[157] R. G. Parr, 'Quantum Theory of Molecular Electronic Structure', W. A. Benjamin, New York, 1963.
[158] R. McWeeny, *Rev. Mod. Phys.*, 1960, **32**, 335.
[159] W. J. Van der Hart, *Mol. Phys.*, 1971, **20**, 385; 1971, **20**, 399.
[160] W. J. Van der Hart, *Mol. Phys.*, 1971, **20**, 407.
[161] N. Trinajstic, *Chem. Phys. Letters*, 1971, **10**, 172.
[162] M. J. S. Dewar and G. L. Gleicher, *J. Amer. Chem. Soc.*, 1965, **87**, 685.
[163] M. J. S. Dewar, 'The Molecular Orbital Theory of Organic Chemistry', McGraw-Hill, New York, 1969.
[164] N. Trinajstic, *Bull. Chem. Soc. Japan*, 1971, **44**, 3208.
[165] N. Trinajstic, *Spectroscopy Letters*, 1971, **4**, 85.
[166] A. D. McLachlan, *Mol. Phys.*, 1960, **3**, 333.
[167] C. L. Honeybourne, *Tetrahedron Letters*, 1971, 4927.
[168] C. L. Honeybourne, *Mol. Phys.*, 1971, **21**, 1057.
[169] J. C. Bickerton and R. E. Moss, *Mol. Phys.*, 1971, **21**, 341.

discussed this question, and have shown in fact that the spin densities calculated by the McLachlan method should be identical in corresponding anions and cations, in agreement with the pairing theorem. Applications of the modified method to conjugated ligands bonded to paramagnetic centres have been described.[170] Another modification has also been described and applied to benzo[*cd*]pyrenyl.[171, 172] Extended Hückel (EHT) calculations cannot give the negative spin densities which are frequently observed, but EHT orbitals used in a McLachlan formulation give reasonable negative spin densities in vinyl and in hydrocarbon systems.[173] EHT spin densities in alkoxyl and alkylperoxyl free radicals have been reported.[174]

The UHF method using P–P–P approximations, and annihilation of the quartet, continues to be used, and calculations have been reported on the anions of styrene, 4-vinylpyridine, and *p*-fluorostyrene.[175] There is good agreement with the experimental values. Various other radical ions derived from methylpyrene,[176] α-methyl- and *p*-fluoromethyl-styrene,[177] and various nitroaromatic compounds have all been studied.[178] Conclusions regarding the ^{14}N σ–π parameters are similar to those cited above. Various hydrazine anion- and cation-radicals have been studied by Russian workers using the UHF method.[179, 180] In an earlier paper, Tino [181] examined the use of ten different methods for π-spin density calculations. There seems to be relatively little to choose between the various methods in this survey, which examined the McConnell equation and also the Colpa–Bolton equation.

In a similar series of calculations on some diazine and naphthyridine anions, Cavalieri D'Oro *et al.*[182] computed spin densities by various procedures. McLachlan spin densities give better agreement than RHF + CI or UHF, and Hückel spin densities were also reliable in these cases. McLachlan-type calculations have also been useful in interpreting the spin distribution in phenyl-substituted imidazyl and pyrryl radicals.[183]

McLachlan calculations of the biacetyl semidione radical using an adjustable β_{CO}(hydroxyl) to account for orientation changes and hence

170 C. L. Honeybourne, *Inorg. Nuclear. Chem. Letters*, 1971, **7**, 1121.
171 J. Nowakowski, *Chem. Phys. Letters*, 1972, **13**, 589.
172 J. Nowakowski, *Theor. Chim. Acta*, 1970, **18**, 133.
173 R. J. Fitzgerald and D. H. Dugre, *Theor. Chim. Acta*, 1971, **22**, 361.
174 O. Yamamoto, H. Kato, and T. Yonezawa, *Nippon Kagaku Zasshi*, 1970, **91**, 907.
175 N. K. Ray and K. K. Sharma, *Theor. Chim. Acta*, 1971, **22**, 403.
176 N. K. Ray and K. K. Sharma, *Indian J. Chem.*, 1971, **9**, 1163.
177 N. K. Ray and K. K. Sharma, *Indian J. Chem.*, 1971, **9**, 1004.
178 D. N. Nanda, J. Subramanion, and P. T. Narasimhan, *Theor. Chim. Acta*, 1971, **22**, 369.
179 V. Gubanov, A. K. Chirkov, and R. O. Matevosyan, *Zhur. strukt. Khim.*, 1971, **12**, 538.
180 V. Gubanov, V. I. Koryakov, A. K. Chirkov, and R. O. Matevosyan, *Zhur. strukt. Khim.*, 1971, **11**, 1006.
181 J. Tino, *Coll. Czech. Chem. Comm.*, 1971, **36**, 3382.
182 P. Cavalieri D'Oro, G. Macagnoni, G. F. Pedulli, and P. Palmieri, *Mol. Phys.*, 1971, **20**, 365.
183 R. D. Allendoerfer and A. S. Pollock, *Mol. Phys.*, 1971, **22**, 661.

temperature dependence of the splittings give a reasonable account of their variation in these systems.[184]

In other recent papers the spin redistribution during inversion of the energy levels in monosubstituted benzene anions has been considered,[185] and a comparison made of open-shell SCF calculations on a variety of molecules[186] using either the Roothaan[187] or Longuet-Higgins–Pople procedures.[188] The latter, when supplemented by configuration interaction, gave the most accurate spin densities.

5 Miscellaneous Theoretical Calculations

This section deals with a few of the many e.s.r. papers in which theoretical calculations have been used to help in the interpretation of e.s.r. observation.

Preparation of the radical-cation of *p*-quaterphenyl[189] has been reported, and McLachlan-type spin density calculations give good agreement with experiment (for which some splitting constants in the cation are less than those at the corresponding position in the anion) only if the *para*-bonded rings are not in the same plane. A series of cycloalkylbenzene anions has been studied by Gerson *et al.*[190] and compared with the alkylbenzene anions. The preferred conformation has the alkyl or cycloalkyl β-proton near to the nodal plane of the π-system.

A detailed study of the sign and magnitude of the metal spin density in alkali-metal radical ion pairs has appeared for the specific case of the $Na^+C_{10}H_8^{\overline{\cdot}}$ pair. The anion is considered as a planar π-system and the interaction treated by first-order perturbation theory.[191] Zeroth- and first-order contributions to the metal spin density are comparable, but have different signs. The sign of the total spin density on the Na nucleus varies with the position of the Na in the ion pair. There were some difficulties associated with calculating the zeroth-order spin density. The metal hfcc has a tendency to become negative with increasing atomic number and this was related to the polarizing power of the ion.

An interesting observation on hyperconjugation arises from the results for the cation of (13) in which the dihedral angle θ between the C—H bonds and the sulphur $3p_z$ orbitals is readily determined.[192] The proton coupling constant is very small, and no simple correlation with θ is found, indicating that the coupling constants of alkyl protons bonded to a π^+ system by bivalent sulphur do not arise predominantly from a hyper-

[184] P. B. Ayscough and R. E. D. McLung, *Mol. Phys.*, 1971, **20**, 35, 47.
[185] V. Lyubinov, S. P. Ionov, and G. V. Ionov, *Zhur. fiz. Khim.*, 1970, **44**, 2433.
[186] P. Carsky and R. Zahradnik, *Coll. Czech. Chem. Comm.*, 1971, **36**, 961.
[187] C. C. J. Roothaan, *Rev. Mod. Phys.*, 1960, **32**, 179.
[188] H. C. Longuet-Higgins and J. A. Pople, *Proc. Phys. Soc.*, 1955, **A68**, 591.
[189] M. K. Carter, *J. Phys. Chem.*, 1971, **75**, 902.
[190] F. Gerson, G. Moshuk, and M. Schiryzer, *Helv. Chim. Acta*, 1971, **54**, 361.
[191] G. W. Cauterns, C. Corvaja, and E. de Boer, *J. Chem. Phys.*, 1971, **54**, 3026.
[192] J. Q. Chambers, N. D. Caufield, D. R. Williams, and D. L. Coffen, *Mol. Phys.*, 1970, **19**, 581.

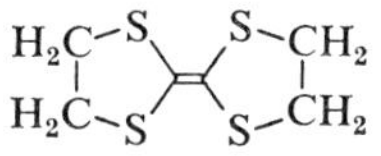

(13)

conjugative interaction, but are consistent with a spin polarization mechanism.

Hyperfine interactions in dimer cations give information concerning molecular interactions, and most studies have assumed a symmetrical sandwich structure for the dimer. Observations of the monomer and dimer cations of 1,4-, 1,5-, and 1,8-dimethylnaphthalenes have been reported, and it is observed that, unlike in most other investigations, the hfcc in the dimer differed considerably from one-half of that in the monomer.[193] Calculations of the structure of the dimer using Hückel theory, and including resonance integrals between upper and lower molecules for different sandwich arrangements, were carried out, and for the 1,4-molecule SCF–MO calculations were also carried out. The calculated spin distribution was consistent with the two possible structures put forward.

Various radicals derived from simple amines and amino-acids have been prepared[194] and the experimental results interpreted with the aid of the Q parameters calculated previously.[138] The NH_2 group turns out to be non-coplanar with the nodal plane of the π-system. A problem which could not be explained was the temperature dependence of the amino-group proton splittings.

Aliphatic nitrosamines have been reduced to the anion-radicals (14),[195] and it is found that the odd electron is predominantly on the nitroso-nitrogen,

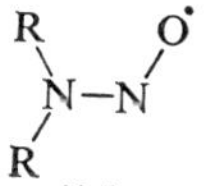

(14)

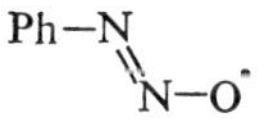

(15)

a conclusion in agreement with INDO calculations. On the other hand the radical (15) is a σ-radical.[196] Several theoretical studies on various nitroxides have been reported.[197–199]

193 H. Yoshimi and K. Kuwata, *Mol. Phys.*, 1972, **23**, 297.
194 R. Poupko, A. Loewenstein, and B. L. Silver, *J. Amer. Chem. Soc.*, 1971, **93**, 580.
195 G. R. Stevenson and C. J. Colón, *J. Phys. Chem.*, 1971, **75**, 2704.
196 R. M. Paton, J. I. G. Cadogan, and C. Thomson, *J. Chem. Soc.* (*B*), 1971, 583.
197 R. Briere, G. Chapelet-Letourneux, H. Lemaire, and A. Rassat, *Mol. Phys.*, 1971, **20**, 211.
198 C. Morat and A. Rassat, *Bull. Soc. chim. France*, 1971, 893.
199 Y. Ellinger, A. Rassat and R. Subra, *J. Chim. phys.*, 1971, **68**, 730.

2
ENDOR and ELDOR

BY N. M. ATHERTON

1 Introduction

ENDOR and ELDOR are branches of double-resonance spectroscopy. In a double-resonance experiment one monitors a particular transition of interest while simultaneously exciting another transition: it is convenient to label them as the *observed* and the *pumped* transitions. The pumped transition need not, in general, have a level in common with that being observed.

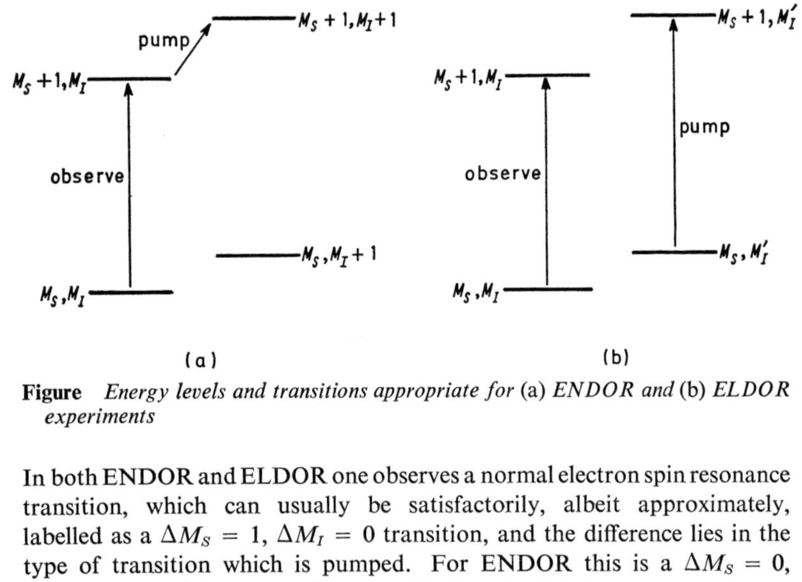

Figure *Energy levels and transitions appropriate for* (a) *ENDOR and* (b) *ELDOR experiments*

In both ENDOR and ELDOR one observes a normal electron spin resonance transition, which can usually be satisfactorily, albeit approximately, labelled as a $\Delta M_S = 1$, $\Delta M_I = 0$ transition, and the difference lies in the type of transition which is pumped. For ENDOR this is a $\Delta M_S = 0$, $\Delta M_I = 1$ transition, hence Electron Nuclear Double Resonance, whereas in ELDOR it is a $\Delta M_S = 1$, $\Delta M_I' = 0$ transition, where the prime indicates a different M_I from that labelling the observed transition, and so the acronym derives from Electron–Electron Double Resonance. The Figure shows energy-level schemes appropriate for ENDOR and ELDOR experiments.

The terms of reference of this Report are to cover the literature for 1971 and the first part of 1972, but in fact rather more references are included in order to achieve a balanced picture and to make the Report more or less self-contained. ENDOR is a relatively old technique but has been little used by chemists and is probably unfamiliar to them so the review of recent work is set against a sketch of the historical development. ELDOR is treated differently. It is the youngest member of the family of magnetic double-resonance experiments and has a very small literature: comprehensive coverage of this has been attempted.

2 ENDOR

Introduction.—The first ENDOR experiments were carried out by Feher in 1956.[1] These experiments were concerned with donors in silicon, and a very high proportion of subsequent ENDOR work has been in the field of solid-state physics. Although work by Cederquist[2] has often been quoted, it was the work of Hyde and Maki[3] in 1964 which led to the first published report of the observation of ENDOR from a radical in solution. The shorter relaxation times in solution mean that high r.f. power is required for such experiments and the associated technical difficulties are far from trivial. Hyde and Maki's breakthrough has hardly led to an explosion of interest in ENDOR among chemists but there have been several elegant applications to problems of chemical interest and one can anticipate that there will be a steady increase of these.

Although a number of earlier textbooks had given very brief introductions to ENDOR,[4–7] it was not until 1970, with the appearance of Abragam and Bleaney's massive treatise,[8] that a really thorough unified treatment became available. Despite the limitation that the discussion is confined to ENDOR in solids, this treatment should be compulsory reading for any beginner seriously interested in the technique. Of essentially equal value, though in a slightly different sense, is Kwiram's review article, published in 1971.[9] This gives comprehensive coverage of the literature through 1970 while at the same time providing an excellent introduction to the subject. An earlier article by Hyde,[10] although not strictly a review, also gives a good, brief survey of the subject and is valuable introductory reading.

ENDOR in Solids.—Some of the most striking illustrations of the power of ENDOR for achieving high resolution have come from the studies of

1 G. Feher, *Phys. Rev.*, 1956, **103**, 834.
2 A. L. Cederquist, Ph.D. Thesis, Washington University, 1963.
3 J. S. Hyde and A. H. Maki, *J. Chem. Phys.*, 1964, **40**, 3117.
4 *E.g.* D. J. E. Ingram, 'Free Radicals as Studied by e.s.r.', Butterworth, London, 1958.
5 *E.g.* G. E. Pake, 'Paramagnetic Resonance', Benjamin, New York, 1962.
6 *E.g.* A. Carrington and A. D. McLachlan, 'Introduction to Magnetic Resonance', Harper and Row, New York, 1967.
7 *E.g.* P. B. Ayscough, 'Electron Spin Resonance in Chemistry', Methuen, London, 1967.
8 A. Abragam and B. Bleaney, 'Electron Paramagnetic Resonance of Transition Ions', Oxford University Press, 1970.
9 A. L. Kwiram, *Ann. Rev. Phys. Chem.*, 1971, **22**, 133.
10 J. S. Hyde, in 'Magnetic Resonance in Biological Systems', ed. A. Ehrenberg, B. Malmström, and T. Vänngard, Pergamon Press, Oxford, 1967, p. 63.

impurity and defect centres, and in spite of the volume of earlier work these systems have continued to receive attention. The data for the shallow donors As, P, and Sb in silicon have been re-analysed by Castner,[11] who finds that the agreement between theory and experiment can be improved by admitting configurational admixture of components of the conduction band into the ground-state wavefunction. F-centres in NaCl, KBr, KCl, NaF, and LiF have been studied by Grachev *et al.*[12] Isotropic hyperfine coupling constants at 20 K and 77 K are reported and the data analysed to give information about the band structures. Studies of centres which do not yet appear to have been fully recorded in the literature are those of O_2^- in KCl, KBr, and KI,[13] of the F-centre in MgO,[14] of an F-centre at high pressure,[15] and of electric field effects in the ENDOR of F-centres.[16]

The second traditional area of application of ENDOR has been the study of metal ions. A particularly striking example of the power of the ENDOR method is the work of Reddy *et al.* on Yb^{3+} in CaF_2.[17] ENDOR of ^{19}F, ^{1}H, and ^{17}O nuclei was observed and it was possible to identify three centres T_1, T_2, and T_3 as $YbFO_4$, YbF_7O, and $YbHO_4$, occupying CaF_8 sites. The details of the structures of each centre were determined: in each case the Yb^{3+} ion has complete second, third, and fourth co-ordination shells but an incomplete first shell. An actinide which has been studied in CaF_2 is Pu^{3+}.[18] The fluorine ligand hyperfine splitting parameters were measured and the signs of the isotropic and anisotropic components measured in a triple-resonance experiment using two radiofrequencies. The analysis of the hyperfine couplings shows that Pu^{3+} is more covalently bonded than other actinides in this lattice. The 'self-ENDOR', *i.e.* the Pu^{3+} ENDOR, was also studied and the apparent nuclear *g*-factor measured and compared with the free-atom value, which was available from beam experiments. The deviation, which arises from a second-order cross-term between the hyperfine interaction and the electron Zeeman energy, was consistent with an earlier estimate of the crystal field splitting.

Ligand hyperfine interactions have been studied for several transition-metal ions. For Ti^{3+} in SrF_2 the dipolar hyperfine coupling to the fluorines in the second, third, and fourth co-ordination shells is in good agreement with that predicted using the point-dipole approximation.[19] Ti^{3+} and V^{3+} have also been studied in CdF_2, CaF_2, and SrF_2 using triple resonance

[11] T. G. Castner, *Phys. Rev.* (*B*), 1970, **4**, 911.
[12] V. G. Grachev, M. F. Deigen, G. I. Neimark, and S. I. Pekar, *Phys. Status Solidi* (*B*), 1971, **43**, K93.
[13] A. M. Doyle, *Diss. Abs.* (*B*), 1970, **31**, 2216.
[14] L. E. Halliburton, *Diss. Abs.* (*B*), 1971, **32**, 1789.
[15] A. B. Wolbarst, *Diss. Abs.* (*B*), 1971, **32**, 2361.
[16] Z. Usmani, *Diss. Abs.* (*B*), 1971, **32**, 4287.
[17] T. Rs. Reddy, E. R. Davies, J. M. Baker, D. N. Chambers, R. C. Newman, and B. Ozbay, *Phys. Letters* (*A*), 1971, **36**, 231.
[18] W. Kolbe and N. Edelstein, *Phys. Rev.* (*B*), 1971, **4**, 2869.
[19] Yu. F. Mitrofanov, Yu. E. Pol'skii, and M. L. Falin, *Fiz. tverd. Tela*, 1971, **13**, 1830.

(electron–nucleus–nucleus).[20] These experiments show that there are strong interactions between the spins of nuclei in the same co-ordination sphere. The hyperfine interaction with ^{17}O for Mn^{2+} in ^{17}O-enriched MgO has been studied by Freund *et al.*[21] and the coupling expressed in terms of the usual bonding parameters. An interesting point is that no significant quadrupole interaction could be detected in the case of Mn^{2+} though it is indicated that this is not so for Cr^{3+} or Fe^{3+}. Saxena and Vincent [22] have also studied Mn^{2+} in MgO. Ziegler [23] has made a very detailed study of Cr^{+} in NaF, NaCl, and NaBr. The ligand magnetic hyperfine and, where appropriate, quadrupole interactions are discussed in terms of two models for the bonding. A covalency model is somewhat more satisfactory than an overlap model but still does not give a satisfactory account of the small anisotropy of the interaction with the nearest-neighbour ligands.

Over the past twelve years or so there has been an enormous amount of work done on trapped radicals in single crystals, and it is surprising that ENDOR has been used so little on these systems. For the period covered by Kwiram [9] the publication rate averages about two papers per annum and that average is maintained. Kwiram himself has made a detailed study of glutaric acid.[24] Of the two radicals formed on irradiation, one was removed by u.v. irradiation so that the remaining $HOOC\cdot\dot{C}H\cdot CH_2\cdot CH_2\cdot COOH$ could be studied without interference. The anisotropy of the β-proton couplings can be studied much more precisely using ENDOR than if one is restricted to ordinary e.s.r. and so the relative orientations of the α- and β-protons can be determined with some confidence from the tensor axes. Box has also continued work in this area, the latest reported studies being on the radical-pairs in irradiated potassium hydrogen malonate.[25] The hyperfine couplings were found to be one-half the values for isolated $^{-}OOC\cdot\dot{C}H\cdot COO^{-}$ radicals, as expected for strong exchange. A study of the radical (1), formed in γ-irradiated 1,2,4-triazole, has been reported by Gloux.[26] The spin Hamiltonian is discussed in some detail in this paper and particular attention is paid to a cross-term between two hyperfine tensors which appears when the energy is calculated to second order

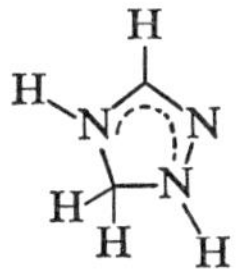

(1)

[20] Yu. F. Mitrofanov, Yu. E. Pol'skii, and M. L. Falin, *Zhur. eksp. teor. Fiz.*, 1971, **61**, 1486.
[21] P. Freund, B. F. Hann, and J. Owen, *J. Phys.* (*C*), 1971, **4**, L296.
[22] G. P. Saxena and C. Vincent, *Canad. J. Phys.*, 1971, **49**, 2276.
[23] H. Ziegler, *Phys. Status Solidi*, 1972, **49**, 367.
[24] A. L. Kwiram, *J. Chem. Phys.*, 1971, **55**, 2484.
[25] H. C. Box, *J. Phys. Chem.*, 1971, **75**, 3426.
[26] P. Gloux, *Mol. Phys.*, 1971, **21**, 829.

retaining the *x* and *y* components of the spin operators. This indirect internuclear coupling accounts for a non-crossing phenomenon which is observed in the angular dependence of the ENDOR frequencies.

Another area where ENDOR remains comparatively little used is the study of triplet states. The value of the technique is impressively illustrated by Hutchison's work and a valuable survey of it was provided by Hutchison himself some years ago.[27] The latest system studied by the Chicago group is anthracene in phenazine.[28] The anisotropic proton couplings are used to derive spin populations and these are compared with the results of calculations and with values in the radical-cation and radical-anion, which should be the same as those in the triplet if the pairing theorem holds. It is concluded that the assumption of a constant Q^{H}_{CH} for all positions of a triplet hydrocarbon in a solid is not accurate.

All the work described so far has been on samples oriented in single crystals. Clearly there would be a considerable saving in labour if the ENDOR information could be obtained from randomly oriented samples and one suspects that chemists would be more anxious to use ENDOR if this were the case. In fact it is, and several papers have described how useful information has been obtained from such studies. The possibilities were pointed out by Hyde some time ago.[10] In the ENDOR experiment one saturates only part of the spectrum, and for a polycrystalline sample this will correspond to molecules in a very limited range of orientations. Only these molecules contribute to the ENDOR spectrum, so one expects that this should remain relatively sharp. Studies by Hyde and his collaborators, and by Kwiram, illustrated the point for copper complexes [29] and for α- and β-protons in π-radicals,[30] while Ehrenberg, Eriksson, and Hyde have also recorded considerable successes with studies of flavoproteins.[31–33] The notable application of powder ENDOR in the period under review here has been to ferredoxin.[34] The results showed that the $S = \frac{1}{2}$ system comprises two inequivalent iron ions at the active site, and hyperfine tensors for both sites were determined (^{57}Fe-labelled ferredoxin was used). The results support the view that the pairs of iron ions comprise one each of high-spin Fe^{3+} and Fe^{2+} antiferromagnetically coupled to give a total spin of one-half. There have been one or two powder ENDOR studies of non-biological systems. Iwasaki *et al.* have reported

[27] C. A. Hutchison, in 'The Triplet State', ed. A. B. Zahlan, Cambridge University Press, 1967, p. 63.
[28] R. H. Clarke and C. A. Hutchison, *J. Chem. Phys.*, 1971, **54**, 2962.
[29] G. H. Rist and J. S. Hyde, *J. Chem. Phys.*, 1968, **49**, 2449; 1970, **52**, 4633.
[30] J. S. Hyde, G. H. Rist, and L. E. G. Eriksson, *J. Phys. Chem.*, 1968, **72**, 4269; A. L. Kwiram, *J. Chem. Phys.*, 1968, **49**, 2860.
[31] A. Ehrenberg, L. E. G. Eriksson, and J. S. Hyde, *Biochim. Biophys. Acta*, 1968, **167**, 482.
[32] L. E. G. Eriksson, J. S. Hyde, and A. Ehrenberg, *Biochim. Biophys. Acta*, 1969, **192**, 211.
[33] L. E. G. Eriksson, A. Ehrenberg, and J. S. Hyde, *European J. Biochem.*, 1970, **17**, 539.
[34] J. Fritz, R. Anderson, J. Fee, G. Palmer, R. H. Sands, J. C. M. Tsibris, I. C. Gunsalus, W. H. Orme-Johnson, and H. Beinert, *Biochim. Biophys. Acta*, 1971, **253**, 110.

briefly the spectrum of trapped electrons in irradiated 2-methyltetrahydrofuran.[35] A detailed understanding of the powder ENDOR lineshape is desirable if the maximum amount of information is to be obtained from the spectra. Leniart, Hyde, and Vedrine[36] have discussed the problem of the powder lineshape, in particular the way in which motion of electrons can induce nuclear relaxation by modulating the dipolar interaction. The same authors have also studied proton powder ENDOR in γ-irradiated zeolites:[37] the spectra indicate that hydrogen atoms are trapped in the centres of the sodalite cages, hyperfine interaction with four different types of lattice proton being observed.

ENDOR in Solution.—The ENDOR spectra of radicals in solution tend to be very solvent- and temperature-dependent. It is generally easier to obtain spectra of neutral radicals than of radical-ions, at least partly because one can use non-polar solvents and accordingly much bigger sample volumes. A number of different types of radical have been studied during the period under review.

Allendoerfer and Pollock's work[38] on phenyl-substituted imidazyl (2) and pyrryl (3) radicals exemplifies work in this field. Values of the proton

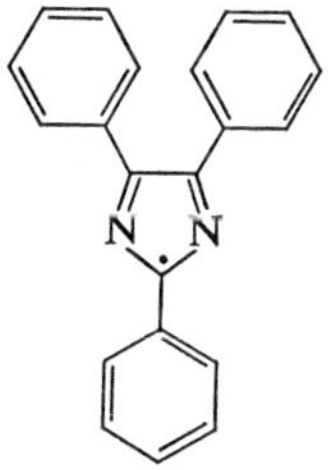

(2)

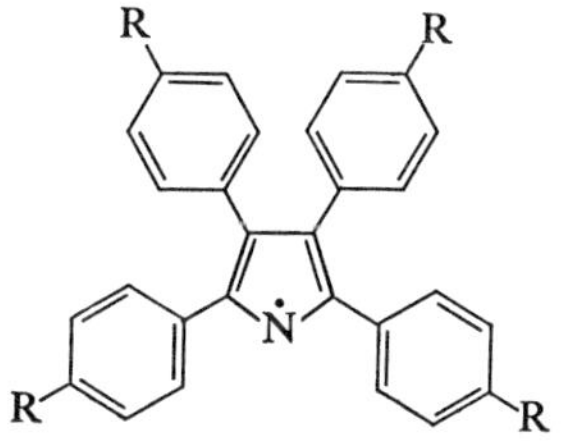

(3)

coupling constants were readily obtained from the ENDOR spectra, and the values confirmed by computer simulation of the e.s.r. spectra, which also gave the nitrogen couplings. The patterns of spin populations reflected by the coupling constants were reproduced quite well by McLachlan calculations which took account of the non-coplanarity of the phenyl and five-membered rings and of departures of the phenyl rings from regular hexagonal geometry. Möbius *et al.*[39] have studied pentaphenylcyclopentadienyl and several of its methyl-substituted derivatives. The coupling constants in the methyl derivatives are temperature-dependent and this was interpreted in terms of thermal mixing of the symmetric and antisymmetric states which would be degenerate in the parent radical but whose degeneracy

[35] M. Iwasaki, H. Muto, B. Eda, and K. Nunome, *J. Chem. Phys.*, 1972, **56**, 3166.
[36] D. S. Leniart, J. S. Hyde, and J. C. Vedrine, *J. Phys. Chem.*, 1972, **76**, 2079.
[37] J. C. Vedrine, J. S. Hyde, and D. S. Leniart, *J. Phys. Chem.*, 1972, **76**, 2087.
[38] R. D. Allendoerfer and A. S. Pollock, *Mol. Phys.*, 1971, **22**, 661.
[39] K. Möbius, H. van Willigen, and A. H. Maki, *Mol. Phys.*, 1971, **20**, 289.

was taken to be lifted on substitution. The symmetries of the upper and lower states were determined from interpreting the patterns of methyl proton splittings, it being assumed that the methyl group could be accounted for solely in terms of the inductive effect, and the separations were estimated from the temperature dependence of the coupling constants.

Stable phenoxyl radicals have always been favourites for study by solution ENDOR and further work has been reported. Allendoerfer and Eustace[40] have studied the 4-formyl-2,6-di-t-butylphenoxyl radical: considerable care is needed to obtain this radical, for the PbO_2 oxidation of 4-hydroxy-3,5-di-t-butylbenzaldehyde leads rapidly to the formation of a secondary radical. In the formyl radical the coupling to the formyl proton is temperature-dependent and this is ascribed to torsional oscillations of the formyl group. Allendoerfer and Eustace discuss at some length the question of relative intensities of the ENDOR signals from the various classes of proton in this radical, and indeed this problem is raised in several of the papers discussed in this section. Comment on this aspect of solution ENDOR is deferred until the next section of this Report. A series of phenoxyl radicals has been studied by Atherton *et al.*[41] It was observed that the methylene proton coupling in the ethoxy-group in 4-ethoxy-2,6-di-t-butylphenoxyl radical was quite temperature-dependent whereas that of the methyl protons in the methoxy-homologue was not. The difference was ascribed to a slowing down of the free rotation in the ethoxy-compound. It seems reasonable that the rotation of alkoxy-substituents in these neutral radicals should be faster than it is in alkoxy-substituted radical-cations,[42] but there is an unexplained contrast with the 4-methoxy-2,6-dimethylphenoxyl radical. The e.s.r. spectrum of this radical shows linewidth alternation, indicating that there is much slower flipping of the methoxy-group.[43]

Nitroxides form another class of stable radicals which have been very widely studied by magnetic resonance and which are of particular interest because of their use as spin-labels and paramagnetic probes. Allendoerfer and Engelmann[44] have studied the proton ENDOR of bis-(*p*-methoxyphenyl)nitroxide (4), and obtained precise values of the coupling constants. Contrasting results have been obtained for alicyclic nitroxides: Leniart

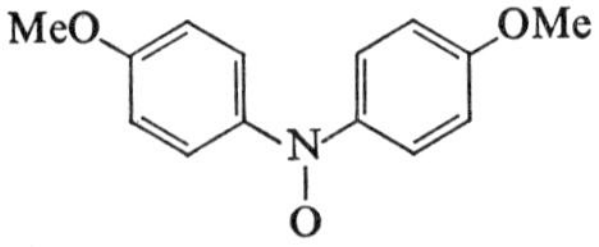

(4)

[40] R. D. Allendoerfer and D. J. Eustace, *J. Phys. Chem.*, 1971, **75**, 2765.
[41] N. M. Atherton, A. J. Blackhurst, and I. P. Cook, *Trans. Faraday Soc.*, 1971, **67**, 2510.
[42] P. D. Sullivan, *J. Phys. Chem.*, 1970, **74**, 2563.
[43] W. J. van den Hoek, W. G. B. Huysmans, and M. J. C. van Gemert, *J. Magn. Resonance*, 1970, **3**, 137.
[44] R. D. Allendoerfer and J. H. Engelmann, *Mol. Phys.*, 1971, **20**, 569.

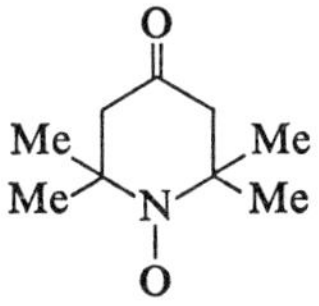

(5)

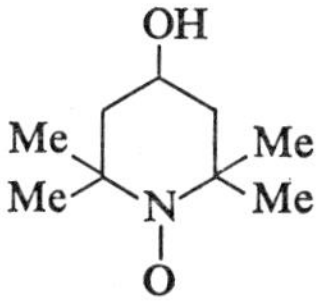

(6)

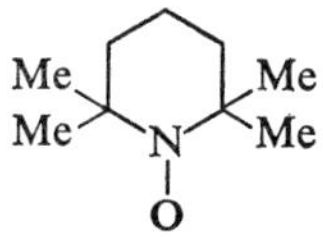

(7)

et al.[45] have studied 1-oxyl-2,2,6,6-tetramethyl-4-piperidine [tanone (5)], 1-oxyl-2,2,6,6-tetramethyl-4-hydroxypiperidine [tanol (6)], and 1-oxyl-2,2,-6,6-tetramethylpiperidine [tempo (7)], and observed the ^{14}N ENDOR. These spectra illustrate a number of interesting aspects of ENDOR. First, there is the r.f. enhancement effect,[8, 46] which makes it possible to observe reasonably intense signals from nuclei with small magnetic moments; secondly, there is the problem of how to treat properly the second-order hyperfine shift for a radical in solution;[47] and thirdly, there are the effects of the relaxation mechanisms on the ENDOR intensities. This third point is raised again in the next section. The most recent ENDOR study of nitroxides is that by Möbius and his collaborators who have observed the ^{14}N spectra of (6) in the liquid-crystal, 'Phase IV'.[48] The quadrupole coupling contribution does not vanish from the ENDOR frequencies of the partially oriented radicals, and so by comparing the spectra in the isotropic and nematic phases the quadrupole coupling tensor can be determined if axial symmetry is assumed. Interpretation of the data suggests that the electric field gradient is determined essentially by the unpaired electron. This work is important because it pioneers a technique which makes quadrupole couplings relatively accessible and it would not be surprising to see several applications of it in future.

Möbius and his group have also studied proton ENDOR in liquid crystals, and have reported the spectra of phenalenyl and 2,4,6-tri-t-butylphenoxyl in Phase IV.[49] The same type of information is obtained as in conventional e.s.r. experiments in liquid crystals, although of course one is better off by virtue of the better resolution and accuracy of ENDOR. Certainly ENDOR offers considerable advantages over e.s.r. for the study of radicals of low symmetry in these solvents.

In ether solvents, in which radical-ions produced by alkali-metal reduction are generally obtained, it seems that ENDOR spectra are usually only observed at temperatures close to the freezing point of the solvent. Notwithstanding this problem there is a steady trickle of reports of spectra

[45] D. S. Leniart, J. C. Vedrine, and J. S. Hyde, *Chem. Phys. Letters*, 1970, **6**, 637.
[46] D. H. Whiffen, *Mol. Phys.*, 1966, **10**, 595; S. Geschwind in 'Hyperfine Interactions', ed. A. J. Freeman and R. B. Frankel, Academic Press, New York, 1967, p. 225.
[47] G. K. Fraenkel, *J. Chem. Phys.*, 1965, **42**, 4275; *J. Phys. Chem.*, 1967, **71**, 139.
[48] K. P. Dinse, K. Möbius, M. Plato, R. Biehl, and H. Haustein, *Chem. Phys. Letters*, 1972, **14**, 196.
[49] K. P. Dinse, R. Biehl, K. Möbius, and H. Haustein, *Chem. Phys. Letters*, 1971, **12**, 399.

from these systems. A dramatic illustration of the use of ENDOR for resolution enhancement is Allendoerfer and Chang's study of hexahelicene (8) radical-anion,[50] prepared by potassium reduction in dimethoxyethane. Simple symmetry considerations suggest that there should be eight pairs of equivalent protons but the ENDOR spectrum shows that there are ten coupling constants, and the observed e.s.r. is reproduced very well by a computer simulation based on these couplings. It is suggested that the departure from the molecular symmetry is due to ion-pairing. The

(8)

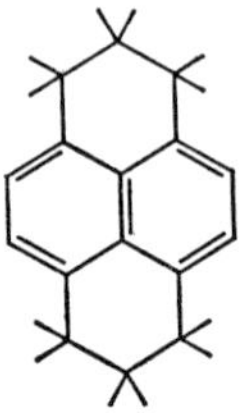

(9)

terphenyl anions also have very complicated e.s.r. spectra, and accurate coupling constants have been obtained from an ENDOR study.[51] The interest in these radicals is in the possibility of σ–π delocalization but it appears that this may not be important even though the radicals are non-planar. Shain[52] has obtained a liquid-phase ENDOR spectrum of the 1,2,3,6,7,8-hexahydropyrene (9) radical anion: the spectrum confirms that the axial and equatorial γ-protons have different coupling constants and the measured value of the smallest coupling constant agrees with one suggested earlier on the basis of computer-simulated e.s.r. spectra. It is interesting that Shain was able to observe an ENDOR spectrum only when lithium was used as the reducing agent, and not when sodium or potassium were used. Spectra of other hydrocarbon radical anions have been observed with Li, Na, or K as reducing agent.[51, 53, 54]

More positive comment on ion-pairing is provided by Allendoerfer and Papez' study of durosemiquinone, reduced with potassium in dimethoxy-ethane.[55] They were able to distinguish the ENDOR spectra of both an ion pair and the free anion. Allendoerfer and Papez have also found the technique of ENDOR-induced e.p.r. useful in their studies of ion pairs.[55, 56] In this experiment, which was originally introduced by Hyde,[57] one observes the change in the total e.s.r. spectrum due to excitation of one ENDOR transition. It would be interesting to see further applications of the method.

[50] R. D. Allendoerfer and R. Chang, *J. Magn. Resonance*, 1971, **5**, 273.
[51] R. Biehl, K. P. Dinse, and K. Möbius, *Chem. Phys. Letters*, 1971, **10**, 605.
[52] A. L. Shain, *Mol. Phys.*, 1971, **22**, 733.
[53] A. Lagendijk, N. F. M. Tromp, M. Glasbeek, and J. D. W. van Voorst, *Chem. Phys. Letters*, 1970, **6**, 152.
[54] K. P. Dinse, R. Biehl, K. Möbius, and M. Plato, *J. Magn. Resonance*, 1972, **6**, 444.
[55] R. D. Allendoerfer and R. J. Papez, *J. Amer. Chem. Soc.*, 1970, **92**, 6971.
[56] R. D. Allendoerfer and R. J. Papez, *J. Phys. Chem.*, 1972, **76**, 1012.
[57] J. S. Hyde, *J. Chem. Phys.*, 1965, **43**, 1806.

Ethanol is quite a satisfactory solvent for solution ENDOR, and the spectra of several semiquinone radical anions in it have been reported by Atherton and Blackhurst.[58] An interesting feature of this work is that again it illustrates the application of ENDOR to sort out overlapping e.s.r. signals from two different radicals, in this case the primary radicals from *p*-benzoquinone or 2,5-dimethyl-*p*-benzoquinone and secondary radicals resulting from substitution of alkoxy-groups from the solvent, methanol or ethanol.

ENDOR Mechanisms.—The simplest picture of the ENDOR experiment is that one saturates or partially saturates an e.s.r. transition and then increases the population difference between the two levels by driving the nuclear spin transitions. The increase of population of the lower level of the e.s.r. transition at the ENDOR frequency, or the decrease of population of the upper, may occur by either absorption or induced emission by the nuclear spins. A picture of this sort, which essentially corresponds to Feher's original spin-packet shifting mechanism,[59] gives some understanding of why ENDOR transitions occur and, if one is only interested in measuring and interpreting the frequencies, could possibly be regarded as adequate. However, it does not explain the relative intensities of ENDOR signals, and since practically all the papers dealing with solution ENDOR comment on whether or not the relative signal intensities are proportional to the numbers of contributing protons, it is clear that deeper insight should be sought.

Quite generally, the magnitude of the ENDOR response depends critically on the relaxation rates, and only if these are identical can two inequivalent nuclei be expected to give signals of the same intensity. The relaxation problem can be formulated phenomenologically by taking the ENDOR signal to correspond to a change in T_1, the electron spin–lattice relaxation time. This method is discussed by Abragam and Bleaney [8] and has independently been elaborated upon in the context of solution ENDOR by Allendoerfer and Maki.[60] A particularly useful result of this work is an equation which enables one to correct the ENDOR intensities for the overlap of e.s.r. lines: for small couplings one may be saturating more than one e.s.r. transition, and the signals from these nuclei have diminished intensity. The effect is striking in the spectra of the stable phenoxyl radicals: the relatively enormous number of t-butyl protons give relatively weak signals. The correction has been applied in several cases and found to work well in many of them.[40, 44, 58, 60]

If all the protons are of the same type, for example if they are all α-protons in a π-radical, then experience suggests that the relative intensities will reflect the relative numbers of contributing protons fairly well, particularly if the Allendoerfer and Maki correction [60] is applied. This

[58] N. M. Atherton and A. J. Blackhurst, *J. C. S. Faraday II*, 1972, **68**, 470.
[59] G. Feher, *Phys. Rev.*, 1959, **114**, 1219.
[60] R. D. Allendoerfer and A. H. Maki, *J. Magn. Resonance*, 1970, **3**, 396.

seems reasonable in the sense that all the protons can be expected to be subject to the same sorts of relaxation processes. For ENDOR of α-protons the most important relaxation mechanism appears to be the electron–nuclear dipolar (END) interaction. This is established both experimentally and theoretically by Freed, who has carried out very detailed analyses of e.s.r. saturation and ENDOR effects using Redfield theory.[61, 62] The predictions of the theory were tested by comparison with Hyde's results and confirm the importance of the END mechanism. The intensities in the ^{14}N ENDOR spectra studied by Leniart *et al.*[45] also indicate that the END mechanism dominates for that nucleus.

Severe departures from proportionality to the numbers of nuclei have been observed for the relative intensities of the ENDOR signals arising from different types of proton, for example α- and β-protons in a π-radical.[39, 58] It is obvious that the END mechanism must be much less effective for a β-proton than for an α-proton with the same isotropic coupling, for the anisotropy of the hyperfine interaction must be much less. It has been suggested[63] that modulation of the isotropic hyperfine interaction by internal motions (internal rotation or torsional oscillation) may provide the important ENDOR mechanism for β-protons.

The ENDOR spectra of solids often show signals at the free nuclear frequency that are due to nuclei relatively remote from the paramagnetic centre. This distant ENDOR or polarization ENDOR, which was first studied and explained by Lambe *et al.*,[64] is distinguishable from the normal ENDOR in several ways. For example, the effect manifests itself differently in the absorption and dispersion spectra, and the recovery time of signals is characterized by the nuclear rather than the electron-spin relaxation time. The distant ENDOR mechanism is that polarization of nuclei near the electron spin occurs through forbidden 'spin-flip' transitions and that this polarization diffuses to the more remote nuclei. Recently a different type of distant ENDOR has been studied. This is described by Wenckebach *et al.*,[65] who discovered it while following up nuclear polarization studies of copper Tutton salts. Here the polarization of the nuclei is effected by saturating the allowed $\Delta M_S = 1$ transitions. The theory of the effect has been fully worked out for both steady-state and transient conditions and there is excellent agreement with experiment. Feher and Isaacson[66] have recently demonstrated the effectiveness of a different ENDOR mechanism, applicable for homogeneously broadened lines. Polarization of nuclei causes a shift in the position of a homogeneously broadened e.s.r. line, and the trick is to depolarize the nuclei by r.f. irradiation so that the line shifts

61 J. H. Freed, *J. Chem. Phys.*, 1965, **43**, 2312.
62 J. H. Freed, *J. Phys. Chem.*, 1967, **71**, 38.
63 N. M. Atherton and B. Day, *Chem. Phys. Letters*, 1972, **15**, 428.
64 J. Lambe, N. Laurance, E. C. McIrvine, and R. W. Terhune, *Phys. Rev.*, 1961, **122**, 1161.
65 W. Th. Wenckebach, L. A. H. Schreurs, and H. Hoogstraate, *Physica*, 1971, **52**, 455.
66 G. Feher and R. A. Isaacson, *J. Magn. Resonance*, 1972, **7**, 111.

back and there is a change in the intensity of absorption at a particular field. (The theory of the shift is analogous to that for contact shifts in n.m.r.[67]) The original polarization of the nuclei may be effected either by saturating the e.s.r. signal or by working at very low temperature. In the latter case it is not necessary to saturate the e.s.r. line, and it is suggested that the method may prove useful for obtaining ENDOR signals from systems which are hard to saturate. The effects are demonstrated by spectra obtained from colloidal Li in LiF.

The radiation fields used in ENDOR experiments are not weak, and a variety of effects can arise from this. For example, in calculating transition probabilities one should go to a higher order than first in time-dependent perturbation theory, and then the possibility of multiple-quantum transitions arises. Two types of double-quantum transition can be distinguished, those in which a nuclear- and an electron-spin transition occur, and those in which two nuclear-spin transitions occur. The theory has been fully worked out by Freed *et al.*[68] and agrees well with experiment. The second type of transition is perhaps of most interest because of its possible application in assigning the numbers of protons giving rise to ENDOR signals. The theory predicts that if one saturates a degenerate transition arising from a group of equivalent nuclei (or the $M_I = 0$ transition for $I = 1$) then the ENDOR line should split at high r.f. power, whereas if a non-degenerate line is saturated no such splitting should occur. The theory was tested and shown to be correct in the original work on coherence effects [68] and has recently been used by Möbius and his collaborators to assign hyperfine couplings in the spectra of the radical anions of 1-phenyl- and 2-phenyl-naphthalene.[54] A somewhat different effect which can occur because the radiation fields are strong arises because the energy levels can themselves be altered. Again this possibility was originally discussed by Freed *et al.*[68] The physics is comparable to that which obtains in the spin-tickling and spin-decoupling experiments so familiar in n.m.r.,[69] and the analogy has been expanded upon to some extent.[70] However, the suggestion that a particularly dramatic manifestation of the effect has been observed [70] should be regarded with caution.

It is clear that there is ample scope for further studies of the ENDOR mechanism and indeed such studies may be expected. The writer subscribes to Kwiram's view that 'the full potential of the [ENDOR] technique cannot be realized until a [comprehensive] theory [of the ENDOR mechanism] is developed'.[9]

ENDOR Instrumentation.—Poole's book [71] gives some consideration to the factors entering into the design of spectrometers and reviews some of the

[67] See *e.g.* ref. 6, p. 221.
[68] J. H. Freed, D. S. Leniart, and J. S. Hyde, *J. Chem. Phys.*, 1967, **47**, 2762.
[69] See *e.g.* R. A. Hoffmann and S. Forsen, *Progr. N.M.R. Spectroscopy*, 1966, **1**, 15.
[70] N. M. Atherton, A. J. Blackhurst, and I. P. Cook, *Chem. Phys. Letters*, 1971, **8**, 187.
[71] C. P. Poole, 'Electron Spin Resonance: A Comprehensive Treatise on Experimental Techniques,' Interscience, New York, 1967.

early instruments built for solid-state experiments. Hyde[57] has given a detailed description of the high-power pulsed r.f. instrument developed for solution experiments and which is the basis of a commercially available spectrometer. An instrument using continuous r.f. has been developed by Maki and his collaborators,[72] and Allendoerfer[40] has more recently described a descendant of it. Gothe[73] has discussed the building of both e.s.r. and ENDOR spectrometers.

One or two points of detail have been discussed at more length. Fourrier-Lamer and Lenoir[74] find better matrix ENDOR signals using a resonant helix than they do with a cavity. Hyde *et al.*[75] have pointed out the disadvantages of using magnetic field modulation in powder ENDOR studies and suggest an alternative scheme which they call 'reference phase modulation'. Yagi *et al.*[76, 77] have described a wire-walled cavity for ENDOR at V-band frequencies.

3 ELDOR

The usual ELDOR response is a decrease in the intensity of e.s.r. absorption. This is in contrast to ENDOR, where the signal corresponds to an increase in e.s.r. absorption. A simple view of the ELDOR experiment is that the reduction arises because of the transfer of saturation from the pumped to the observed transition. It is clear from the Figure that the response will depend critically on the nuclear relaxation rates.

The earliest ELDOR experiments were performed as a result of interests in cross-relaxation in solids. In the first, Sorokin, Lasher, and Gelles studied nitrogen centres in diamond.[78] The results showed clearly the importance of cross-relaxation processes in which the total Zeeman energy is conserved. The next work was carried out by Unruh and Culvahouse,[79] who used the method to analyse the relaxation processes for Co^{2+} in $La_2Zn_3(NO_3)_{12},24H_2O$. The high spin of ^{59}Co ($I = \frac{7}{2}$) meant that analysis of the data was complicated, but a very detailed comparison of theory and experiment was made possible by this feature, and in fact the data could be explained very satisfactorily. The third of the early papers was by Moran,[80] and dealt with F-centres in KCl. An interesting result of this work was that it showed that 'forbidden' transitions (those commonly called 'spin-flip' transitions, in which an electron- and a nuclear-spin turn over) made a

[72] A. H. Maki, R. D. Allendoerfer, J. C. Danner, and R. T. Keys, *J. Amer. Chem. Soc.*, 1968, **90**, 4225.
[73] K. H. Gothe, *Wiss. Z. tech. Hochsch. Ilmenau*, 1970, **16**, 67.
[74] A. Fourrier-Lamer and J. Lenoir, *Compt. rend.*, 1971, **272**, *B*, 98.
[75] J. S. Hyde, T. Astlind, L. E. G. Eriksson, and A. Ehrenberg, *Rev. Sci. Instr.*, 1970, **41**, 1598.
[76] H. Yagi, M. Inoue, T. Tatukawa, O. Yaeguchi, and S. Kato, *Jap. J. Appl. Phys.*, 1970, **9**, 1534.
[77] M. Inoue and H. Yagi, *Fukui Daigaku Kogakubu Kenkyu Hokoku*, 1971, **19**, 101.
[78] P. P. Sorokin, G. J. Lasher, and I. L. Gelles, *Phys. Rev.*, 1960, **118**, 939.
[79] W. P. Unruh and J. W. Culvahouse, *Phys. Rev.*, 1963, **129**, 2441.
[80] P. R. Moran, *Phys. Rev. (A)*, 1964, **135**, 247.

considerable contribution to the double-resonance intensity, and it was suggested that the nuclear relaxation rates were so high that it was a questionable approximation to treat the inhomogeneous line-broadening in these systems as a static problem.

The first time chemists became involved with ELDOR was with the publication of a paper by Hyde, Chien, and Freed in 1968.[81] This paper gives a detailed description of an ELDOR spectrometer and describes the observations for several radicals in fluid solution. A theory of ELDOR is developed, following the same general pattern as that proposed earlier by Freed for solution ENDOR,[61, 62] and the effects of the END interaction and of Heisenberg exchange (HE) as ELDOR mechanisms are discussed in detail. The END mechanism dominates at low temperature and low concentration of radical but at higher temperature and concentration the HE mechanism dominates. In later work Freed and his collaborators have again demonstrated the value of ELDOR for studying HE.[82]

Hyde's first experiments [81] were of the field-sweep variety. Initial setting of the observing and pumping frequencies is required so that both become resonant with the required transition frequencies simultaneously as the field is swept. A more attractive way of performing the experiment is to operate in the frequency-swept mode: one observes the change in intensity of absorption at a particular point as a function of pumping frequency. (The frequency-swept experiment is analogous to ENDOR.) Frequency-swept experiments were first carried out on γ-irradiated malonic acid crystals:[83] they confirmed that the three radicals present were $\dot{C}H(COOH)_2$ and $\dot{C}H_2COOH$ in two spectroscopically distinct sites. Again these experiments illustrate the power of double-resonance techniques for sorting out overlapping spectra and obtaining precise data from them. A detailed description of the frequency-swept instrumentation is contained in a later paper which deals with experiments on DPPH in both fluid and frozen solutions.[84] The small difference between the nitrogen coupling constants, which is not easy to measure from conventional e.s.r., was obtained to high precision. In more recent work Nechtschein and Hyde have described frequency-swept experiments in which the pump-power was pulsed.[85] Interesting transient effects, which are not readily accounted for, were observed.

In parallel with the exploration of the ELDOR technique by Hyde and his collaborators has been the work by a group in the U.S.S.R. Their first results were described in 1968.[86] This group has been concerned with

81 J. S. Hyde, J. C. W. Chien, and J. H. Freed, *J. Chem. Phys.*, 1968, **48**, 4211.
82 M. P. Eastman, G. V. Bruno, and J. H. Freed, *J. Chem. Phys.*, 1970, **52**, 321.
83 J. S. Hyde, L. D. Kispert, R. C. Sneed, and J. C. W. Chien, *J. Chem. Phys.*, 1968, **48**, 3824.
84 J. S. Hyde, R. C. Sneed, and G. H. Rist, *J. Chem. Phys.*, 1969, **51**, 1404.
85 M. Nechtschein and J. S. Hyde, *Phys. Rev. Letters*, 1970, **24**, 672.
86 V. A. Benderskii, L. A. Blumenfeld, P. A. Stunzas, and E. A. Sokolov, *Nature*, 1968, **220**, 365.

studying spin exchange and has looked at TCNQ salts in some detail. The experimental observations on the pure salts are described in two papers [86, 87] and the spin-exchange frequencies were measured. A third paper [88] reports a study of the interactions between triplet excitons and paramagnetic ($S = \frac{1}{2}$) impurities in TCNQ salts. The spectrometer used in this work has been described:[89] it incorporates a helical resonator, rather than a bimodal cavity as described by Hyde and earlier workers. The latest publication from this Russian group is a theoretical one: using a density matrix formalism and a pulse description of the modulation of the exchange interaction, the ELDOR response expected for interaction of pairs of $S = \frac{1}{2}$ systems, pairs of $S = 1$ systems, and of an $S = \frac{1}{2}$ and an $S = 1$ system is discussed.[90]

Hausser and his co-workers have also described an ELDOR spectrometer which uses a helix and a cavity.[91] With this arrangement they obtain very good signal : noise ratios at room temperature for 2,2,5,5-tetramethyl-3-carbamoyl-1-pyrrolidyloxyl (10) in solution in benzene.

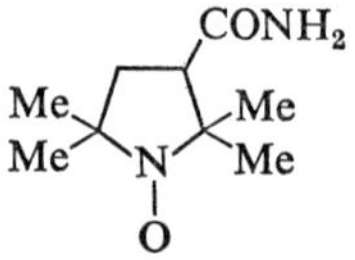

(10)

[87] P. A. Stunzas, V. A. Benderskii, L. A. Blumenfeld, and E. A. Sokolov, *Optika i Spektroskopiya*, 1970, **28**, 278.
[88] P. A. Stunzas, V. A. Benderskii, and A. A. Sokolov, *Optika i Spektroskopiya*, 1970, **28**, 487.
[89] E. A. Sokolov and V. A. Benderskii, *Pribory Tekh. Eksp.*, 1969, **3**, 232.
[90] P. A. Stunzas and V. A. Benderskii, *Optika i Spektroskopiya*, 1971, **30**, 1041.
[91] H. M. Vieth, H. Brunner, and K. H. Hausser, *Z. Naturforsch.*, 1971, **26a**, 167.

3
Relaxation Processes and Linewidths

BY P. W. ATKINS

1 Introduction

An important event of the year was the NATO Advanced Study Institute on Electron Spin Relaxation in Liquids: the lectures presented on this course have been collected into a published volume,[1] and although this Reporter must declare his interest in the volume (as one of its two editors) it is probably fair to say that it is the first extended review of the theories of electron spin relaxation processes in solution. The book deals with the general formal background of spin relaxation theories: superoperators, time-ordering, density operators (Muus), stochastic processes (Pedersen), stochastic theories of lineshape (Jensen), projection operators (Platz), cumulants (Nielsen), linear response (Oppenheim), and then investigates the structure of spin relaxation theories (Deutch, Freed). This is followed by a review of applications (Kivelson), spin-rotation (Atkins), ^6S ions (Luckhurst), and five contributions on slowly tumbling molecules and liquid crystals (Gordon and Messenger, Lynden-Bell, Freed, Luckhurst, Kaplan). The book concludes with an account of the lineshape of triplet excitons in disordered systems (Lemaistre and Kottis) and saturation and double resonance in liquids (Freed). Some of the individual chapters will be referred to in later parts of this Report. Several other books on magnetic resonance have appeared during the year: one that deals explicitly with magnetic relaxation is that by Poole and Farach,[2] with chapters on the Bloch equations, saturation and pulse methods, relaxation in liquids and solids, Kubo–Tomita and Redfield theories, the special problems that arise in solids, and specific accounts of electron, nuclear, ultrasonic, and dielectric relaxation and the Mössbauer effect. The book constitutes a helpful review of the theory and applications without too much mathematical detail. Sections on relaxation processes and linewidths will also be found in the books by Bolton and Wertz[3] and by Dixon.[4] A lengthy review article on the theories of electron spin relaxation in solution has also

1 'Electron Spin Relaxation in Liquids', ed. L. T. Muus and P. W. Atkins, Plenum Press, London and New York, 1972.
2 C. P. Poole and H. A. Farach, 'Magnetic Relaxation', Academic Press, New York, 1971.
3 J. R. Bolton and J. E. Wertz, 'Electron Spin Resonance', Wiley, New York, 1972.
4 W. T. Dixon, 'The Theory and Interpretation of Magnetic Resonance Spectra', Plenum Press, London and New York, 1972.

appeared:[5] it describes in detail the Kubo–Tomita and Redfield theories and describes their interconnection, and discusses a variety of molecular motions that lead to relaxation.

The period under review has seen considerable progress in the discussion of the lineshape of slowly tumbling molecules and this Report will reflect the tendency. The theoretical structure of the Redfield theory has been elucidated in a notable paper.[6, 7] Developments in experimental procedures have included spin-echo studies of electron spin relaxation in liquids[8] (relaxation of solutions of 2,4,6-tri-t-butylphenoxyl) and direct observation of spectral decay from an initial chemically induced spin polarization,[9] which promises to be the basis of a method for the determination of T_1 and T_2.

2 Basic Theory

The basis of the Deutch–Albers theory[6, 7] of the structure of the Redfield equations is the difference of the time scales of the spin relaxation process (which is slow) and the molecular motion (which is relatively rapid). Three consequences of this difference are that the spin system may be described by a master equation (which to the lowest order of approximation is the Redfield equation), that the relaxation follows a linear regression law, and that the relaxation times may be expressed as Fourier transforms of time-correlation functions (which in lowest order have a time dependence determined by the fast lattice motion alone). The approach devised by Albers and Deutch[6, 7] is to note that the Redfield equation

$$(\partial/\partial t)\,\sigma(t)_{\alpha\alpha'} = -\,i\omega_{\alpha\alpha'}\,\sigma(t)_{\alpha\alpha'} + \sum_{\beta,\beta'} R_{\alpha\alpha'\beta\beta'}\,\sigma(t)_{\beta\beta'} \tag{1}$$

where $\omega_{\alpha\alpha'} = E_\alpha - E_{\alpha'}$ and σ is the spin-density matrix, is an expression in terms of an average quantity (σ) because the lattice components have been eliminated; they then seek a more fundamental microscopic quantity $G_{\alpha\alpha'}$ which when suitably averaged generates σ:

$$\sigma(t)_{\alpha\alpha'} = \mathrm{tr}\rho(0)\,G_{\alpha\alpha'} \tag{2}$$

where $\rho(0)$ is the initial density matrix of the system. All the quantities of interest in an e.s.r. experiment are known if $G_{\alpha\alpha'}$ is known, because if one can assume initial factorization of ρ into a lattice and a spin component it is simple to show that the equation of motion of σ is the same as the equation of motion of $\langle G_{\alpha\alpha'}\rangle$ (the angle brackets denoting a lattice average), and that the magnetization correlation function is a matrix function of the correlation function $\langle G_{\alpha\alpha'}(t)G_{\beta\beta'}(0)\rangle$. It is shown that $G_{\alpha\alpha'}(t)$ satisfies an equation

[5] P. W. Atkins, *Adv. Mol. Relaxation Processes*, 1972, **2**, 121.
[6] J. Albers and J. M. Deutch, *J. Chem. Phys.*, 1971, **55**, 2613.
[7] J. M. Deutch, in ref. 1, p. 127.
[8] A. D. Milov, K. M. Salikhov, and Yu. D. Tvetsko, *Chem. Phys. Letters*, 1971, **8**, 523.
[9] P. W. Atkins and K. A. McLauchlan, Proceedings of the Fifth Symposium on Molecular Spectroscopy, Institute of Petroleum, 1971, p. 171.

of motion of the Redfield type with the fast lattice motion effects appearing as an additional rapidly fluctuating term, and that the magnetization follows an equation of the Bloch type with an analogous fast fluctuation term. Lattice averaging eliminates these additional terms and leads to the Redfield and Bloch equations.

The derivation of the equation of motion of G depends on the introduction of the projection operator P, which projects in the sense

$$P(\cdots) = \langle\cdots\rangle = \mathrm{tr}\rho_l(\cdots) \tag{3}$$

where ρ_t is the lattice density matrix. The (exact) equation derived in this way is

$$\dot{G}(t)_{\alpha\alpha'} = -\mathrm{i}\omega_{\alpha\alpha'}G_{\alpha\alpha'}(t) + K(t)_{\alpha\alpha'} + \sum_{\beta\beta'}\int_0^t \mathrm{d}\tau\langle\beta' \mid F(\tau)_{\alpha\alpha'} \mid \beta\rangle G(t-\tau)_{\beta\beta'} \tag{4}$$

where

$$\langle\beta' \mid F(\tau)_{\alpha\alpha'} \mid \beta\rangle = \langle\beta' \mid \langle[H_1, K(\tau)_{\alpha\alpha'}] \mid \beta\rangle \tag{5}$$

with H_1 as the spin–lattice interaction hamiltonian and $K_{\alpha\alpha'}$ as a random force term which satisfies $\langle K_{\alpha\alpha'}\rangle = 0$. The similarity of equation (4) to a generalized Langevin equation illustrates the scheme of the calculation, and the effect of the lattice is contained in $K_{\alpha\alpha'}$. The corresponding equation of motion for the spin-density matrix is therefore

$$\dot{\sigma}(t)_{\alpha\alpha'} = -\mathrm{i}\omega_{\alpha\alpha'}\sigma(t)_{\alpha\alpha'} + \sum_{\beta\beta'}\int_0^t \mathrm{d}\tau\langle\beta' \mid F(\tau)_{\alpha\alpha'} \mid \beta\rangle\, \sigma(t-\tau)_{\beta\beta'} \tag{6}$$

which is similarly exact except for the introduction of the initial condition $\rho = \rho_l\sigma$. The first term on the right of equation (4) is eliminated by transferring into the interaction picture with $G(t)^*_{\alpha\alpha'} = G_{\alpha\alpha'}(t)\exp(\mathrm{i}\omega_{\alpha\alpha'}t)$ and since then

$$\dot{G}(t)^*_{\alpha\alpha'} = K(t)^*_{\alpha\alpha'} + \sum_{\beta\beta'}\int_0^t \mathrm{d}\tau F(\tau)_{\alpha\alpha'\beta\beta'}\exp(\mathrm{i}\omega_{\beta\beta'}\tau) \times G(t-\tau)^*_{\beta\beta'}\exp[\mathrm{i}(\omega_{\alpha\alpha'} - \omega_{\beta\beta'})t] \tag{7}$$

the crucial step in the analysis must be the investigation of $F_{\alpha\alpha'\beta\beta'}$:

$$F(\tau)_{\alpha\alpha'\beta\beta'} = \langle\beta' \mid F(\tau)_{\alpha\alpha'} \mid \beta\rangle \exp(\mathrm{i}\omega_{\beta\beta'}\tau). \tag{8}$$

If the lattice hamiltonian is of order λ ($\lambda \ll 1$) only the lowest order form of this expression need be obtained, and if the interaction hamiltonian is of the form

$$H_1 = \sum_q F_q S_q \tag{9}$$

the appropriate lowest order approximation to F is

$$\begin{aligned} F(\tau)^{(0)}_{\alpha\alpha'\beta\beta'} = \sum_{qq'}\{ & S_{q'\alpha\beta}S_{q\beta'\alpha'}[g_{qq'}(\tau)\exp(\mathrm{i}\omega_{\beta\alpha}\tau) + g_{qq'}(-\tau)\exp(-\mathrm{i}\omega_{\beta'\alpha'}t)] \\ & - \delta_{\alpha'\beta'}\sum_{\gamma} S_{q\alpha\gamma}S_{q'\gamma\beta}g_{qq'}(\tau)\exp(\mathrm{i}\omega_{\beta\gamma}\tau) \\ & - \delta_{\alpha\beta}\sum_{\gamma} S_{q\beta'\gamma}S_{q'\gamma\alpha'}g_{qq'}(-\tau)\exp(-\mathrm{i}\omega_{\beta'\gamma}\tau)\} \end{aligned} \tag{10}$$

where

$$g_{qq'}(t) = \langle F_q(t)\, F_{q'}(0)\rangle \tag{11}$$

Carrying through the insertion of this equation into equation (7) to lowest order in H_1 yields

$$\dot{G}(t)_{\alpha\alpha'} = -\, i\omega_{\alpha\alpha'}\, G(t)_{\alpha\alpha'} + \sum_{\beta\beta'} R_{\alpha\alpha'\beta\beta'}\, G(t)_{\beta\beta'} + K(t)^{(0)}_{\alpha\alpha'} \tag{12}$$

where

$$R_{\alpha\alpha'\beta\beta'} = \int_0^\infty d\tau F(\tau)^{(0)}_{\alpha\alpha'\beta\beta'} \tag{13}$$

which is the usual Redfield matrix. This expression differs from the conventional analysis in the presence of $K(t)^{(0)}_{\alpha\alpha'}$, which is the lowest order approximation to the rapidly fluctuating force due to the lattice; this term disappears when averaging is used to move to equation (6), and so one arrives at the Redfield equation. Since the analysis has invoked approximations about the strength of the perturbation and the rapidity of the lattice fluctuations one may conclude [6, 7] that the Redfield equations are valid when $t \gg \tau_c$ and $\tau_c \ll T_2$, τ_c being a lattice correlation time.

A similar analysis can be extended to the Bloch equations [6, 7] and both calculations may be extended to higher order: in general one deduces that

$$\dot{\sigma}(t)_{\alpha\alpha'} = -\, i\omega_{\alpha\alpha'}\, \sigma(t)_{\alpha\alpha'} + \sum_{n=0}^{\infty} R^{(n)}_{\alpha\alpha'\beta\beta'}\, \sigma(t)_{\beta\beta'} \tag{14}$$

with

$$R^{(n)}_{\alpha\alpha'\beta\beta'} = \int_0^\infty d\tau F(\tau)^{(n)}_{\alpha\alpha'\beta\beta'} \tag{15}$$

where $F^{(n)}$ is the nth order function derived from the random force. This equation has the same form as the generalized Redfield equation derived by Freed by the cumulant method.[10]

A theory employing projection operator techniques has also been described [11] and it is such that the Kubo theory of the response of a system emerges as the lowest order approximation. The development of the paper consists of replacing an inhomogeneous equation by a linear homogeneous integrodifferential equation.

Another general approach to the lineshape problem in magnetic resonance has been described [12] and although it refers principally to n.m.r. its method naturally is also applicable to e.s.r. In the calculation, the magnetization of the sample is first reduced to the calculation of two autocorrelation functions: beginning with the general problem of determining the nine components of the relaxation tensor it was found that, for a particular direction of the field and a particular symmetry of the hamiltonian for the isolated system, only three components of the tensor need be

[10] J. H. Freed, *J. Chem. Phys.*, 1968, **49**, 376; also ref. 1, p. 165.
[11] V. M. Kenbre, *Phys. Rev.* (*A*), 1971, **5**, 2327.
[12] F. Lado, J. D. Memory, and G. W. Parker, *Phys. Rev.* (*B*), 1971, **4**, 1406.

calculated. The corresponding three components of the magnetization autocorrelation function then required could be reduced to the task of finding only two. The next section of the paper is devoted to a general discussion of such autocorrelation functions in terms of orthogonal expansion of the dynamical quantity, and the approach is akin to the Zwanzig [13] and Mori [14] formalisms. The dynamical evolution of the magnetic moment is expressed in terms of an infinite set of time-independent vectors in an appropriate space and chosen so that the desired correlations are just the projections of the evolving magnetic moment operator on to the first of the orthogonal vectors. The equations of motion of the coefficients of the expansion are obtained and solved as Laplace transforms, and the transform of the correlation functions are obtained as the ratios of infinite order determinants. The method is related to the moment expansion method, and in principle it too requires a knowledge of all the moments of the corresponding spectral function; nevertheless it has the advantage of making more efficient use of the few moments that are in practice available, and it is expressed in terms of a form that can be parameterized in a lucid fashion. A section is devoted to calculations based on various models of memory functions, and a perturbation expansion, and the method is applied to a few examples.

Clauser and Blume [15] have described a stochastic theory of lineshapes which although directed towards Mössbauer spectra has some features that are also applicable to e.s.r. (and n.m.r.). The theory is constructed quantum-mechanically throughout so that off-diagonal terms in the hyperfine interaction may be taken into account in a proper fashion. They aim to extend the theory which treats the hyperfine modulation as a classical modulation modelled by an external field (and which therefore can have no off-diagonal elements).

Spherical tensors, because of their transformation properties, are a particularly helpful way of expressing a relaxation perturbation when rotational modulation is the mechanism of relaxation; several papers describing particular aspects of their application have appeared,[5, 16–18] and the technique is now common in work on spin relaxation.

3 Slowly Rotating Molecules

The motion of rotating radicals has received particular attention during the year, and it is now possible to predict the spectral lineshape for virtually all speeds and modes of rotation.

[13] R. Zwanzig, 'Lectures in Theoretical Physics', ed. W. E. Britten, B. W. Downs, and J. Downs, Interscience, New York, 1961, vol. 3, p. 135.
[14] H. Mori, *Progr. Theor. Phys.* (*Kyoto*), 1965, **34**, 399.
[15] M. J. Clauser and M. Blume, *Phys. Rev.* (*B*), 1971, **3**, 583.
[16] S. Sykora, *J. Chem. Phys.*, 1971, **54**, 2469.
[17] N. C. Pyper, *Mol. Phys.*, 1971, **21**, 1.
[18] P. W. Atkins, *Mol. Phys.*, 1971, **21**, 97.

A general analysis of fast and slow motional effects in spectra has been given by Clauser [19] by determining linewidths directly as eigenvalues of superoperators. Results for slow and fast motion have been given, and the structure of the theory simplifies the problem by revealing how a particular term in the hamiltonian affects the appearance of the spectrum. When the relaxation is slow it is found that the broadening is proportional to the relaxation rate, and the shift is proportional to the square of the latter; when the relaxation is fast the hyperfine positions are determined primarily by the thermally averaged effective hamiltonian and the extent to which the lines are broadened (or narrowed) is inversely proportional to the relaxation rate. If the hamiltonian contains terms off-diagonal in the electronic states, a line-broadening mechanism is available which is most effective when the relaxation rate corresponds to the electronic level spacing.

Sillescu [20] has extended the Debye model and the rotational jump model in order to allow for the time fluctuation of the rotational diffusion constant and the jump rate constant which arises because of the rapid local fluctuations of the environment of a particular radical:[21, 22] fluctuations that decay rapidly compared with the rotational correlation time lead to an exponential correlation time, whereas slow fluctuations lead to a distribution of correlation times. The paper introduces higher approximations to the molecular reorientation process in a systematic manner: in the first approximation the reorientation is described by a one-dimensional Markov process and the exponentially decaying correlation functions by one parameter, such as the rotational diffusion constant. In the second approximation this parameter is allowed to fluctuate, and these fluctuations are themselves described by a one-dimensional Markov process so that the overall process is now two-dimensional and is described by non-exponential correlation functions. Correlation functions for spin relaxation are discussed in detail, and the e.s.r. lineshape is calculated for a case where the rotation is so slow that the rotational correlation time is comparable to the spin relaxation time. When this is so the Debye model and the jump model (an activated process) lead to markedly different shapes. This arises because $3\cos^2\theta - 1$ has its maximum value at $\theta = 0$ and its maximum slope at $\theta = \pi/4$: if reorientation occurs through small angular steps ($\delta\theta \ll \pi/4$) a more efficient relaxation mechanism is provided in the vicinity of $\theta \sim \pi/4$ than at $\theta \sim 0$. This gives rise to a bump in the lineshape in the high-frequency region of the spectrum which is averaged out when reorientation occurs through arbitrarily large angles as in the jump model. These differences disappear when the rotation is fast.

Gordon and Messenger [23] have presented a treatment of a lineshape calculation based on the rotational diffusion equation but which avoids

19 M. J. Clauser, *Phys. Rev.* (*B*), 1971, **3**, 3748.
20 H. Sillescu, *J. Chem. Phys.*, 1971, **54**, 2110.
21 W. Kauzmann, *Rev. Mod. Phys.*, 1942, **14**, 12.
22 J. E. Anderson and R. Ullmann, *J. Chem. Phys.*, 1967, **47**, 2178.
23 R. G. Gordon and T. Messenger, ref. 1, p. 341.

both adiabatic and perturbation approximations. The paper analyses the equation of motion

$$(\partial/\partial t)\{\mu_{if}(\Omega, t)\, P(\Omega, t)\} = (i/\hbar)\, [H(\Omega), \mu]_{if}\, P(\Omega, t) + \mu_{if}\, D\nabla^2_{\Omega}P(\Omega, t) \quad (16)$$

where μ_{if} is a magnetic moment matrix element and $P(\Omega, t)$ is the probability that the molecule has the orientation Ω at the time t: the first term on the right is the quantum-mechanical evolution of the moment and the second is the classical diffusive evolution of the molecular orientation. In matrix notation the Fourier transform of equation (16) after integration by parts is

$$\{\omega\mathbf{1} - [L(\Omega) - i\mathbf{1}D\nabla^2_{\Omega}]\}\cdot f(\Omega, \omega) = \mu(0)\, P(\Omega, 0) \quad (17)$$

f being the one-sided transform of $\mu(\Omega, t)P(\Omega, t)$ and L the Liouville operator $(1/\hbar)\,[H(\Omega), \ldots]$. The problem is to solve equation (17), for its solutions immediately give the spectral intensity through the relation

$$I(\omega) = (1/\pi)\, im\langle\mu(0)\cdot f(\Omega, \omega)\rangle. \quad (18)$$

Gordon and Messenger describe two approaches: a solution by a finite difference method and a solution in terms of an expansion of orthogonal functions. The latter is helpful if the hamiltonian can be written in terms of a few of the orthogonal polynomials, for the matrix elements are then easily evaluated, but the finite difference method is more flexible because any spin hamiltonian may be used. Gordon and Messenger describe a transformation of equation (17) which enables the frequency dependence to be dealt with after a diagonalization procedure: this greatly reduces the labour (or time) for the calculation, for the lineshape can be calculated from a diagonalized matrix and the transformation matrix, and the latter need be calculated only once for the particular system rather than at each point of the spectrum. Lynden-Bell [24] has shown how the transformation properties of the Liouville operator $L(\Omega, t)$ may be used to simplify the calculation procedure even further. The diagonalization procedure [25] (a sequence of Jacobi rotations to complex, symmetrical, tridiagonal form, followed by a modification of a QR procedure [26]) is described in detail, and the whole computer program is published. Some applications to tumbling triplet molecules and nitroxyl radicals are described: the agreement between theory and experiment is excellent, and rotational correlation times can be determined with some precision. The results emphasize the importance of including at least pseudo-secular terms in the hamiltonian, for they have a marked effect on the appearance of the spectrum (electron flipping non-secular terms are only important at fast tumbling rates, and this region can be dealt with by perturbation procedures, but see ref. 27).

[24] R. M. Lynden-Bell, ref. 1, p. 383.
[25] H. Rutishauser, *Proc. Amer. Math. Soc. Symp. Appl. Maths.*, 1963, **15**, 219.
[26] J. G. F. Francis, *Computer J.*, 1962, **4**, 265; 332.
[27] S. A. Goldman, G. V. Bruno, C. F. Polnaszek, and J. H. Freed, *J. Chem. Phys.*, 1972, **56**, 716.

Freed, Bruno, and Polnaszek[28, 29] have treated a similar problem for slowly tumbling doublets and triplets by a method based on Kubo's stochastic Liouville approach[30] (like Gordon and Messenger[23]) and an expansion in orthogonal polynomials. For example, when the stochastic Liouville operator corresponds to the rotational diffusion operator the appropriate polynomials are the Wigner rotation matrices. In this way are generated an infinite set of coupled differential equations for the coefficients of the polynomials, but because of the finite rank of the relaxing perturbation (typically 2) only a few equations are directly coupled to a particular equation (for example the equation for the coefficient C_L may be directly coupled only to the equations for C_{L+2} and C_{L-2}). The equations are decoupled at some point by setting $C_L = 0$ for an appropriately high value of L. A typical result for a spin-$\frac{1}{2}$ system in the absence of saturation and in which the relaxation perturbation is the secular effect of an axially symmetric g-tensor is that as $|(g_{\parallel} - g_{\perp})/D|$, with the diffusion constant D in appropriate units, approaches unity the line broadens; where the ratio lies between 1 and 10 the resonance peak shifts downfield and becomes asymmetric; between 10 and 100 a new high-field peak appears; at 1000 the features of the solid-like spectrum sharpen and the powder spectrum emerges. Significant differences of lineshape are calculated in the case of a nitroxyl radical when different reorientational models are used, and the importance of including pseudo-secular terms from the nuclear dipolar hyperfine interaction is demonstrated (thus again illustrating the inadequacy of adiabatic models in systems of these dynamic characteristics). Saturation can be treated, and by an appropriate modification (the elimination of the high-field approximation when the zero-field splitting is large) triplet lineshapes can also be calculated.[28, 29] For a theory of triplet lineshapes a start is made from linear response theory and the lineshape is expressed in terms of the magnetization correlation function. The susceptibility, in terms of the transform

$$S_q(s) = \int_0^\infty \mathrm{d}t S_q(t) \exp(-st) \tag{19}$$

may be written

$$\chi''(\omega) = (\gamma_e^2\omega/2NkT)\,\mathrm{tr}\,\{[\langle S_q(\mathrm{i}\omega)\rangle + \langle S_q(-\mathrm{i}\omega)\rangle]\,S_q\} \tag{20}$$

where the trace is over spin degrees of freedom and the angle brackets indicate a classical orientational average. The equation for $\langle S_q(s)\rangle$ is obtained from the stochastic Liouville equation in the same form as equation (17) and the solutions expanded in terms of orthogonal polynomials; the hamiltonian for the spin system now contains the zero-field

[28] J. H. Freed, G. V. Bruno, and C. F. Polnaszek, *J. Chem. Phys.*, 1971, **55**, 5270.
[29] J. H. Freed, ref. 1, p. 387.
[30] R. Kubo, *Adv. Chem. Phys.*, 1969, **16**, 101; *J. Phys. Soc. Japan*, 1969, **26**, Suppl. 1.

terms. The procedure once again yields a set of coupled differential equations, which when truncated for $L > n$ is of dimension $2 + 9n/2$. Examples of high-, low-, and zero-field spectra are presented for a range of rotational correlation times. As an example of a result in the high-field limit, as the rotational motion increases from the rigid limit both the $\Delta m = 1$ and $\Delta m = 2$ lines broaden initially, but fast motion narrows the former until it becomes Lorentzian. The authors [28] point out that since the rotational motion has quite different effects on different types of triplet e.s.r. spectra one should expect to be able to obtain detailed information about the nature of rotational motion. For example, in the high-field cases the $\Delta m = 1$ transition is largely quantized in the molecular frame and broadening occurs until $D_{\text{spin-spin}}/D_{\text{diffusion}} \sim 6$ when narrowing sets in; but the $\Delta m = 2$ transition continues to broaden with a width contribution of $6D_{\text{diffusion}}$ because of the orientation dependence of its transition moment. At zero field, where the spins are quantized in the molecular frame, the slow tumbling broadens the lines by $4D_{\text{diffusion}}$.

This approach has been applied to a study of anisotropic rotational reorientation and slow tumbling of peroxylamine disulphonate in liquid and frozen media.[27] The range 10^{-11}—10^{-6} s in rotational correlation time was covered by varying the solvent (the rapid reorientation in ice suggests the formation of a clathrate) and activation energies for rotation were determined from the linewidth data. For correlation times shorter than about 10^{-9} s the normal relaxation theories are suitable, and the results could be interpreted in terms of axially symmetric rotational diffusion with the axis in the plane of the N, O, and S atoms and parallel to the S—S direction. When the rotational correlation time exceeded 10^{-9} s the stochastic Liouville approach,[28, 29, 31] extended to accommodate completely anisotropic *g*- and hyperfine tensors, was found to give quite good agreement with observation, and the agreement was improved by incorporating axially symmetric diffusion. Various types of diffusional process were considered, and some improvement over the results with Brownian diffusion were obtained by permitting large jumps to occur. Some evidence is presented to show that the non secular contributions are less then expected: this might indicate the importance of the short-time behaviour of the reorientation, which could deviate markedly from its long-time Brownian nature.

A number of other approaches to the slow tumbling lineshapes of e.s.r. spectra may be referred to. Kaplan [32] has treated slow tumbling by a modification of his theory of hindered motion for n.m.r.:[33] his equations are related to those found by Gordon and Messenger,[23] but are incomplete because only secular terms are retained. In the same contribution he has investigated the lineshape to be expected from radicals in thin films: he

[31] J. H. Freed, G. V. Bruno, and C. F. Polnaszek, *J. Phys. Chem.*, 1971, **75**, 3385.
[32] J. I. Kaplan, ref. 1, p. 443.
[33] J. I. Kaplan, *J. Chem. Phys.*, 1958, **28**, 278.

shows that for magnetic fields up to some critical value $\pi(K/\chi_a)^{\frac{1}{2}}/d$, where K is an elastic constant, χ_a the diamagnetic susceptibility, and d the thickness of the film, the lineshape does not vary, but above this value the spectrum broadens inhomogeneously. In the case of biradicals, Luckhurst and Pedulli [34] have shown that at high viscosities the modulation of the scalar electron–electron exchange in flexible molecules (which at high temperatures is a dominant relaxation mechanism) is quenched and modulation of the anisotropic magnetic interactions becomes dominant. Their paper gives an expression for the broadening arising from this process and applies the theory to the determination of configurations, electron–electron separations, and the sign of the exchange coupling.

The lineshape of a triplet molecule which is simultaneously interacting with the molecular environment and an externally applied field has also been calculated by Roberts and Lynden-Bell,[35] and because they are able to deal with adiabatic and non-adiabatic situations and for all rates of motion it is applied by them to the cases where the field is zero, weak, intermediate, and strong, and where the motion (which is assumed to be due to strong collisions) is slow or intermediate. The strong collision model supposes that for each collision the orientation of the molecule is changed; therefore the effect on a density matrix $\rho(\Omega)$ for molecules in an orientation Ω is

$$(\partial\rho(\Omega)/\partial t)_{\text{collision}} = -\,k\{\rho(\Omega) - \langle\rho(\Omega_1)\rangle_\Omega\} \tag{21}$$

where k is the collision frequency and $\langle\rho(\Omega_1)\rangle_\Omega$ is the average spin density matrix of all those molecules arriving at Ω after a collision: for strong collisions in an isotropic, homogeneous system this is equal to ρ itself (the ensemble average density matrix). Consequently, allowing for relaxation to an equilibrium state, the equation of motion employed is

$$\text{i}(\text{d}/\text{d}t)\,\rho(\Omega) = \{L_Z + L_M(\Omega) + L_{\text{r.f.}}\}\,\rho(\Omega) - \text{i}k\rho(\Omega) + \text{i}k(\rho - \rho_{\text{eq}}) \tag{22}$$

where the Liouville operators correspond to the Zeeman, the molecular, and the r.f. field. The formal solution of this equation is easily obtained in terms of the inverse operator $\{L_Z + L_M(\Omega) - (\omega + \text{i}k)\}^{-1}$ and so in principle the magnetization (and hence the lineshape) can be determined.

The application to a tumbling triplet was then made by numerical integration over the molecular orientations. The authors conclude that in the high-field case the lineshapes of a tumbling triplet are not very sensitive to the choice of a diffusional or a strong collision model for the reorientational process: this justifies the use of the latter when the applied field is not much larger than the molecular field, but also implies that the lineshapes are not particularly helpful in determining the mode of reorientation. In

[34] G. R. Luckhurst and G. F. Pedulli, *Mol. Phys.*, 1971, **20**, 1043.
[35] J. Roberts and R. M. Lynden-Bell, *Mol. Phys.*, 1971, **21**, 689.

the intermediate and slow motion regions the behaviour is strongly dependent on the relative magnitudes of the molecular field and the Zeeman field. In the low- and zero-field regions the powder spectrum is rapidly broadened even by slow motion, and soon disappears.

In a contribution on partially oriented systems, Nordio and Busolin [36] attempt to overcome defects of earlier models of reorientation in an anisotropic medium (which suffer either because they are impractical for quantitative calculation or because their assumptions are too restrictive). The authors employ a conventional Wigner rotation matrix expansion in the solution of the diffusion equation but employ a simplified model potential to represent the environment: this is chosen so as to make the coupled equations for the coefficients tractable (but not without truncation). This procedure leads immediately to the evaluation of the spectral densities: the components of the dipole and *g*-tensors relax with a spectrum of characteristic times which are functions of an alignment parameter, and all contributions to the linewidth vanish in the limit of complete ordering. In a following paper [37] the approach is generalized: both prolate and oblate radicals can be treated, and the restraining potential is less restrictive (and expressed as a Legendre polynomial expansion). The lineshape change on ordering is calculated for vanadyl complexes and nitroxyl radicals: for the former the shape seems to be determined almost completely by the quadratic term in the potential, in agreement with earlier work.[38, 39]

Two final papers on slow motion can be mentioned. Nitroxyl radical spectra have been calculated as a function of the rotational correlation time by numerical solution of the coupled Bloch equations.[40] For a dilute solution of spin-labelled oxyhaemoglobin the rotational correlation time was found to be about $26 \pm 2 \times 10^{-9}$ s, and it increased as sucrose was added. Lazarev and Strynkov [41] have monitored slow rotation and diffusion of the iminoxyl radical in viscous media and polymers (polythene) and note that the motion of a small radical in a polymer is due to migration of microscopic defects into the amorphous portion of the polymer, the migration rate being determined by the molecular motion of the polymeric molecule.

4 Miscellaneous Solids

The lineshape of powders has received some attention. Owen and Vincow [42] have reported a computer simulation of the powder spectra from radicals with extensively delocalized electron systems (naphthalene,

[36] P. L. Nordio and P. Busolin, *J. Chem. Phys.*, 1971, **55**, 5485.
[37] P. L. Nordio, G. Rigatti, and U. Segre, *J. Chem. Phys.*, 1972, **56**, 2117.
[38] P. G. James and G. R. Luckhurst, *Mol. Phys.*, 1970, **19**, 489.
[39] G. R. Luckhurst, ref. 1, p. 411.
[40] R. C. McCalley, E. J. Shimshick, and H. M. McConnell, *Chem. Phys. Letters*, 1972, **13**, 115.
[41] A. V. Lazarev and V. B. Strynkov, *Doklady Akad. Nauk. S.S.S.R.*, 1971, **197**, 627.
[42] G. S. Owen and G. Vincow, *J. Chem. Phys.*, 1971, **54**, 368.

anthracene, and perylene radical cations). Canters and Johnson[43] have reported calculations on the numerical evaluation of moments and shapes of magnetic resonance lines for crystals and powders. They evaluated the van Vleck moment equations numerically and obtained analytical expressions for the second and fourth moments of dipole broadened lines for cubic crystals and powders as a function of the spin concentration. They also discuss practical limits on the enhancement of resolution.

Other reports for solids of various kinds that have chemical significance have included the report by Acrivos and Azebu[44] of an extremely narrow fast-passage absorption in the amorphous blue and grey precipitates (and the solution) of the system Na–Pb–NH_3 ($T_1 \gg 10^{-2}$ s $\gg T_2 \geqslant 10^{-5}$ s) and they propose a metal–nonmetal transition induced by the formation of highly negatively charged polyanions (*e.g.* Pb_7^{4-}; Pb_9^{4-}) which reduce the electron-pair correlation time. Of some interest may be the report by Pifer and Magno[45] of the temperature dependence of the lineshape of the spectrum from lithium foil, and the observation (made possible by the use of a more sensitive apparatus) that Dyson's theory applies at high temperatures but deviations occur for temperatures less than 120 K. The temperature dependence of conduction electrons in bulk sodium metal, the relaxation being dominated by the weak spin–orbit coupling in combination with a strong electron–phonon interaction, is in agreement with theory,[46] but below 10 K impurities dominate. From $(T_1T_2)^{\frac{1}{2}}$ measurements, Lin and Kevan[47] have studied the spatial distribution of electrons trapped in γ-irradiated organic glasses (3-MP, TEA, and MTHF). They find that electrons are trapped in clusters at low doses; spur radii are found to be greater than 13.0 nm in 3-MP, 10.1 nm in TEA, and 6.3 nm in MTHF. The radius increases as the polarity decreases, and an indication of the distance travelled before the electrons are trapped can be obtained. As a change from trapped electrons, trapped hydrogen and deuterium atoms in γ-irradiated acidic glasses show anomalous differential saturation of the hyperfine components, which is ascribed to cross relaxation induced by paramagnetic metal ions.[48] In a similar vein to the electron clustering observation, it has been demonstrated that hydrogen atoms trapped in γ-irradiated calcium phosphates are also generated in an initially non-uniform distribution.[49] The possible concentration-dependent relaxation time of N_2^- in *X*-irradiated sodium azide at 4.2 K has been studied and found to be anisotropic and independent of concentration;[50] the relaxation is ascribed to an intramolecular process.

43 G. W. Canters and C. S. Johnson, *J. Magn. Resonance*, 1972, **6**, 1.
44 J. V. Acrivos and J. Azebu, *J. Magn. Resonance*, 1971, **4**, 1.
45 J. H. Pifer and R. Magno, *Phys. Rev.* (*B*), 1971, **3**, 663.
46 W. Kolbe, *Phys. Rev.* (*B*), 1971, **3**, 320.
47 D. P. Lin and L. Kevan, *J. Chem. Phys.*, 1971, **55**, 2629.
48 S. K. Wong and J. K. S. Wan, *J. Chem. Phys.*, 1971, **55**, 4940.
49 Y. P. Virmani, J. D. Zimbrick, and E. J. Zeller, *J. Phys. Chem.*, 1971, **75**, 1936.
50 B. Ford and E. Gelerinter, *J. Chem. Phys.*, 1971, **55**, 3660.

Although the van Vleck method for second and fourth moments is widely used for lineshape analysis in solids, one paper[51] draws attention to the fact that the method is erroneous in the presence of both a rapidly modulated (that is, narrowed) interaction with unlike spin species, and a broadening from like spins which exceeds the broadening arising from unlike spins. The determination of Debye temperatures by e.s.r. seems to be a more speculative hope than had been realized.[52]

5 Exchange Processes

The study of spin exchange in solids has continued: an example is the α(tetragonal) → β(monoclinic) polymorph transformation in some crystalline solid solutions of a metal phthalocyanine with a metal-free phthalocyanine.[53] Another example is a study of the angular dependence of the width of an exchange-narrowed line in a single crystal of TMPD:[54] the width follows the second moment as given by the secular part of the dipole–dipole interaction, a behaviour expected for spatially uncorrelated spin-excitations. As a curiosity from solid-state studies of exchange, attention may be drawn to a study of antisymmetric exchange effects in exchange-narrowed lines.[55] The Dzialoshinskii–Moriya antisymmetric exchange (which is of the form $\mathscr{D}(\mathrm{i},\mathrm{j})\cdot \boldsymbol{s}_\mathrm{i} \wedge \boldsymbol{s}_\mathrm{j}$) is shown to contribute in the paramagnetic phase in $Cu(CO_2H)_2,4H_2O$, and in a manner similar to the symmetric anisotropic exchange and a dipole–dipole interaction.

Zitzerman[56] has investigated the generalization of the Kaplan–Alexander theory of exchange to high spins, and in particular he treats the case of a spin-S in collision with a spin-$\frac{1}{2}$. He attempts to reduce the complexity of Lynden-Bell's calculation based on the collision of triplet excitons,[57] and criticizes the model on the grounds that the choice of a dimer density matrix is artificial because the instantaneous dimer species do not contribute to the absorption signal. He preserves a one-particle description throughout and assumes that the solution is dilute, the exchange integral is constant during the collision time, the exchange hamiltonian is of the form $j\boldsymbol{s}_1\cdot\boldsymbol{s}_2$, the characteristic splittings in the spectrum (Δ) are short in the sense $\Delta\tau_c \ll 1$, τ_c being the collision duration. By examining the structure of the evolution operator for such a model the explicit forms for the relaxation matrix elements are obtained and examples are discussed, in particular the broadening effect of an arbitrary spin on a spin-$\frac{1}{2}$ is described.

Salikhov *et al.*[58] describe the exchange effects for solutions of free radicals and transition-metal complexes. The point of using such systems

[51] R. E. Walstedt, *Phys. Rev. (B)*, 1972, **5**, 41.
[52] J. S. M. Harvey, *Chem. Phys. Letters*, 1971, **10**, 62.
[53] M. Abkowitz and I. Chen, *J. Chem. Phys.*, 1971, **54**, 811.
[54] T. Z. Huang, R. P. Taylor, and Z. G. Soos, *Phys. Rev. Letters*, 1972, **28**, 1054.
[55] T. Castner and M. S. Seehra, *Phys. Rev. (B)*, 1971, **4**, 38.
[56] V. Yu. Zitzerman, *Mol. Phys.*, 1971, **20**, 1005.
[57] R. M. Lynden-Bell, *Mol. Phys.*, 1964, **8**, 71.
[58] K. M. Salikhov, A. B. Doctorov, Yu. N. Molin, and K. I. Zamaraev, *J. Magn. Resonance*, 1971, **5**, 189.

is that the strength of the exchange interaction may be varied over a wide range by changing the ligands of the complex. The theory of such systems must take into account the possibility that the spin–lattice relaxation times of the complexes are shorter than the duration of the collisions, and the spins involved may exceed $\frac{1}{2}$. The authors set up a collision matrix for the effect of the exchange and deduce the lineshape from it. A variety of results are obtained for different collisional conditions, and are tabulated. A series of experiments on 2,2,6,6-tetramethyl-4-oxopiperidine 1-oxyl, αα-diphenyl-β-picrylhydrazyl (DPPH), and the bisdiphenylchromium cation with complexes based on vanadyl, Co^{II}, and Ni^{II} support the theoretical conclusions. One result predicted and supported is that when $T_1 \gg \tau_c$ the collision efficiency for a typical experimental condition $\Delta\tau_c \ll 1$ should be smaller than unity even when $J\tau_c \gg 1$.

In the case of chemical exchange, Leigh[59] has shown how the Bloch equations may be solved exactly for a two-site system and investigated by a variety of approximations. In the absence of exchange one naturally obtains two relaxation times (T_{1A} and T_{1B}), when one seeks to describe the effect of electron and nuclear relaxation. For $T_{1A} \gg T_{1B}$ and $\tau_A \gg \tau_B$ one obtains

$$1/\bar{T}_1 = 1/T_{1A} + N_A/N_B(\tau_B + T_{1B}) \tag{23}$$

which is a familiar equation, but derived with less restriction. In the case of T_2 under the conditions $T_{2A} \gg T_{2B}$ and $\tau_A \gg \tau_B$ one obtains the equation originally derived by Swift and Connick.[60]

Lang and Freed[61] have made a careful study of spin exchange in a binary liquid solution near the critical point. Mixtures of 2,2,4-trimethylpentane and perfluoro-n-heptane (which are known to have a large critical anomaly in the kinematic viscosity) were studied by introducing a small amount of di-t-butyl nitroxide and concentrating on the spin exchange, which depends on the translational diffusion of the radicals. The difference in the linewidths in the presence (δ_m) and absence [$\delta_m(0)$] of exchange is given by

$$\delta_m - \delta_m(0) = A_m^{-1}\, C_{T_0}\{T/\nu(T)\} \tag{24}$$

where

$$A_m = 3\sqrt{3} \times 10^3 f_m \mid \gamma_e \mid \rho(T_0)\, a/4dfN_A\, k \tag{25}$$

where m is the spectral index of the line, f_m a statistical factor for each hyperfine line, ρ the density, a the molecular radius, d the encounter distance of two radicals undergoing exchange, f a factor taking into account the charge of the species ($f = 1$ for uncharged radicals), and C the molar concentration, with $C(T) = C_{T_0}\{\rho(T)/\rho(T_0)\}$. The factor A_m is independent of temperature and the experiments are based on the dependence on ν

[59] J. S. Leigh, *J. Magn. Resonance*, 1971, **4**, 308.
[60] T. J. Swift and R. E. Connick, *J. Chem. Phys.*, 1962, **37**, 307.
[61] J. C. Lang and J. H. Freed, *J. Chem. Phys.*, 1972, **56**, 4103.

of the linewidth as given by equation (24). The linewidth was found to satisfy an equation of this form in the non-critical region, except for the existence of a puzzling non-zero intercept in some cases. Within the critical region it was found that the linewidth contribution is not linear in T/ν, but when a quantity $\nu' = \nu - \Delta\nu$ is defined ($\Delta\nu$ being the contribution from critical effects) then the contribution is linear in T/ν' both inside and outside the critical region: this shows that the molecular diffusion is unaffected by the contribution to the kinematic viscosity due to long-range correlations of concentration fluctuations. Furthermore, it may be noted that the critical fluctuations (with correlation time $\sim 10^{-3}$ s) are essentially static in comparison with the exchange frequency ($1/\omega_{HE} \sim 10^{-8}$ s). The paper also includes an account of the corrections necessary in the presence of unresolved hyperfine structure.

Highly concentrated aqueous solutions of manganese nitrate have been studied [62] and the collapse of the structure analysed by a computer simulation using component linewidths and positions calculated from Anderson's theory, in which transitions occur only between adjacent hyperfine components.

Hughes and Soos [63] have drawn attention to the similarity in one-dimensional spin dynamics between the inorganic system Me_4NMnCl_3 and a variety of organic charge-transfer and free-radical crystals which have been analysed as Heisenberg linear antiferromagnetics. The theoretical analysis developed for organic systems is also applicable to inorganic systems and both provide good examples of linear antiferromagnetics.

6 Spin–Rotation Relaxation

Relaxation processes involving spin–rotation interactions in radicals continue to be found in a variety of systems. One example is the fluorosulphate radical (SO_3F) where a study of the temperature dependence of the linewidth (which increased with temperature) enabled most of the alternative processes to be eliminated, leaving spin–rotation as the favoured mechanism.[64] Likewise, a study of chlorine dioxide (a favourite in this matter) has shown a not unexpected deviation of the motional correlation time from the modified Debye theory, and spin–rotation interaction is considered.[65]

A theoretical account of the spin–rotation relaxation process has been given [66] as an illustration of the application of the formulation of relaxation theories. One particular problem encountered is the presence of a bivariate correlation function when the spin–rotation coupling tensor is anisotropic: the relaxation depends on the joint orientation and angular momentum

[62] E. F. Strother, H. A. Farach, and C. P. Poole, *Phys. Rev.* (*A*), 1971, **5**, 2079.
[63] R. C. Hughes and Z. G. Soos, *Phys. Rev.* (*B*), 1971, **4**, 3253.
[64] P. M. Nutkowsky and G. Vincow, *J. Phys. Chem.*, 1971, **75**, 712.
[65] L. Burlamacchi and M. Romanelli, *Chem. Phys. Letters*, 1971, **10**, 59.
[66] P. W. Atkins, ref. 1, p. 279.

correlations (such a correlation function has been classified[5] as of type A_2) and when the correlation times for the reorientation and angular momentum are very different it has been the practice[67, 68] to factorize the correlation function into functions for the two modes separately (or $A_2 \rightarrow A_1'A_1''$ in the classification scheme, A_1' denoting a correlation function of one orientation variable, and A_1'' denoting a correlation function for one momentum variable). There appear to be two procedures depending on where one decouples the angular momentum and interaction tensors:[66] one may first decouple $\boldsymbol{C}^{(2)}$ and $\boldsymbol{J}^{(1)}$ and then refer $\boldsymbol{C}^{(2)}$ to the rotating molecular axes (in which we know its components), or one may refer the contraction $(\boldsymbol{C}^{(2)}\boldsymbol{J}^{(1)})^{(1)}$ to the molecular axes and only then effect the decoupling. The overall correlation function can be written as an exponential with a time constant τ_{J2} (with $\tau_{J2}^{-1} = \tau_J^{-1} + \tau_2^{-1}$) for the first scheme and τ_{J1} (with $\tau_{J1}^{-1} = \tau_J^{-1} + \tau_1^{-1}$) for the second: τ_1 and τ_2 are the correlation times for first- and second-rank spherical harmonics, respectively. In order to assess which coupling procedure is more correct one must assess the degree of correlation of the decoupled components: in the first case one must suppose that the angular momentum components in the laboratory frame are uncorrelated with the orientation, and in the latter one must suppose that the angular momentum components in the rotating molecular frame are uncorrelated with orientation. It is suggested[66] that the latter decoupling is more nearly complete: if this is so then it is probably the first example of a first-rank spherical harmonic correlation function (which, of course, are of principal importance in dielectric relaxation). Two approaches to the solution of this problem may be by considering McClung's theory of diffusion[69] where, to a particular approximation, it is unnecessary to decouple the modes, or to deal with the classical bivariate correlation function which can be determined for Brownian motion.[70] A step towards this has been taken by Freed[71] who has adapted Chandrasekhar's solution for rotation in a plane.[70]

Spin–rotation processes also dominate the spin relaxation above 320 K in diethyl phosphate- and chloro-complexes of iron in diethyl hydrogen phosphate,[72] but the linewidths of copper acetate in acetic acid solutions can be accounted for without invoking the process.[73]

7 Molecular Rotation

It is difficult to decide which papers one ought to include in this section: the determination of the characteristics of molecular motion is the main object of many e.s.r. studies, yet the papers that aid interpretation are often

[67] P. S. Hubbard, *Phys. Rev.*, 1963, **131**, 1155.
[68] P. W. Atkins and D. Kivelson, *J. Chem. Phys.*, 1966, **44**, 169.
[69] R. E. D. McClung, *J. Chem. Phys.*, 1969, **51**, 3842.
[70] S. Chandrasekhar, *Rev. Mod. Phys.*, 1943, **15**, 1.
[71] J. H. Freed, *J. Chem. Phys.*, 1972, **56**, 1407.
[72] G. P. Vishnevkaya, A. F. Karimova, B. M. Kozyrev, and A. A. Muratova, *Doklady Akad. Nauk S.S.S.R.*, 1971, **196**, 377.
[73] G. Nyberg, *J. Phys. Chem.*, 1971, **75**, 2228.

concerned solely with statistical mechanics. Here too e.s.r. merges most successfully with n.m.r. A few pages that seem to have particular pertinence to spin–relaxation studies are selected for comment.

A classification of molecular motion has been described[5] in which Class A motions are those that involve the rotation or translation of the radicals or the motion (such as jumps or internal rotation) of groups of nuclei or individual nuclei within the radicals and Class B motions are those that involve the vibrational or electronic excitation of the radical, but do not involve motions of the type of Class A. The former class contains all the 'conventional' relaxation processes in fluids, and the latter contains processes that are of dominating importance in solids but which also appear to be of some significance in liquids in order to account for contributions to the linewidth that do not seem to be explicable in terms of rotational processes. These classes are divided into A_1 and A_2 depending on whether the motion depends on one dynamical variable or two (and so gives rise to a bivariate correlation function), and into B_r and B_v depending on whether the excitation process is real (as in an Orbach process) or virtual (as in a Raman process). A_1 motions are further subdivided into A_1' and A_1'', where the former denotes a motion that depends on the position variable, and the latter that the motion depends on a momentum variable. Correlation functions relating to a particular class of motion are classified in the same way as the motion itself. An example of this classification scheme is given in the section on spin–rotation theory.

Berne[74] has described a hydrodynamic theory of the angular velocity autocorrelation function, which he computes by using a frequency-dependent version of Stokes' law and which takes into account the fact that, because of molecular collisions, the angular velocity of a rotating sphere depends on the time. The drag torque on a sphere rotating non-uniformly about one of its diameters is considered, and it is shown that the autocorrelation function for the angular velocity decays asymptotically as $t^{-5/2}$. This long, persistent tail has consequences for orientational correlation functions. For example, the Debye model is based on the assumption that the angular momentum decays very rapidly, but the $t^{-5/2}$ behaviour throws this assumption into doubt. In an approximate calculation it is shown that the asymptotic behaviour of the orientational correlation function is

$$D_l(t) \rightarrow \tfrac{3}{4}(2/\pi)^{\frac{1}{2}}\,\{l(l+1)\,D_{\mathrm{R}}\}^{-1}\,D_{\mathrm{R}}\,\lambda^3\,t^{-5/2} \quad (t \rightarrow \infty)$$

where $\lambda^2 = a^2\rho/2\eta$, a the radius, ρ the density, η the viscosity, and $D_{\mathrm{R}} = kT/8\pi\eta a^3$. Thus a $t^{-5/2}$ tail is predicted for reorientation.

The modified Debye theory of Kivelson *et al.*,[75, 76] in which the hydrodynamic radius is modified by a parameter κ which is related to the anisotropy of the solute–solvent interaction, has received some support in a study

[74] B. J. Berne, *J. Chem. Phys.*, 1972, **56**, 2164.
[75] D. Kivelson, M. G. Kivelson, and I. Oppenheim, *J. Chem. Phys.*, 1970, **52**, 1810.
[76] D. Kivelson and R. E. D. McClung, *J. Chem. Phys.*, 1968, **49**, 3380.

of vanadyl complexes in a variety of alcohols.[77] The theory requires $0 < \kappa \leqslant 1$, and the fitted values of κ all lie within this range. Activation energies for rotational diffusion are reported, and correlated with those for viscosity: the unusual results for methanol are mentioned. A discrepancy is observed between the calculated and observed residual linewidths (after allowing for spin–rotational effects) at low temperatures (although reasonable at high temperatures) which appears to be unrelated to solvent properties, and it is speculated that it might be due to electric-field fluctuation effects.[78, 79] Mention has been made in another section of Sillescu's analysis of the extension of the Debye theory to allow for time fluctuation in the rotational diffusion constant.[20] Nir and Stein [80] have investigated two modes of diffusion in liquids: they propose that a flow mechanism, based on the movement of solvent molecules, is the only one available to large diffusants with a molecular weight greater than ten times that of the solvent; smaller diffusants can migrate in this way, but they can also move by activated diffusion within the lattice of the solvent matrix. The role of the size of the diffusant is discussed. Further evidence for incipient gas-like jumps through angles approaching 1 rad has been obtained from n.m.r. and i.r. studies [81] (and, incidentally, that translation involves motion incorporating jumps of about 0.5 nm), and O'Reilly and Peterson [82] have analysed the self-diffusion and rotational correlation times in toluene. They demonstrate that the microviscosity theory of Gierer and Wirtz [83] is inapplicable to most liquids and find as a general rule that molecules rotating rapidly and isotropically just below the melting point have a dynamically coherent rotation in the liquid (the barrier for rotation is much less than kT at a liquid lattice site). In most liquids the barrier is high and free rotation not possible: rotation then occurs by a hard collision upon translation into an interstitial site, and is quenched when the molecule returns to a more typical environment.

8 Alternating Linewidths

One of the reasons why the Kubo–Tomita–Kivelson theory of e.s.r. linewidths has been displaced by theories more overtly based on density matrices has been the belief that it was incapable of accounting for alternating linewidths.[84, 85] But as the theory is a general framework for relaxation processes it would be surprising if the belief were valid: in fact it is not valid, for it has been shown [18] that the secular contributions to the width

[77] N. S. Angerman and R. B. Jordan, *J. Chem. Phys.*, 1971, **54**, 837.
[78] D. Kivelson, *J. Chem. Phys.*, 1966, **45**, 1324.
[79] P. W. Atkins, *Mol. Phys.*, 1967, **12**, 201.
[80] S. Nir and W. D. Stein, *J. Chem. Phys.*, 1971, **55**, 1598.
[81] W. G. Rothschild, *J. Chem. Phys.*, 1971, **55**, 1402; 1972, **56**, 1402.
[82] D. E. O'Reilly and E. M. Peterson, *J. Chem. Phys.*, 1972, **56**, 2262.
[83] T. Gierer and K. Wirtz, *Z. Naturforsch.*, 1953, **8b**, 532.
[84] D. Kivelson, *J. Chem. Phys.*, 1964, **41**, 1904.
[85] J. H. Freed and G. K. Fraenkel, *J. Chem. Phys.*, 1963, **39**, 326.

do lead to an alternation, and Kivelson[86] has extended the argument to include non-secular terms. The theoretical difficulty had lain in the exponentiation procedure of the Kubo–Tomita–Kivelson theory.

The theory is in good shape, and so too now are the lines, for alternation effects continue to be reported and interpreted in terms of molecular dynamic processes. Subramanian and Narasimhan[87] report on the spectra of electrogenerated radical ions in the range from + 20 to − 80 °C, and in particular sterically hindered 2,2′-dinitrobiphenyl and 1,8-dinitronaphthalene, and account for the lineshape in terms of a two-site jump model with out-of-phase correlation. van den Hoek *et al.*[88] have made a comparative study of a series of 4-aryloxyphenoxyl and 4-methoxyphenoxyl radicals, and discover that the *ortho* couplings are inequivalent at 20 °C, but at higher temperatures the symmetrical linewidths indicate that fast internal motion is occurring. Claridge *et al.*[89] find it possible to explain the alternating linewidths in perinaphthalene, pleidane, 1,2,3,6,7,8-hexahydropyrene, and dypleiadene in terms of two- and four-jump models. Pasimeni *et al.*[90] monitored durosemiquinone and alkali-metal ion-pairs at different temperatures, and from a two-jump model determine an energy barrier for K, Rb, Cs migration and a likely position for the cation relative to the ring. The linewidths support the assignment of a positive sign to the metal hyperfine interaction. Chen *et al.*[91] have reported on the two-jump model as applied to ketyl ion-pairs in ethereal and polar solvents, and find it satisfactory: it is shown that ketyls never exist as solvent-separated ion-pairs under the conditions employed. Elson *et al.*[92] have observed an alternating linewidth effect in the spectrum from cyclohex-2-enone radical anion and its simple alkyl derivatives; they find that the conformational lifetime is about 7×10^{-8} s in the parent molecule and 3×10^{-8} s in the 5,5-dimethyl derivative. Jones and Symons[93] have noted that the spectra of *m*-dinitrobenzene anions are characterized by fast exchange (appearing as linewidth alternation) in alcoholic media and slow exchange in water: they sought to employ this difference to determine how water is modified by co-solvents, in particular t-butanol. The mean lifetime of the asymmetrically solvated units is found to increase from 8×10^{-7} s at 291 K to 4.5×10^{-6} s at 282 K. Addition of t-butanol in low concentration slightly increases the lifetime, but when the concentration exceeds 0.04 mol fraction the lifetime decreases rapidly. Dohrmann *et al.*[94] have investigated the effect of pH on the alternation effect in alloxan, parabanic acid, and related compounds,

[86] D. Kivelson, ref. 1, p. 213.
[87] J. Subramanian and P. T. Narasimhan, *J. Chem. Phys.*, 1972, **56**, 2572.
[88] W. J. van den Hoek, J. F. Th. de Winter, and J. Smidt, *J. Magn. Resonance*, 1972, **6**, 15.
[89] R. F. C. Claridge, B. M. Peake, and R. M. Golding, *J. Magn. Resonance*, 1972, **6**, 29.
[90] L. Pasimeni, M. Brulstolon, and C. Corvaja, *J. C. S. Faraday II*, 1972, **68**, 223.
[91] K. S. Chen, S. W. Mao, K. Nakamura, and N. Hirota, *J. Amer. Chem. Soc.*, 1971, **93**, 6004.
[92] I. H. Elson, T. J. Kemp, and T. J. Stone, *J. Amer. Chem. Soc.*, 1971, **93**, 7091.
[93] D. Jones and M. C. R. Symons, *Trans. Faraday Soc.*, 1971, **67**, 961.
[94] J. K. Dohrmann, R. Livingston, and H. Zeldes, *J. Amer. Chem. Soc.*, 1971, **93**, 3343.

and have also observed [95] alternation in radicals from uracil, orotic acid, iso-orotic acid, and thymine. Ion-pairs of sodium or potassium naphthalenide in mixed solvents over a wide temperature range are the subject of a report by Lee *et al.*,[96] and alternation in nitrosamine radical ions has also been reported.[97] These effects are dealt with more fully in the chapter on dynamical processes.[98]

9 Miscellaneous Effects

Values of T_1 and T_2 for the benzene radical anion have been obtained over a range of concentrations.[99] A least-squares fit of the data gave (in seconds)

$$T_2 = (6.22 \pm 0.12) \times 10^{-5}/T - (1.47 \pm 0.11) \times 10^{-7} \quad (173\text{ K} < T < 243\text{ K})$$

$$T_2 = -(5.11 \pm 0.26) \times 10^{-5}/T + (5.04 \pm 0.30) \times 10^{-7} \quad (153\text{ K} < T < 173\text{ K})$$

$$T_1 = -(3.46 \pm 0.62) \times 10^{-5}/T + (4.01 \pm 0.80) \times 10^{-7} \quad (153\text{ K} < T < 213\text{ K})$$

although there was considerable scatter on the T_1 temperature dependence. The radical is of particular interest because it is an example of an orbitally degenerate species, and there is considerable evidence that the radical is ion-paired. An analysis of the effect of unresolved metal hyperfine structure,[100] on the assumption that the lifetime of the ion-pair is long compared to the inverse metal hyperfine frequency, led to the observation that, as the ratio of the hyperfine splitting to the component linewidth is increased for Lorentzian components, there is increased decay of amplitude in the wings of the spectrum and the observed peak-to-peak linewidth increases to approximately three times the individual component width just before the components begin to be resolved. The experimental results show a reasonably good fit to these predictions out to about three half-widths. This work is extended in a later paper [101] which shows that the source of the hyperfine structure is the potassium ion and that the inhomogeneity of the radical lines introduces an error into the measurement of spin–lattice relaxation times by the saturation technique. This error can be reduced if the correct T_2 is used in the determination of T_1, and the paper reports improved values of the latter:

$$T_1 = -2.00 \times 10^{-5}/T + (3.53 \pm 0.21) \times 10^{-7} \quad (154\text{ K} < T < 219\text{ K})$$

[95] J. K. Dohrmann and R. Livingston, *J. Amer. Chem. Soc.*, 1971, **93**, 5363.
[96] L. Lee, R. Adams, J. Jagur-Grodzinski, and M. Swarc, *J. Amer. Chem Soc.*, 1971, **93**, 4149.
[97] G. R. Stevenson and C. J. Colón, *J. Phys. Chem.*, 1971, **75**, 2704.
[98] R. F. Adams, this volume, Chapter 5.
[99] R. D. Rataiczak and M. T. Jones, *J. Chem. Phys.*, 1972, **56**, 3898.
[100] M. T. Jones, M. Komarijnsky, and R. D. Rataiczak, *J. Phys. Chem.*, 1971, **75**, 2769.
[101] M. T. Jones and M. Komarijnsky, *J. Chem. Phys.*, 1972, **56**, 4404.

Evidence continues to accumulate that the relaxation in a variety of high-spin complexes is due to the collisional modulation of the zero-field interaction. In a study of solvated iron, manganese, and chromium ions, Rubinstein *et al.*[102] have derived theoretical expressions for T_1 and T_2 and have discovered that the mean lifetime between relaxing collisions lies in the range 4—9 $\times$ 10^{-12} s. Reed *et al.*[103] have examined the frequency and ligand dependence of manganous complexes in aqueous solution and they too find modulation correlation times in the range 3—9 $\times$ 10^{-12} s at 300 K. Two zero-field splitting contributions could be distinguished, the static and the transient, and no connection between their magnitudes could be discerned. Burlamacchi *et al.*[104, 105] have also contributed to this problem by studying manganous and ferric ions in a variety of solvents chosen because they can attain high viscosities before they solidify. It is concluded that the modulation of the zero-field interaction is not the only effective process in most cases. Another contribution is due to Reuben,[106] who has studied aqueous solutions of gadolinium(III) aquo, cacodylate, and bovine serum albumin complexes, and ascribes the relaxation to the modulation of the transient zero-field splitting of Gd^{III}. He makes his analysis quantitative by using the ^8S-state relaxation theory of Hudson and Lewis[107] and diagonalizing the relaxation matrix numerically.

Linewidths in gases have not been the subject of many reports. Uttee[108] has studied the line-broadening cross-sections of F and I resonances and concludes that they exceed the geometrical cross-sections determined from transport properties. This is apparently generally true for non-*S*-state atoms where coupling between orbital and spin momenta can occur.

10 Lineshape and Relaxation Techniques

The general basis of techniques employed to analyse relaxation has been described in the book by Poole and Farach.[2] Apart from a brief mention of an initial study of the relaxation by a direct observation of the magnetization decay[9] and the spin-echo technique,[8] a detailed description of the saturation technique has been given by Rataiczak and Jones.[99] Zhidkov *et al.*[109] have described a method for the extraction of relaxation parameters from saturation studies of non-uniformly broadened lines, and consider a model based on non-interacting spin pockets. Buckmaster and Skirrow have continued the study of the shape of modulation broadened unsaturated Lorentzian lines, and in ref. 110 consider the dispersion mode. Exact

102 M. Rubinstein, A. Barum, and Z. Luz, *Mol. Phys.*, 1971, **20**, 67.
103 G. H. Reed, J. S. Leigh, and J. E. Pearson, *J. Chem. Phys.*, 1971, **55**, 3311.
104 L. Burlamacchi, *J. Chem. Phys.*, 1971, **55**, 1205.
105 L. Burlamacchi and M. Romanelli, *J. Chem. Phys.*, 1971, **55**, 979.
106 J. Reuben, *J. Phys. Chem.*, 1971, **75**, 3164.
107 A. Hudson and J. W. E. Lewis, *Trans. Faraday Soc.*, 1970, **66**, 1297.
108 C. J. Uttee, *J. Chem. Phys.*, 1971, **54**, 5437.
109 O. P. Zhidkov, Ya. S. Lebedev, A. F. Mikhailov, B. N. Protorov, *Theor. and Exp. Chem.*, 1970, **3**, 135 (*Teor. i eksp. Khim.*, 1967, **3**, 240).
110 H. A. Buckmaster and J. D. Skirrow, *J. Magn. Resonance*, 1971, **5**, 285.

formulae are found which relate the location and amplitude of the extrema, the extreme amplitude extrema ratio, the location and amplitude of the slope extrema, the slope extrema ratio, and the location of the zeros of the observed and true half-widths; however, this Reporter wonders who is likely to make use of such an elaborate analysis.

4
Triplets and Biradicals

BY A. HUDSON

There has been a significant increase in the number of papers concerned with photo-excited triplet states during the past two years. This is mainly attributable to the rapid development of microwave-optical double-resonance (MODR) techniques which are more sensitive than the conventional e.s.r. experiment. However, many interesting results have been reported using standard methods and we shall discuss these first. Systems containing exchange-coupled transition-metal ions will not be considered in this chapter.

1 Ground and Low-lying Excited State Triplets

The e.s.r. of triplet nitrenes has been reviewed.[1] Photolysis of *p*-diazidobenzene produces initially[2] the mononitrene (1) ($D^* = 0.932\ cm^{-1}$) which reacts slowly with oxygen at 77 K to give a second triplet species (2) ($D^* = 0.110\ cm^{-1}$). The structures of the two species observed on further photolysis of *p*-diazidobenzene have been reassigned.[3] The spectrum originally attributed[4] to (3) is probably due to (4) with $D^* = 0.059\ cm^{-1}$; however, (3) is also observed with $D^* = 0.171\ cm^{-1}$. *p*-Nitrophenylnitrene

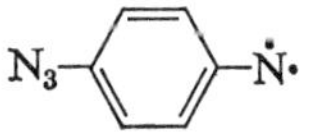

(1)

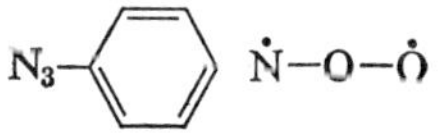

(2)

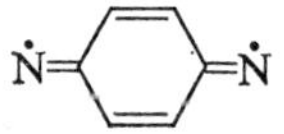

(3)

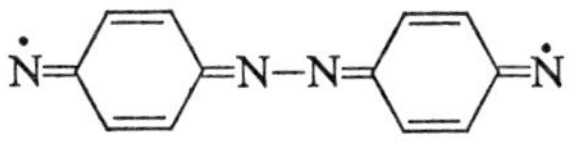

(4)

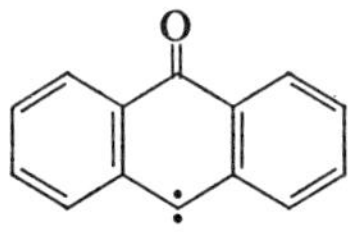

(5)

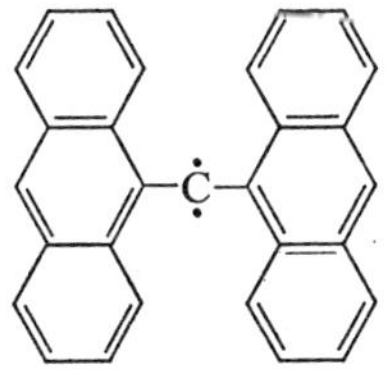

(6)

[1] E. Wasserman, *Progr. Phys. Org. Chem.*, 1971, **8**, 319.
[2] J. S. Brinen and B. Singh, *J. Amer. Chem. Soc.*, 1971, **93**, 6623.
[3] B. Singh and J. S. Brinen, *J. Amer. Chem. Soc.*, 1971, **93**, 542.
[4] A. M. Trozzolo, R. W. Murray, G. Smolinsky, W. A. Yager, and E. Wasserman, *J. Amer. Chem. Soc.*, 1963, **85**, 2526.

has been investigated in an oriented system.[5] U.v. irradiation of diazoanthrone in a benzophenone matrix at 77 K gives [6] the carbene anthronylidene (5) with $D = 0.365\ cm^{-1}$ and $E = 0.0177\ cm^{-1}$. Unlike other aromatic methylenes, 9,9′-dianthrylmethylene (6) is linear with the aromatic rings perpendicular to one another.[7] It is exceptionally stable and has been observed from 4 to 480 K. D Decreases linearly with temperature to a value of $0.113\ cm^{-1}$ at 4 K; E is essentially zero. 2-Isopropylidenecyclopentane-1,3-diyl (7) has [8] $D = 0.027$ and $E = 0.0023\ cm^{-1}$, compared

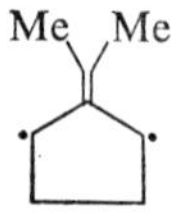

(7)

with values of $D = 0.024\ cm^{-1}$ and $E = 0$ for trimethylenemethane. The latter has been the subject of further theoretical work.[9]

Following initial reports in 1970,[10–12] there has been continued interest [13–15] in the simplest methylene, CH_2. A re-analysis of the electronic spectrum [16] has confirmed that methylene is bent rather than linear as previously supposed. A bond angle of about 136° has been deduced from studies [13, 14] of the ^{13}C hyperfine interaction, in good agreement with theoretical predictions.[17] The first observations were made in a xenon matrix;[10–14] investigations in SF_6 and octafluorocyclobutane [15] have shown that the zero-field parameters are increased by about 50%, owing to intermolecular spin–orbit interactions with Xe. This perturbation is effectively removed in the SF_6 matrix and the parameters obtained should then be

5 J. A. Hebden and C. A. McDowell, *J. Magn. Resonance*, 1971, **5**, 115.
6 P. Devolder, P. Bourlet, C. Dupret, and D. Dessaux, *Chem. Phys. Letters*, 1972, **14**, 57.
7 E. Wasserman, V. J. Kuck, W. A. Yager, R. S. Hutton, E. D. Greene, V. P. Abegg, and N. M. Weinshenker, *J. Amer. Chem. Soc.*, 1971, **93**, 6335.
8 J. A. Berson, R. J. Bushby, J. M. McBride, and M. Tremelling, *J. Amer. Chem. Soc.*, 1971, **93**, 1544.
9 P. J. Hay, W. J. Hunt, and W. A. Goddard, *J. Amer. Chem. Soc.*, 1972, **94**, 638.
10 E. Wasserman, W. A. Yager, and V. J. Kuck, *Chem. Phys. Letters*, 1970, **7**, 409.
11 R. A. Bernheim, H. W. Bernard, P. S. Wang, L. S. Wood, and P. S. Skell, *J. Chem. Phys.*, 1970, **53**, 1280.
12 E. Wasserman, V. J. Kuck, R. S. Hutton, and W. A. Yager, *J. Amer. Chem. Soc.*, 1970, **92**, 7491.
13 R. A. Bernheim, H. W. Bernard, P. S. Wang, L. S. Wood, and P. S. Skell, *J. Chem. Phys.*, 1971, **54**, 3223.
14 E. Wasserman, V. J. Kuck, R. S. Hutton, E. D. Anderson, and W. A. Yager, *J. Chem. Phys.*, 1971, **54**, 4120.
15 E. Wasserman, R. S. Hutton, V. J. Kuck, and W. A. Yager, *J. Chem. Phys.*, 1971, **55**, 2593.
16 G. Herzberg and J. W. C. Johns, *J. Chem. Phys.*, 1971, **54**, 2276.
17 J. M. Foster and S. F. Boys, *Rev. Mod. Phys.*, 1960, **32**, 305; S. V. O'Neil, H. F. Schaefer, and C. F. Bender, *J. Chem. Phys.*, 1971, **55**, 162, and refs. therein; P. J. Hay, W. J. Hunt, and W. A. Goddard, *Chem. Phys. Letters*, 1972, **13**, 30; J. E. Del Bene, *ibid.*, 1971, **9**, 68; S. Y. Chu, A. K. Q. Siu, and E. F. Hayes, *J. Amer. Chem. Soc.*, 1972, **94**, 2969.

close to those for gas-phase methylene. Wasserman and his co-workers conclude that $D = 0.76 \pm 0.02$ and $E = 0.052 \pm 0.017$ cm^{-1}. These values are in excellent agreement with those calculated by Harrison [18] using *ab initio* methods. He finds a dipolar contribution to D of 0.70 cm^{-1}, which on adding an intramolecular spin–orbit contribution [19] of *ca.* 0.05 cm^{-1} gives $D = 0.75$ cm^{-1}. The calculated value for E is 0.052 cm^{-1}. The effects of deviations from planarity on the zero-field splitting in methylenes has also been investigated theoretically.[20]

Applications of e.s.r. and ENDOR to the study of chemically reacting triplet-state species in single crystals have been reviewed [21] and a detailed account has been given of the reactions of diphenyldiazomethane and diphenylmethylene in single 1,1-diphenylethylene crystals.[22]

The dianion of 1,3,5-triphenylbenzene has a triplet ground state irrespective of solvent, counter ion, or temperature, but the dianions of 2,4,6-triphenyl- and 2,4,6-tritolyl-*sym*-triazine have singlet ground states with thermally accessible triplets in poorly solvating media and triplet ground states in better solvating systems.[23] The triazine dianions have much larger D values than the benzene derivative. E.s.r. has been used [24] to study the photo-oxidation of the triplet dianions of triphenylbenzene and triphenylene. Triplets have also been observed in frozen solutions of nitroaromatic radical-anions and alkali-metal cations.[25]

Irradiation of single crystals of carbazide gives a triplet species which is either $\dot{N}$HNHCONH$\dot{N}$H or, possibly, a radical pair.[26] The zero-field splitting corresponds to an interelectronic distance of 4.37 Å. There have been several studies of radiation-produced radical pairs [27] in organic crystals.

A thermally excited triplet state has been detected [28] in the yellow-orange modification of Fremy's salt with $D = \pm 0.746$ and $E = \pm 0.0044$ cm^{-1}. The singlet–triplet excitation energy is 0.27 eV. Triplets have also been observed in Wurster's Blue perchlorate [29] and donor–acceptor complexes of TCNQ.[30] A detailed e.s.r. and *X*-ray diffraction study has been made [31]

[18] J. F. Harrison, *J. Chem. Phys.*, 1971, **54**, 5413; *Internat. J. Quantum Chem.*, 1971, **S5**, 285.
[19] S. H. Glarum, *J. Chem. Phys.*, 1963, **39**, 3141.
[20] J. Higuchi, *Bull. Chem. Soc. Japan*, 1971, **44**, 2634.
[21] C. A. Hutchison, jun., *Pure Appl. Chem.*, 1971, **27**, 327.
[22] D. C. Doetschman and C. A. Hutchison, jun., *J. Chem. Phys.*, 1972, **56**, 3964.
[23] J. A. M. van Broekhoven, J. L. Sommerdijk, and E. de Boer, *Mol. Phys.*, 1971, **20**, 993.
[24] H. van Willigen, *Chem. Phys. Letters*, 1971, **11**, 294.
[25] C. A. McDowell and F. Nakano, *J. Phys. Chem.*, 1971, **75**, 1205.
[26] K. Reiss and W. Gordy, *J. Chem. Phys.*, 1971, **55**, 5329.
[27] M. Iwasaki, K. Minakata, and K. Toriyama, *J. Chem. Phys.*, 1971, **55**, 1472; 1971, **54**, 3225; H. C. Box, *J. Phys. Chem.*, 1971, **75**, 4326; J. Wilske and H. Heusinger, *Internat. J. Radiation Phys. and Chem.*, 1970, **2**, 131.
[28] B. D. Perlson and D. B. Russell, *J. C. S. Chem. Comm.*, 1972, 69.
[29] M. Itoh, *Bull. Chem. Soc. Japan*, 1971, **44**, 2886.
[30] A. A. Alikin, V. S. Grechishkin, R. V. Grechishkina, and V. M. Gusarov, *Optika i Spektroskopiya*, 1971, **31**, 249.
[31] M. J. Hove, B. M. Hoffmann, and J. A. Ibers, *J. Chem. Phys.*, 1972, **56**, 3490.

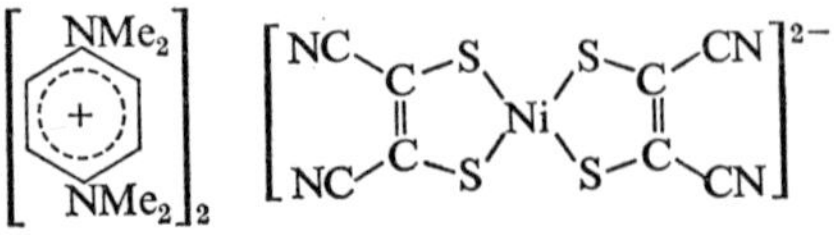

(8)

of the radical salt $[TMPD^+]_2[Ni(mnt)_2]^{2-}$ (8) which exhibits triplet exciton spectra characteristic of an alternating chain of spins. The exciton creation energy is 0.240 eV, $|D| = 215$ G, and $|E| = 28$ G.

2 Photoexcited Triplet States

The excited triplet states of a number of charge-transfer complexes of tetracyanobenzene and aromatic hydrocarbons have been observed [32] in oriented liquid-crystal glasses. The triplet-state e.s.r. spectrum of triphenylene in a cooled oriented nematic phase has been analysed to provide information about the distribution function of the solute molecules.[33]

Phosphorescent dibenzothiophen has been investigated [34] in a dibenzofuran single crystal and shown to be very similar to the biphenyl triplet which has been studied [35] in the same single-crystal host. Biphenyl itself has been used as a host for acridine;[36] the D value of 0.07366 cm^{-1} is intermediate between those of phenazine (0.07440 cm^{-1}) and anthracene (0.07156 cm^{-1}). Triplet excitons have been observed in acridine orange.[37] Spectra obtained from zone-refined biphenyl have been attributed to impurities trapped at crystal-defect sites.[38]

The zero-field splitting parameters of thiopyronine (9) have been measured in glassy solvents [39] and found to be concentration dependent.

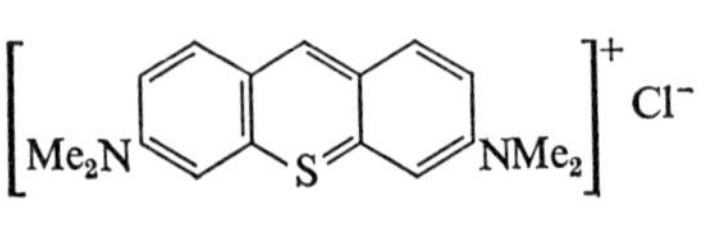

(9)

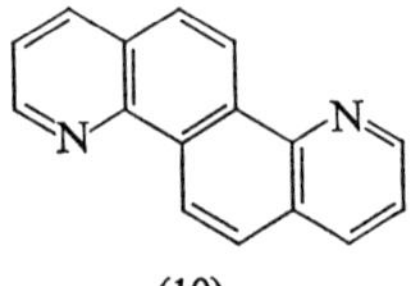

(10)

For a 3×10^{-3} mol l^{-1} solution in phosphate buffer–glycol, $D = 0.0410$ and $E = 0.0075$ cm^{-1}. Phosphorescent 4,10-diazachrysene (10) has been studied both as the free base and as its mono- and di-protonated forms.[40] The D values are 0.0901, 0.0804, and 0.0765 cm^{-1} and the E values 0.0265, 0.0204, and 0.0161 cm^{-1} for B, BH^+, and BH_2^{2+}, respectively.

32 P. Krebs, E. Sackmann, and J. Schwarz, *Chem. Phys. Letters*, 1971, **8**, 417.
33 P. Krebs and E. Sackmann, *Mol. Phys.*, 1972, **23**, 437.
34 M. Baiwir, *Chem. Phys. Letters*, 1971, **9**, 482; *Bull. Soc. roy. Sci. Liège*, 1971, **40**, 162.
35 J. Mispelter, *Chem. Phys. Letters*, 1971, **10**, 539.
36 J.-Ph. Grivet, *Chem. Phys. Letters*, 1971, **11**, 267.
37 H. Schmidt, *Z. Naturforsch.*, 1971, **26a**, 763.
38 R. M. Haines, A. Pryce, and L. Shields, *J. Chem. Soc.* (*B*), 1971, 1420.
39 S. Lalitha and A. Haug, *J. Amer. Chem. Soc.*, 1971, **93**, 3340.
40 H. Bulska, A. Chodkowska, and A. Grabowska, *Chem. Phys. Letters*, 1972, **12**, 508.

The triplet state of chrysene itself has been the subject of an extensive study[41] using low-field e.s.r. In a *p*-terphenyl host, resonances were observed from seven distinct species. These multiple absorptions arise from an orientational inequivalence among the guest molecules which produces small variations in D and E. Similar effects have been found in studies of triphenylene in dodecahydrotriphenylene.[42] Guest rearrangement in the host was detected and it is likely that E is attributable to the molecular crystal field. The orientation of the principal axes of the zero-field splitting tensor for chrysene and related species has been considered theoretically.[43] The half-field transitions of triplet phenylacetylene, phenylmethylacetylene, and diphenylacetylene have been reported;[44] D^* lies in the range 0.145—0.152 cm^{-1}. The zero-field splittings of some cyanine dyes show[45] a regular dependence on the chain length of the conjugated π-system. The π–π^* triplet-state e.s.r. spectra of substituted methyl benzoates indicate that a considerable spin density is distributed on to the carboxy-group.[46] Substituent effects on the zero-field splittings have been discussed in terms of contributions from diallyl radical and *para*-quinoidal structures of the triplet.

For a series of dibenzoylmethane chelates the e.s.r. spectra indicate[47] that the triplet exciton is localized on one of the ligands; D and E are *ca.* 0.100 and 0.027 cm^{-1}, respectively. The kinetics of the formation and decay of the phenanthrene triplet dissolved in benzophenone have been investigated[48] at 77 K. Illumination of the crystal at 750 nm depopulates the excited state; absorption spectra detected at this wavelength may arise from a triplet phenanthrene–benzophenone exciplex.

The kinetics of population and depopulation in triplet states have mostly been studied at liquid-helium temperatures. Transient effects have been detected at 77 K using light-intensity modulation with a phase-sensitive detector locked to the modulation frequency. The latter can be varied with a rotating sector or by employing a pulsed light source. Only light-dependent signals with lifetimes or decay times shorter than the light period are observed; radiation damage centres are suppressed. The triplet state of phenanthrene in benzophenone has fast and slow decay components and spin polarization has been reported.[49, 50] This technique is readily accessible with conventional e.s.r. spectrometers and is likely to find increasing application.

[41] R. E. Gerkin and A. M. Winer, *J. Chem. Phys.*, 1972, **56**, 1359.
[42] R. E. Gerkin and P. Szerenyi, *J. Chem. Phys.*, 1971, **55**, 2575.
[43] L. I. Epstein, *J. Chem. Phys.*, 1972, **56**, 1779.
[44] C. G. Wade and S. E. Webber, *J. Chem. Phys.*, 1972, **56**, 1619.
[45] R. A. Pierce and R. A. Berg, *J. Chem. Phys.*, 1972, **56**, 5087.
[46] D. R. Arnold, J. R. Bolton, and J. A. Pedersen, *J. Amer. Chem. Soc.*, 1972, **94**, 2872.
[47] H. Levanon, *Chem. Phys. Letters*, 1971, **9**, 257.
[48] H. Levanon and S. I. Weissman, *Chem. Phys. Letters*, 1971, **10**, 25.
[49] H. Levanon and S. I. Weissman, *J. Amer. Chem. Soc.*, 1971, **93**, 4309.
[50] H. Levanon, *Varian Instrument Applications*, 1971, **5**, (4), 6.

The effects of methyl substitution on the naphthalene triplet have been reported;[51] triplet energy transfer from β-methylnaphthalene has been studied in naphthalene single crystals.[52] The anthracene triplet has been investigated by ENDOR in a phenazine host.[53] The sign of D for the triplet state of 1,5-diazanaphthalene in durene has been shown to be positive, from an analysis of weak simultaneous electron-spin–nuclear-spin transitions.[54] 1,8-Diazanaphthalene has also been investigated[55] and 1,3-, 1,5-, and 1,8-diazanaphthalene have been studied by phosphorescence–microwave double resonance.[56] The lowest triplet states of chromone (11)

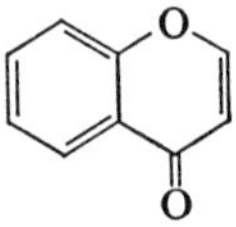

(11)

and 3,4-dimethylacetophenone are π,π^* in ethanol glasses and are detectable by e.s.r. In 3-methylpentane or 2-methyltetrahydrofuran, re-ordering of the triplets occurs, but the n,π^* triplets were not observed by e.s.r.[57] Energy transfer has been studied in aromatic single crystals.[58a] Selective excitation of the singlet states of a donor guest molecule (phenazine or acridine) in a biphenyl or fluorene host populates the lowest triplet of an acceptor guest molecule (naphthalene or pyrene).

Other studies in mixed organic crystals have been concerned with spin–phonon interactions[58b] and the influence of deuteriation.[58c] Energy delocalization in molecular crystals has also been discussed by Hanson,[59] who emphasizes the importance of both coherent and incoherent processes in exciton spectra. Spin–lattice relaxation rates for triplet excitons in molecular crystals have been estimated on the assumption that modulation of the spin–orbit interaction by the orbit–lattice interaction leads to a direct mixing of the spin levels.[60] The spin–lattice relaxation times for the triplet states of naphthalene, quinoxaline, and tetramethylpyrazine in durene and biphenyl hosts have been determined[61] by the pulse-recovery method for the temperature range 1.25—4.23 K. The temperature dependence is characteristic of Raman and direct, *i.e.* one and two phonon, processes. A

[51] B. S. Yamanashi and K. W. Bowers, *J. Magn. Resonance*, 1971, **5**, 109.
[52] N. A. Ashford and C. A. Hutchison, jun., *J. Chem. Phys.*, 1971, **54**, 1165.
[53] R. H. Clarke and C. A. Hutchison, jun., *J. Chem. Phys.*, 1971, **54**, 2962.
[54] J. S. Vincent, *Chem. Phys. Letters*, 1971, **8**, 37; *J. Chem. Phys.*, 1971, **54**, 2237.
[55] A. M. Nishimura and J. S. Vincent, *Chem. Phys. Letters*, 1971, **11**, 609.
[56] A. M. Nishimura, D. S. Tinti, and J. S. Vincent, *Chem. Phys. Letters*, 1971, **12**, 360.
[57] J. B. Gallivan and J. S. Brinen, *Chem. Phys. Letters*, 1971, **10**, 455.
[58a] H. Zimmermann, D. Stehlik, and K. H. Hausser, *Chem. Phys. Letters*, 1971, **11**, 496.
[58b] M. Schwoerer, U. Konzelmann, and D. Klipper, *Chem. Phys. Letters*, 1972, **13**, 272.
[58c] J. Gromer, H. Sixl, and H. C. Wolf, *Chem. Phys. Letters*, 1972, **12**, 574.
[59] D. M. Hanson, *Chem. Phys. Letters*, 1971, **11**, 175.
[60] R. M. Hochstrasser and P. N. Prasad, *Chem. Phys. Letters*, 1971, **9**, 113.
[61] J. P. Wolfe, *Chem. Phys. Letters*, 1971, **10**, 212.

useful account of the effects of intramolecular exciton exchange on triplet-state e.s.r. spectra has been published.[62]

The triplet states of 1-fluoronaphthalene and 2-fluoronaphthalene have been investigated in a durene host.[63] As in fluorine-containing radical ions, the presence of the halogen has little effect on the spin distribution. Hyperfine splittings from ^{19}F have also been resolved in polycrystalline spectra on both $\Delta m = 2$ and $\Delta m = 1$ transitions and have led to the suggestion of a fluorine-labelling method for determining the spin-density distribution in phosphorescent hydrocarbons.[64] The method has been illustrated by an investigation of biphenyl and six fluorinated derivatives.

There has been further interest in the lowest triplet state of benzene. Vibronic interaction of the lower B_{1u} and E_{1u} states has been treated theoretically[65a] and studied by the Zeeman effect;[65b] the strong temperature dependence of the phosphorescent lifetime has been attributed to a pseudo-rotational motion,[66] and PMDR techniques have been used to confirm the $^3B_{1u}$ assignment of D_{2h} triplet benzene.[67]

The lowest triplet state of caffeine has been detected[68] in ethanol at 77 K with $D^* = 0.139\ \text{cm}^{-1}$. Phosphorescent acetophenone, 4-chloroacetophenone, benzonitrile, and phenylacetylene have all been investigated[69] in 1,4-dibromobenzene single crystals. Although the triplet lifetimes are short in this host, strong signals have been obtained. This is because the guest molecules are excited into triplet states by triplet energy transfer from the host. The latter is excited predominantly by S_0–T_1 absorption of the crystal, giving a very high quantum yield for triplet-state formation.

The effects of deuteriation on intersystem crossing in anthracene have been investigated at 1.5 K in a phenazine crystal.[70] The effect of deuterium substitution is largest for the in-plane spin levels, in agreement with theoretical predictions.[71]

3 Optically Detected Magnetic Resonance

Since the first reports in 1967 there has been a rapid growth in the application of double-resonance experiments in which magnetic dipole transitions, induced by a radiofrequency field between triplet-state spin levels, are detected by a change in phosphorescence intensity. These techniques are

[62] J. P. Lemaistre and Ph. Kottis, 'Electron Spin Relaxation in Liquids', Plenum Press, New York, 1972, p. 455.
[63] J. Mispelter, J.-Ph. Grivet, and J.-M. Lhoste, *Mol. Phys.*, 1971, **21**, 999.
[64] J. Mispelter, J.-Ph. Grivet, and J.-M. Lhoste, *Mol. Phys.*, 1971, **21**, 1015.
[65a] J. H. van der Waals, A. M. D. Berghuis, and M. S. de Groot, *Mol. Phys.*, 1971, **21**, 497.
[65b] J. van Egmond, D. M. Burland, and J. H. van der Waals, *Chem. Phys. Letters*, 1971, **12**, 206.
[66] S. Fischer, *Chem. Phys. Letters*, 1971, **10**, 397.
[67] A. A. Gwaiz, M. A. El-Sayed, and D. S. Tinti, *Chem. Phys. Letters*, 1971, **9**, 454.
[68] G. Lancelot and C. Helene, *Chem. Phys. Letters*, 1971, **9**, 327.
[69] S. W. Mao, T. C. Wong, and N. Hirota, *Chem. Phys. Letters*, 1972, **13**, 199.
[70] R. H. Clarke, *Chem. Phys. Letters*, 1971, **12**, 157.
[71] B. R. Henry and W. Siebrand, *J. Chem. Phys.*, 1971, **54**, 1072.

more sensitive than conventional e.s.r. and have been applied in particular to the n,π^* triplet states of carbonyl compounds and nitrogen heterocycles. The experiments can be performed in zero-field and both e.s.r. and ENDOR transitions have been reported. In addition to spin-hamiltonian parameters such as zero-field splittings, hyperfine interactions, and nuclear quadrupole coupling constants, the double-resonance methods yield kinetic information on radiative and non-radiative properties of the excited state. Several variations on the basic experiment have been devized to facilitate these measurements; an introductory account has been given by El-Sayed.[72]

At temperatures below 4.2 K, spin–lattice relaxation between the triplet spin levels is often slower than the phosphorescence from the individual zero-field states. At these temperatures the triplet state retains memory of the pathway by which it was formed. Van der Waals and his co-workers have observed microwave-induced delayed phosphorescence by suddenly sweeping a microwave field through one of the zero-field resonances during phosphorescence decay, and have described applications of the method to quinoline and quinoxaline,[73] pyrazine and pyrimidine,[74] and zinc porphin.[75] It has been shown by Harris [76] that a fast passage through resonance in zero-field may lead to a partial inversion of the populations of the two spin levels. This occurs when a coherent coupling exists between the triplet system and the microwave field and a modulation has been observed in the phosphorescence of quinoxaline when a strong microwave field is resonant with one of the zero-field transitions.[77] A fast-passage method for determining radiative and populating rate constants has been described by Winscom and Maki.[78] In El-Sayed's method [79] the decay of individual vibronic bands is studied by PMDR and gives an insight into the origin of the vibronic bands in terms of the individual spin components. A further technique involves pulsing both the exciting light and the microwave resonance radiation.[80]

All the original experiments in this field were performed at 4.2 K or below, but recently signals from phosphorescent pyrimidine in benzene have been detected at 77 K using strong microwave fields.[81] Cheng and Hirota [82] have made an extensive study of several $^3\pi,\pi^*$ aromatic carbonyls and discuss the influence of experimental factors on signal intensity in PMDR.

72 M. A. El-Sayed, *Accounts Chem. Res.*, 1971, **4**, 23.
73 J. Schmidt, D. A. Antheunis, and J. H. van der Waals, *Mol. Phys.*, 1971, **22**, 1.
74 D. M. Burland and J. Schmidt, *Mol. Phys.*, 1971, **22**, 19.
75 I. Y. Chan, W. G. van Dorp, T. J. Schaafsma, and J. H. van der Waals, *Mol. Phys.*, 1971, **22**, 741, 753.
76 C. B. Harris, *J. Chem. Phys.*, 1971, **54**, 972.
77 J. Schmidt, W. G. van Dorp, and J. H. van der Waals, *Chem. Phys. Letters*, 1971, **8**, 345.
78 C. J. Winscom, and A. H. Maki, *Chem. Phys. Letters*, 1971, **12**, 264.
79 D. S. Tinti and M. A. El-Sayed, *J. Chem. Phys.*, 1971, **54**, 2529; M. A. El-Sayed, *ibid.*, p. 680.
80 M. A. El-Sayed and J. Olmsted, *Chem. Phys. Letters*, 1971, **11**, 568.
81 A. L. Shain, M. Sharnoff, and M. P. Stombler, *J. Chem. Phys.*, 1972, **56**, 2475.
82 T. H. Cheng and N. Hirota, *J. Chem. Phys.*, 1972, **56**, 5019.

The scope of microwave-optical double resonance is well illustrated by the work of Tinti,[83] who has detected the $^3B_1$ state of SO_2 traps in neat SO_2 at $\leqslant$ 4.2 K. He finds $|D| = 0.275$ and $|E| = 0.0157$ cm^{-1}. Two spin states are active in the phosphorescence with the dominant level having a lifetime of 2 ms.

The lowest triplet state of 1,2,4,5-tetrachlorobenzene in durene has $^3B_{1u}$ symmetry;[84] a study of the intersystem-crossing process suggests that vibronic interactions are important in this system. Francis and Harris have observed coherent triplet excitons in molecular crystals of tetrachlorobenzene[85] and Maki and von Schutz have studied optically induced nuclear alignment of ^{35}Cl in the same system.[86] A detailed investigation of $^3\pi,\pi^*$ *p*-dichlorobenzene by optically detected e.s.r. and ENDOR has provided information on the zero-field splitting, chlorine nuclear quadrupole, and chlorine hyperfine interactions.[87] The chlorine electric field gradient is significantly reduced in the triplet state, compared with the ground state, and the molecule is probably distorted.

The n,π^* triplet state of benzil in a benzophenone host[88] has $|D| = 0.092$ and $|E| = 0.021$ cm^{-1}. The conjugated enone testosterone acetate has[89] $|D| = 0.28$ and $|E| = 0.01$ cm^{-1} in its lowest π,π^* triplet state. Theoretical calculations suggest that the zero-field splitting arises from spin–spin dipolar interactions, spin–orbit coupling of $^3\pi,\pi^*$ with $^3n,\pi^*$ states, and spin–orbit coupling of $^3\pi,\pi^*$ with $^1n,\pi^*$. This complicated situation has led the authors to suggest that it may be difficult to use zero-field splittings to assign the orbital character of ketone triplet states. 1-Indanone, α-tetralone, and 2,4,6-trimethylbenzaldehyde have all been studied[90] in durene host crystals at 35 GHz and 2 K. An investigation of acetophenone and benzaldehyde in a 1,4-dibromobenzene host has shown that the zero-field splitting and dynamic properties are primarily determined by the local symmetry of the carbonyl groups.[91] The influence of substituents on these parameters has been investigated[92] for a series of *para*-substituted benzaldehydes. The magnetic field dependence of spin–lattice relaxation has been studied[93a] for phosphorescent pyrazine at 1.6 K; methods for determining T_1 optically have been described.[93b]

83 D. S. Tinti, *Chem. Phys. Letters*, 1971, **12**, 169.
84 C. R. Chen and M. A. El-Sayed, *Chem. Phys. Letters*, 1971, **10**, 307; M. A. El-Sayed and C. R. Chen, *ibid.*, 1971, **10**, 313.
85 A. H. Francis and C. B. Harris, *Chem. Phys. Letters*, 1971, **9**, 181, 188.
86 A. H. Maki and J. U. von Schutz, *Chem. Phys. Letters*, 1971, **11**, 93.
87 M. J. Buckley and C. B. Harris, *J. Chem. Phys.*, 1972, **56**, 137.
88 H. Hayashi, K. Morigaki, and S. Nagakura, *Chem. Phys. Letters*, 1971, **9**, 119.
89 C. R. Jones, F. Pappano, A. H. Maki, and D. R. Kearns, *Chem. Phys. Letters*, 1972, **13**, 521.
90 A. M. Nishimura and J. S. Vincent, *Chem. Phys. Letters*, 1972, **13**, 89.
91 T. H. Cheng and N. Hirota, *Chem. Phys. Letters*, 1972, **13**, 194.
92 A. M. Nishimura and D. S. Tinti, *Chem. Phys. Letters*, 1972, **13**, 278.
93a L. H. Hall and M. A. El-Sayed, *Mol. Phys.*, 1971, **22**, 361.
93b L. H. Hall and M. A. El-Sayed, *J. Chem. Phys.*, 1971, **54**, 4958.

There is a need for theoretical work on spin-relaxation mechanisms in molecular solids. Methods have been described[94] for detecting deviations from C_{2v}, D_{2h}, or D_2 symmetry in phosphorescent molecules. In n-hexane at 1.6 K, 1,4- but not 2,3-dibromonaphthalene retains its C_{2v} symmetry. Hanson[95] has predicted that it should be possible to observe interference effects in optically detected double resonance and has considered in particular the case of electron–electron double resonance. The adiabatic inversion technique has been further developed by Harris and Hoover[96] and applied to the $^3\pi,\pi^*$ state of 2,3-dichloroquinoxaline. The same authors have also described[97] how the functional dependence of the phase shift for a time-dependent phosphorescence intensity changes relative to a repetitive delta function microwave perturbation. This extension of quadrature measurements to PMDR has been applied to *p*-dichlorobenzene and 1,2,4,5-tetrachlorobenzene.

4 Biradicals

A number of [2,2]paracyclophanylene bisnitroxides have been reported.[98] The e.s.r. spectra of the pseudo-*meta* and pseudo-*para* biradicals (12) and (13) give five-line spectra with a spacing of 20.7 MHz, as expected if $J \gg a_N$. The pseudo-*ortho* compound (14) has a large zero-field splitting and a broad solution spectrum. Measurements in glassy solvents gave D values of 806, 294, and 244 MHz for (14), (12), and (13), corresponding to inter-electronic distances of 4.0, 6.4, and 6.8 Å, respectively. The stable aromatic biradical (15) also exhibits a large dipolar interaction.[99]

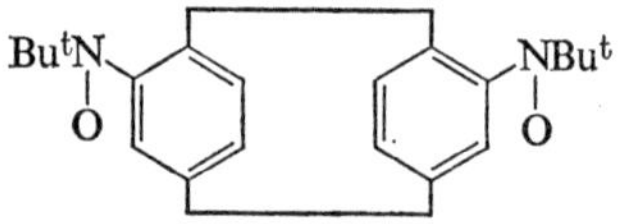

(12)

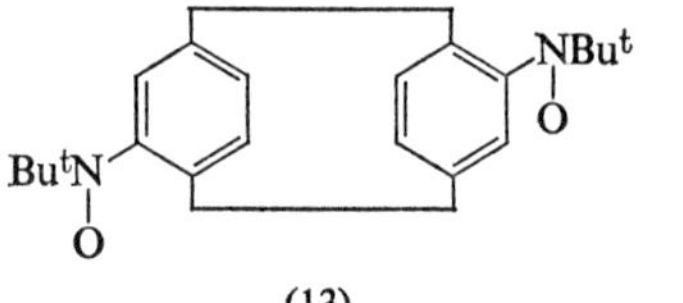

(13)

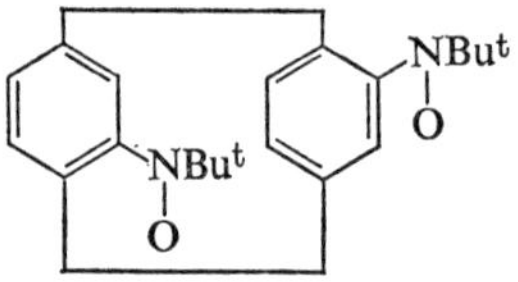

(14)

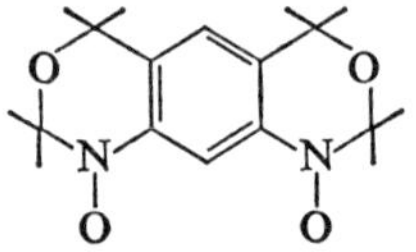

(15)

[94] M. A. El-Sayed, M. Leung, and C. T. Lin, *Chem. Phys. Letters*, 1972, **14**, 329.
[95] D. M. Hanson, *Chem. Phys. Letters*, 1971, **12**, 65.
[96] C. B. Harris and R. J. Hoover, *J. Chem. Phys.*, 1972, **5**, 2199.
[97] C. B. Harris and R. J. Hoover, *Chem. Phys. Letters*, 1971, **12**, 75.
[98] A. R. Forrester and R. Ramasseul, *J. Chem. Soc.* (*B*), 1971, 1638, 1645.
[99] A. Rassat and H. U. Sieveking, *Angew. Chem. Internat. Edn.*, 1972, **11**, 303.

The highly resolved spectra observed in solutions of Schlenk's hydrocarbon (16) are *meta*-substituted triphenylmethyl radicals. However, a triplet species is observed in frozen solutions[100] with $D = 192$ and $E = 17$ MHz. Calculations of D and E suggest that (16) exhibits a marked deviation from planarity. The triplet ground state of Schlenk's hydrocarbon is only *ca.* 1 kJ mol^{-1} more stable than the singlet.[101] The Muller

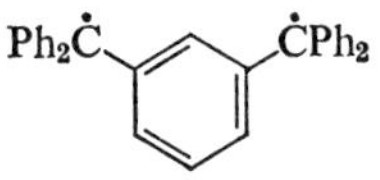

(16)

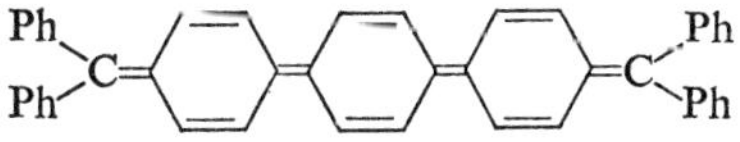

(17)

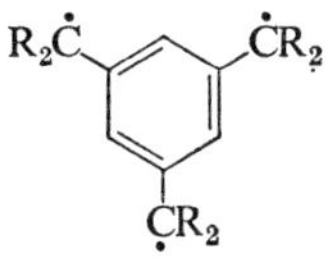

(18) R = *p*-biphenyl

hydrocarbon (17) has a singlet ground state with a triplet ($D = 0.0035$ cm^{-1}) 4 kJ mol^{-1} higher in energy.[101] The novel triradical 1,3,5-benzenetriyl-tris[di(*p*-biphenyl)methyl] (18) has a quartet ground state.[102] In frozen solutions both $\Delta m = 1$ and $\Delta m = 2$ transitions were observed with $D = 128$ MHz.

The biradical (19) has been used to study orientation in phospholipid multilayers and is expected to be a useful spin probe for membrane-structure investigations.[103]

Pyridinyl radicals (20; R = Me or Et) associate to give triplet dimers in a 2-methyltetrahydrofuran glass. The D value (0.0098 cm^{-1}) corresponds

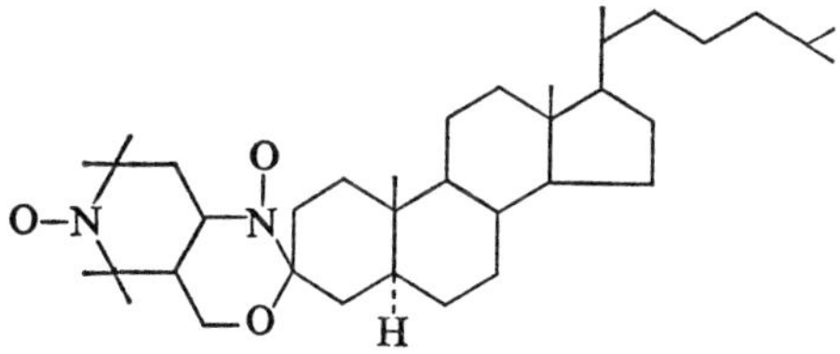

(19)

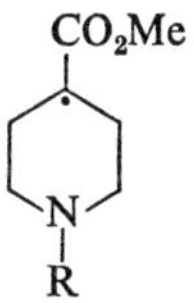

(20)

[100] G. R. Luckhurst, G. F. Pedulli, and M. Tiecco, *J. Chem. Soc. (B)*, 1971, 329.
[101] R. Schmidt and H. D. Brauer, *Angew. Chem. Internat. Edn.*, 1971, **10**, 506.
[102] G. Kothe, E. Ohmes, J. Brickmann, and H. Zimmermann, *Angew. Chem. Internat. Edn.*, 1971, **10**, 938.
[103] J. F. W. Keana and R. J. Dinerstein, *J. Amer. Chem. Soc.*, **93**, 2808.

to an average interelectronic distance of 6.5 Å and there is a small E term ($0.0011\ cm^{-1}$). Irradiation with visible light at 77 K gave a new triplet species with $D = 0.0175\ cm^{-1}$ and $E = 0$. It is suggested[104] that this is due to a more symmetrically arranged dimer with the electrons 5.4 Å apart. Some pyridinyl hetero-biradicals have also been reported.[105] Other systems studied include verdazyl biradicals[106] and nitroxide polyradicals.[107, 108]

A detailed discussion of asymmetric line-broadening in biradical spectra has appeared.[109] The theory is useful in the determination of molecular configurations, electron–electron separations, and the sign of the exchange interaction. The theory of nuclear hyperfine interactions in trimeric species has been discussed[110] for the case $J \gg A$. A useful article[111] on the electronic properties of biradicals contains much of interest to the e.s.r. spectroscopist. Detailed accounts have been given of the calculation of e.s.r. lineshapes for slowly rotating triplet-state species.[112, 113]

Monomer–dimer equilibria have been observed in[114] solutions of 3-carboxy-2,2,5,5-tetramethylpyrrolidine 1-oxyl in non-polar solvents, with heats of dimerization close to 25 kJ mol^{-1}. The temperature dependence of the exchange interaction has also been investigated.

[104] Y. Ikegami, H. Watanabe, and S. Seto, *J. Amer. Chem. Soc.*, 1972, **94**, 3274.
[105] M. Itoh, *J. Amer. Chem. Soc.*, 1972, **94**, 1034.
[106] P. Kopf, K. Morokuma, and R. Kreilick, *J. Chem. Phys.*, 1971, **54**, 105.
[107] I. Yamauchi, T. Fujito, A. Nakajima, H. O. Nishiguchi, and Y. Deguchi, *Bull. Chem. Soc. Japan*, 1971, **44**, 2263.
[108] A. Nakajima, H. O. Nishiguchi, and Y. Deguchi, *Bull. Chem. Soc. Japan*, 1971, **44**, 2120; A. Nakajima, H. O. Nishiguchi, and Y. Deguchi, *Bull. Chem. Soc. Japan*, 1972, **45**, 713; A. I. Kokorin, K. I. Zamaraev, G. L. Grigoryan, V. P. Ivanov, and E. G. Rozantsev, *Biofizika*, 1972, **17**, 34.
[109] G. R. Luckhurst and G. F. Pedulli, *Mol. Phys.*, 1971, **20**, 1043.
[110] A. Hudson, *Mol. Phys.*, 1971, **21**, 61.
[111] L. Salem and C. Rowland, *Angew. Chem. Internat. Edn.*, 1972, **11**, 92.
[112] J. H. Freed, G. V. Bruno, and C. Polnaszek, *J. Chem. Phys.*, 1971, **55**, 5270.
[113] R. M. Lynden-Bell, *Mol. Phys.*, 1971, **22**, 837; J. Roberts and R. M. Lynden-Bell, *ibid.*, 1971, **21**, 689.
[114] J. H. Marshall, *J. Chem. Phys.*, 1971, **54**, 2762.

5
Radical Ions, Ion Pairs, and Dynamic Processes

BY R. F. ADAMS

1 Introduction

Much of the early impetus given to research in the field of e.s.r. spectroscopy was provided by the discovery that well-resolved e.s.r. spectra can often be obtained from stable aromatic radical anions in solution. It was quickly realized that the measurement of nuclear hyperfine splitting constants (hfsc) provides valuable information on the electronic structure of these radical anions and on their environment in solution. For this reason, many new radical ion species have been studied with the aid of e.s.r. spectroscopy over the past fifteen years. Work on the preparation of new radical ion species and measurement of their e.s.r. spectra has continued during the period covered by this Report. However, it has become increasingly more rewarding to carry out more detailed e.s.r. studies of existing species because the behaviour of these radical ion species in different environments provides information on ion solvation, ion-pair structure and equilibria, and dynamic processes involving ion pairs. Thus, radical ions are increasingly being used as probes into their own environment. The sections into which this Report is divided must therefore be regarded not as completely separate areas of activity, but as overlapping and interlocking aspects of the study of the solvation of radical ions.

2 Radical Anions

Solvated Electrons.—The nature and properties of the paramagnetic species obtained on dissolution of alkali metals in ether and amine solvents have been the subject of continuing research and some lively debate. The e.s.r. spectrum of potassium dissolved in tetrahydrofuran (THF) consists of two quartets and a single line, arising from ^{39}K, ^{41}K, and the solvated electron, respectively.[1] Optical spectrophotometric studies show that the paramagnetic species represent only a fraction of the metal dissolved in the THF and that this fraction increases on dilution.

For some time now it has been accepted by most workers in the field that solutions of sodium in amines contain the solvated electron and the Na^+, e^- ion pair. This view has now been questioned[2] on the basis that the

[1] J. Slater and M. C. R. Symons, *J. Chem. Phys.*, 1970, **52**, 1003.
[2] T. R. Tuttle, jun., and P. Graceffa, *J. Phys. Chem.*, 1971, **75**, 843.

e.s.r. spectrum of sodium–amine solutions is similar to that of sodium dissolved in a mixed solvent of THF and [$^2H_8$]naphthalene. It is suggested that the mechanism which operates in the sodium–THF–[$^2H_8$]naphthalene system, *i.e.* fast atom transfer between sodium [$^2H_8$]naphthalenide ion pairs and neutral [$^2H_8$]naphthalene, might also be applied to the sodium–amine solution. This implies that Na^+, $(amine)^{\overline{\cdot}}$ ion pairs are formed and exchange with the neutral amine. There are several difficulties associated with this hypothesis and these have been pointed out by Symons.[3] The behaviour of the two systems when diluted with an inert solvent is not the same in that the e.s.r. spectrum of the sodium–naphthalene solution clearly reveals hyperfine interaction between the nuclei of the naphthalene molecule and the unpaired electron. There is no evidence for similar behaviour in ammonia systems at least. Secondly, ions such as $NH_3^{\overline{\cdot}}$ are likely to be characterized by large positive proton splitting constants, but the available n.m.r. evidence suggests that this is in fact small and negative in these solutions. Although these objections are accepted by the authors of the original hypothesis, a valid point has been raised concerning the nature of sodium–amine solutions and it should be possible to test by volume expansion methods whether these solutions in fact contain amine anions.[4]

Aromatic Hydrocarbon and Heterocyclic Anions.—The radical anion of diphenylfulvene (1) has been prepared in solution[5] and it is the first

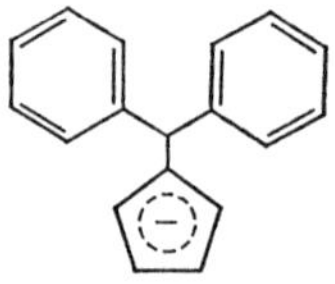

(1)

example of a triarylmethyl radical which bears a formal negative charge on one ring. Compound (1) was prepared from the dianion in 1,2-dimethoxyethane (DME), by shifting the disproportionation equilibrium on addition of the neutral molecule. No metal hf.s.c. are observed in this solvent. The coupling constants are given for the e.s.r. spectra of the anion in dimethyl sulphoxide (DMSO) solution ($a_{\text{H(cyclopentadiene)}}$ 0.179 and 0.200 mT) and agree reasonably well with McLachlan SCF calculations. The e.s.r. spectrum of the radical anion of 1,3,5-triphenylbenzene (TPB) has normal linewidths and *g*-value in liquid ammonia,[6] in spite of the fact that it has an orbitally degenerate ground state and that the dianion of TPB is a ground-state triplet. Radical anions with orbitally degenerate ground states usually

[3] M. C. R. Symons, *J. Phys. Chem.*, 1971, **75**, 3904.
[4] T. R. Tuttle, jun., and P. Graceffa, *J. Phys. Chem.*, 1971, **75**, 3905.
[5] C. M. Camaggi, M. J. Perkins, and P. Ward, *J. Chem. Soc.* (*B*), 1971, 2417.
[6] J. A. M. van Brockhoven, B. M. P. Hendriks, and E. de Boer, *J. Chem. Phys.*, 1971, **54**, 1988.

exhibit very broad e.s.r. spectra because of the effect of the enhanced spin–orbit coupling on the relaxation processes.[7] It is postulated, in the case of $TPB^{\overline{\cdot}}$, that this enhanced spin–orbit coupling may be quenched by out-of-plane bending of the phenyl rings. The authors point out that the spectrum in DME is much less well resolved, and they speculate that this may be due to a more planar configuration of the anion when it is associated with a cation. There is, however, no evidence of a small alkali-metal hfsc from the n.m.r. measurements and also no significant shift in *g*-value. This latter effect may have been offset by the shift in *g*-value caused by the neighbouring cation.

Several new condensed-ring radical anions have been prepared and studied. A combined e.s.r. and ENDOR study of hexahelicene radical ion pairs in DME[8] reveals ten different proton hfsc rather than the eight that a totally symmetric anion should have. The asymmetry is probably caused by ion pairing, and on the basis of Hückel spin densities the cation is predicted to be at one end of the axis of the helix, *i.e.* above the positions of highest charge density. The coupling constants range from 0.266 mT at the positions of high spin density to 0.020 mT, with the three remaining splittings required for an asymmetric anion unresolved. Peake *et al.*[9] have continued their studies of radical anions derived from perinaphthane (2), pleiadane (3), 1,2,3,6,7,8-hexahydropyrene (4), and dipleiadane (5). These

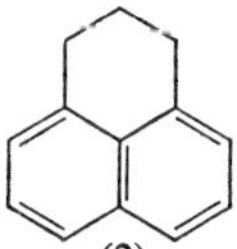
(2)

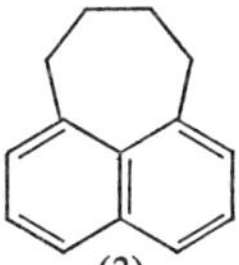
(3)

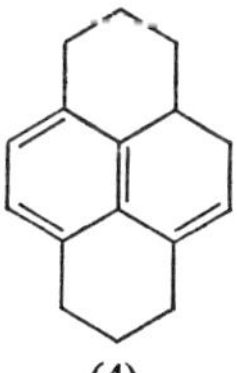
(4)

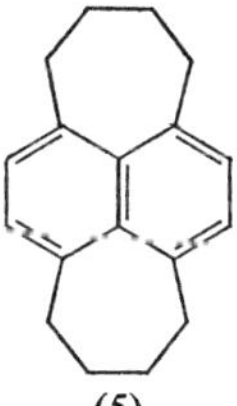
(5)

molecules are non-planar and the aliphatic substituent may move through the plane of the aromatic naphthalene ring [for (4) and (5), this is equivalent to chair–boat type interconversion]. The spectra of the radical anions in DME and THF change drastically between − 95 and + 80 °C, revealing linewidth alternation effects which are attributed to the conformation

[7] M. G. Townsend and S. I. Weissman, *J. Chem. Phys.*, 1958, **32**, 309.
[8] R. D. Allendoerfer and R. Chang, *J. Magn. Resonance*, 1971, **5**, 273.
[9] R. F. C. Claridge, B. M. Peake, and R. M. Goldring, *J. Magn. Resonance*, 1972, **6**, 29.

changes described above. The modified Bloch equations are used to simulate these effects on the basis of a two-jump model for the anions of (2) and (3), and a four-jump model for the anions of (4) and (5). The energies of activation for 1,2,3,6,7,8-hexahydropyrene and dipleiadane are 26 and 33 kJ mol^{-1}, respectively, and the A factors appear to be in the range expected for such a strained ring-flipping process. No accurate kinetic parameters could be obtained for the other two molecules, though E_a is estimated to be *ca.* 30 kJ mol^{-1} in each case.

Another interesting new substituted naphthalene radical anion is that of 1,8-bis(dimethylamino)naphthalene (6).[10] The neutral molecule is strained

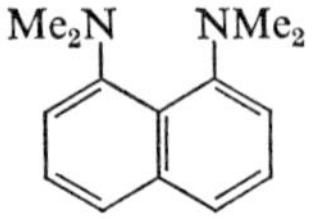

(6)

to the extent that the configuration between the nitrogen lone pairs and the π-system should be considerably reduced. In fact the abnormally high basicity of the molecule (it is known as a 'proton sponge'!) is attributed to the relief of this strain on protonation and/or strong interaction between the nitrogen lone pairs. The spectrum of the radical anion in DME is quite consistent with those of the anions of other 1,8-substituted naphthalenes ($a_{H(4,5)}$ = 0.446 mT, $a_{H(3,6)}$ = 0.177 mT, and $a_{H(2,7)}$ = 0.139 mT) and the assignments have been made by analogy with such compounds. Further confirmation of the assignments comes from Hückel MO calculations, which also reveal that the interaction between the two dimethylamino-substituents is large enough for the nitrogen lone pairs to be significantly displaced with respect to the axis of the π-system.

Several new radical anions have been prepared, with particular interest being directed towards the methylene substituents. The e.s.r. spectra of the radical anions of 9,10-dihydrophenanthrene, 4,5-dihydropyrene, 9,10-dihydroanthracene, and 5,12-dihydrotetracene have been measured and interpreted.[11] Using the n.m.r. method of de Boer and MacLean,[12] it has been possible to obtain the signs as well as the magnitudes of the smaller hfsc. The anion of 9,10-dihydrophenanthrene is non-planar, like its neutral precursor,[13] as evidenced by the non-equivalence of its axial and equatorial proton hfsc. At low temperatures the hfsc are 0.265 and 0.029 mT, respectively. As the temperature is increased, linewidth alternation occurs on the methylene-group lines and at room temperature the methylene proton hfsc are averaged to 0.147 mT. The linewidth effects

[10] F. Gerson, E. Haselbach, and G. Plattner, *Chem. Phys. Letters*, 1971, **12**, 316.
[11] J. Van der Kooij, C. Gooiyer, N. H. Velthorst, and C. MacLean, *Rec. Trav. chim.*, 1971, **90**, 732.
[12] E. de Boer and C. MacLean, *J. Chem. Phys.*, 1966, **44**, 1334.
[13] E. A. Braude and W. F. Forbes, *J. Chem. Soc.*, 1955, 3776.

are due to the ring inversions described above, and the authors have made a complete linewidth analysis of the e.s.r. spectra using Alexander density matrix theory. The spectrum of the anion of 9,10-dihydroanthracene shows a methylene proton hfsc of 0.415 mT but no alkali-metal hfsc. The temperature-dependent linewidths in these spectra have not been fully explained, although the facts are interpretable on the basis of an equilibrium between ion pairs with different methylene-proton hfsc. 4,5-Dihydropyrene anion has a very small (0.0096 mT) methylene-proton hfsc and a lithium splitting of 0.003 mT. In each case the proton coupling constants are assigned by comparison with similar radical anions and by analysis of the n.m.r. spectra of the electron-transfer reaction. The spectrum of the radical anion of the related 5,12-dihydronaphthacene in DME has been reported [14] and the methylene-proton coupling constants are analysed by a configuration interaction model, and by simple Hückel calculations in which adjustments are made to the methylene-carbon resonance-integral.

An interesting radical dianion which contains a methylene group is the 4,5-methylenephenanthrene dianion (7). The radical dianion is obtained by alkali-metal reduction of the monoanion of 4,5-methylenephenanthrene in ethereal solvents.[15] The observed proton hfsc are $a_{H(9,10)} = 0.513$ mT, $a_{H(1,3,6,8)} = 0.313$ mT, $a_{H(2,7)} = 0.056$ mT, and $a_{H(15,methylene)} = 0.044$ mT. Metal splitting is observed for all alkali metals except potassium. However, splitting from only one cation is observed, and it is suggested that the second counterion is in a similar position to that of the counterion of the monoanion because the alkali-metal nucleus does not split the spectrum of the monoanion. The changes

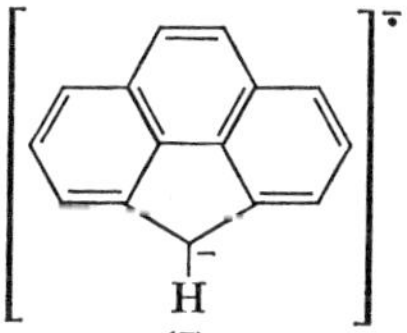

(7)

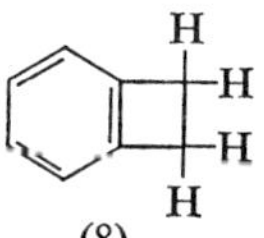

(8)

observed in the dianion metal hfsc with temperature are most easily ascribed to dynamic equilibria between different types (contact and solvent-separated) of ion pair.

A more positive approach to the study of the mechanism of methylene proton hyperfine interaction, and the calculation of observed coupling constants, is taken in a study of the benzocyclobutene (8) radical anion.[16] The problem of calculating the methylene-proton spin densities within the limited framework of Hückel theory is overcome by making the coulomb integrals for the α-carbons consistent with more electronegative carbon

[14] H. Joela, *Suomen Kem.* (*B*), 1971, **44**, 235.
[15] D. Casson and B. J. Tabner, *J. Chem. Soc.* (*B*), 1971, 479.
[16] R. D. Rieke, S. E. Bates, P. M. Hudnall, and C. F. Meares, *J. Amer. Chem. Soc.*, 1971, **93**, 697.

atoms, and by allowing the ring-juncture carbon atoms to become less electronegative. In this way the effect of the strained ring is taken into account because the increased *s*-character of the cyclobutene σ-bond polarizes the σ-electrons away from the α-carbon atoms and makes them more electronegative. With these adjustments, Hückel theory gives reasonable agreement with the observed room-temperature hfsc for the radical anion prepared by reduction with potassium in DME. At 20 °C the methylene coupling constant is 0.530 mT and that for the potassium counterion 0.002 mT. At − 88 °C, however, two different methylene splitting constants (0.575 and 0.525 mT) are observed, which suggests that the ion pair is not symmetrical. Possibly the cation is centred over the benzene ring. The observed linewidth alternation of the methylene-proton lines is therefore the result of the potassium counterion jumping between equivalent positions, above and below the benzene ring. In addition, a dynamic equilibrium is formulated between solvent-separated and contact ion pairs in order to explain the temperature-dependence of the individual methylene-proton hfsc. No attempt has been made, however, to quantify this equilibrium. The changes are only of the order of 10%, which seems to be quite a small change in the average coupling constant over a temperature range of one hundred degrees. Perhaps a progressive re-ordering of solvent molecules only around one type of ion pair would be an equally acceptable explanation. Although it does not apply so much to this particular investigation, in many e.s.r. studies of this nature optical spectroscopic observations would be useful additions to the e.s.r. results. So far, e.s.r. spectroscopists have not made full use of the visible and u.v. spectra of their radical-ion solutions. Obviously, e.s.r. can only detect the paramagnetic species in solution and often cannot distinguish between different types. Diamagnetic ions (especially dianions, carbanions, and dimers), which often affect equilibria involving paramagnetic ions, are most easily detected by optical means.

The decay of 9-alkylfluorene radical anions to the corresponding carbanions has been investigated, by e.s.r. and optical spectrophotometric methods, for a large number of substituents and for different alkali metals.[17] The decay is straightforward in the case of lithium and potassium ion pairs in ethereal solvents, but there is evidence of an equilibrium of the kind shown in equation (1) in the case of sodium reduction.

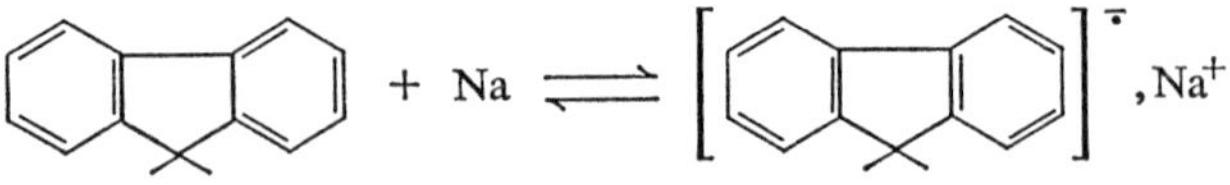

(1)

A series of radical anions of the type (9) have been prepared by electrolytic reduction in acetonitrile and dimethylformamide (DMF).[18] These

[17] B. J. Tabner and T. Walker, *J. C. S. Perkin II*, 1972, 445.
[18] W. M. Gulick, jun., *J. Amer. Chem. Soc.*, 1972, **94**, 29.

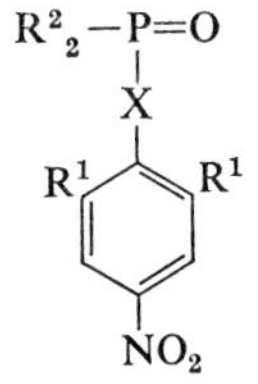

(9)

(10) R^1 = H, R^2 = Ph, X = O
(11) R^1 = Me, R^2 = Ph, X = O
(12) R^1 = Ph, R^2 = Ph, X = O
(13) R^1 = H, R^2 = PhO, X = S
(14) R^1 = Me, R^2 = EtO, X = CH_2

radical anions are a group of compounds from which information about the mechanism of phosphorus hyperfine coupling may be extracted. It has been suggested[19] that the predominant coupling mechanism is hyperconjugation by direct interaction between phosphorus atomic orbitals and the $2p^{\pi}$ orbital of the substituted aromatic ring carbon atom. The spin density and therefore the magnitude of the ^{31}P hfsc depends on the physical proximity of the phosphorus atom to the carbon $2p^{\pi}$ orbital. It should also depend upon the angle θ between the $2p^{\pi}$ axis and the X—P bond (both projected on to a plane perpendicular to the X—C bond). Inspection of the splitting constants in Table 1[18] shows that thera tio of a_P for

Table 1 ^{31}P *and* ^{14}N *hyperfine splitting constants*/mT

Compounds	a_P		a_N	
	CH_3CN	DMF	CH_3CN	DMF
(10)	0.607	0.635	1.040	0.988
(11)	0.932	0.966	1.058	1.000
(12)	1.000	1.049	1.000	0.953
(13)	2.480	2.715	0.848	0.778
(14)	1.914	2.009	1.100	1.029

the 4-nitrophenyl diphenylphosphinate (10) radical anion to a_P for the anion of 2,6-diphenyl-4-nitrophenyl diphenylphosphinate (12) is roughly what is expected for $\langle \cos^2 \theta \rangle$ dependence, (10) being free to rotate about the C—X bond and (12) being totally restricted. The fact that the ratio of the phosphorus splitting constants is not exactly the value expected is interpreted as being due to slightly restricted rotation about the C—X bond in compound (10). It is also possible, of course, that not all significant contributions to the phosphorus hyperfine splitting have been accounted for. As well as ^{31}P and ^{14}N splitting constants, proton splitting constants for all the anions are recorded in both solvents and provide a very useful addition to our knowledge of such systems. This is also true of a study of the radical anion of 1-phenyl-2-trimethylsilylacetylene (15) in THF.[20] On reduction with alkali metal at − 80 °C, the stable anion is formed; no metal splitting

[19] W. M. Gulick, jun., and D. H. Geske, *J. Amer. Chem. Soc.*, 1966, **88**, 2928.
[20] A. G. Evans, J. C. Evans, P. J. Emes, and T. J. Phelan, *J. Chem. Soc.* (*B*), 1971, 315.

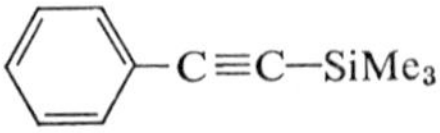

(15)

is observed in the spectrum except with caesium as counterion. The value of $a_{^{29}Si}$ is given as 0.44 mT, and the proton and silicon hfsc do not vary with the alkali metal used. Hückel and McLachlan MO calculations on the anion of (15) have been carried out, treating $SiMe_3$ as a heteroatom, but these calculations are probably too crude because the authors do not obtain reasonable agreement with experiment. The same group of workers has carried out a fairly extensive investigation of the ion pairs formed on reduction of certain azobenzenes and azonaphthalenes with alkali metals.[21] All the compounds studied are easily reduced to the dianion, and in fact the disproportionation equilibrium constant for equation (2) has been measured. The radical anions themselves are prepared by only brief

$$2(RN{=}NR)^{\overline{\cdot}}, Me^{+} \rightleftharpoons (RN{=}NR)^{2-}, 2Me^{+} + RN{=}NR \quad (2)$$

contact with the alkali-metal mirror. Where observed, the alkali-metal splitting constants are temperature-dependent, and in the case of sodium azobenzenide and sodium naphthalene-1-azobenzenide the $M_{Na} = \pm 3/2$ lines are broader than the $M_{Na} = \pm 1/2$ lines. These two observations taken together indicate an equilibrium between ion pairs with different metal splittings.

Radical anions of sulphur and oxygen heterocyclic ring systems are interesting both from the point of view of their chemical properties and (especially the sulphur heterocyclics) because of the need for a better understanding of the heterocyclic atom participation in the spin system. The electrolytic reduction of xanthene (16), thioxanthene (17), and anthrone (18)[22] is an example of the former type of investigation. Electrolysis of (16)—(18) in the presence of a trace of oxygen results in oxidation at the

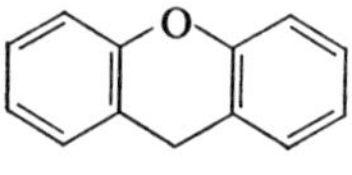

(16)

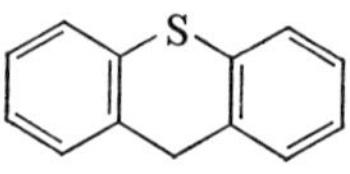

(17)

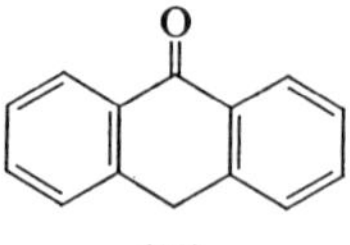

(18)

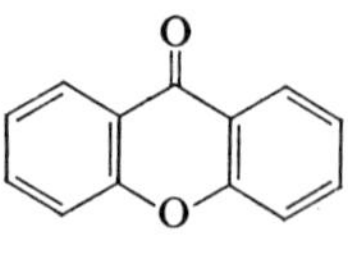

(19)

[21] A. G. Evans, J. C. Evans, P. J. Emes, C. L. James, and J. P. Pomery, *J. Chem. Soc.* (*B*), 1971, 1484.

[22] B. J. Tabner and J. R. Zdysiewicz, *J. Chem. Soc.* (*B*), 1971, 1659.

9-position and subsequent reduction of the resulting 9-one to its radical anion. For example, electrolysis of xanthene at − 40 °C in DMF gives the radical anion of xanthen-9-one (19). The splitting-constant assignments are confirmed by Hückel calculations and by comparison with the spectra of the radicals produced by direct reduction of the 9-ones. Electrolysis of the same compounds under oxygen-free conditions is, by comparison, complex and not fully understood.

The radical anions of the condensed thiophens (20)—(24) are rather short-lived species in solution, even at very low temperatures.[23] In fact the

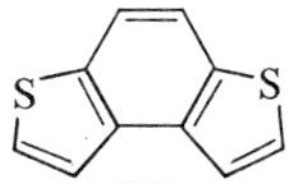
(20)

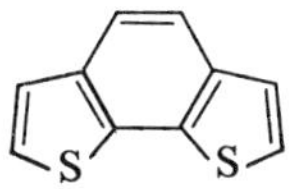
(21)

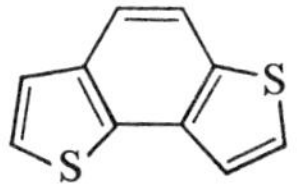
(22)

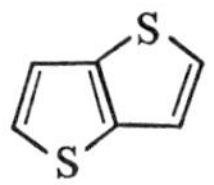
(23)

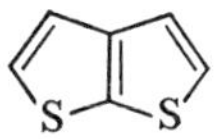
(24)

radical anion of (24) is not detected. Nevertheless, the coupling constants of the observed radical anions have been measured (see Table 2) and

Table 2 *Proton hyperfine splitting constants*/mT *of thiophen derivatives* [24]

Compound	Position 1	2	3	4	5	6	7	8
(20)	0.027	0.511	—	0.415	0.415	—	0.511	0.027
(21)	—	0.572	0.0*	0.052*	0.052*	0.0*	0.572	—
(22)	—	0.140	0.045	0.625	0.032	—	0.601	0.045
(23)	—	0.487	0.052	0.487	0.052	—	—	—

* Assignment is ambiguous.

compared with McLachlan spin densities. An attempt to account for the participation of sulphur *d*-orbitals in the π-system, by introducing parameters to describe possible pd^2 hybridization, was not successful. However, using the sulphur MO parameters of Sullivan,[24] the authors have been able to obtain much closer agreement between theoretical and experimental spin densities. Thiophen derivatives of ethylene, *e.g.* (25)—(29), are much more stable on alkali-metal reduction and give well-resolved e.s.r. spectra.[25] Again, the assignment of the splitting constants is made by comparison

[23] L. Lunazzi, G. Placucci, M. Tiecco, and G. Martelli, *J. Chem. Soc.* (*B*), 1971, 1820.
[24] P. D. Sullivan, *J. Amer. Chem. Soc.*, 1968, **90**, 3618.
[25] L. Lunazzi, A. Mangini, G. Placucci, P. Spagnolo, and M. Tiecco, *J. C. S. Perkin II*, 1972, 192.

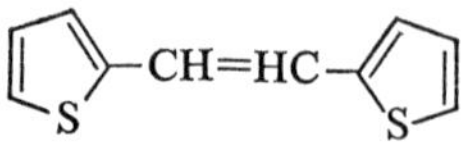

(25)

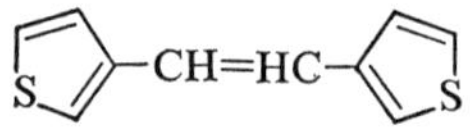

(26)

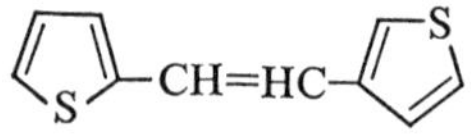

(27)

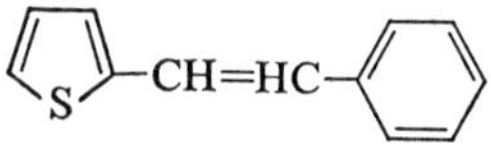

(28)

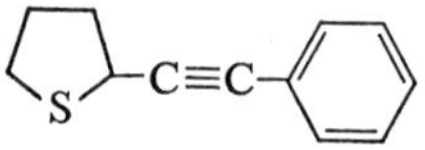

(29)

with McLachlan spin densities and also by selective deuteriation. Once more, it appears that attempts to introduce a measure of sulphur *d*-orbital participation are unsuccessful, and indeed the results are worse than from those calculations which only consider *p*-orbitals. The spectrum of the radical anion of (25) is a mixture of the spectra of two different species – the symmetric *cis,cis-* (or the *trans,trans-*) rotamer and the asymmetric *cis,trans-* rotamer. In addition, the McLachlan calculations reveal that the radical anions of (25)—(29) are twisted along the aryl–ethylene bond. A complete progress report on e.s.r. investigations of thiophen radical ions has been published by the same group.[26]

Ketyls and Semidiones.—Ketyl and semidione radical anions continue to be of great interest to many e.s.r. spectroscopists. As well as the continuing search for novel ketyl and semidione radical anions, there is great interest in the mechanism of hyperfine interaction between the carbonyl-carbon atom and the substituent groups. This is especially true of aliphatic ketyls and semidiones, and this interest has led to great importance being attached to the solution configurations of such ions. It has long been known, for example, that many ketyls are non-planar, and the reasons for

[26] L. Lunazzi, A. Mangini, G. F. Pedulli, and M. Tiecco, *Gazzetta*, 1971, **101**, 10.

this are under active discussion. The ion-pairing properties of ketyls and semidiones are also the subject of a great deal of work.

A particularly comprehensive study of ^{13}C hfsc of a number of semidiones and cyclic ketyls is presented by Russell's group.[27] Continuing their previous interest in aliphatic semidiones,[28] they have attempted to rationalize the large range of carbonyl-carbon splittings as well as those of carbon atoms α and β to the carbonyl group. The ^{13}C hfsc of a number of aliphatic semidiones $R-(C\dot{O})=(CO^-)-R$, where R is methyl, ethyl, isopropyl, and t-butyl, have been measured in DMSO solution at 25 °C. Several trends in the splitting constants are noted. The *cis*- and *trans*-semidiones of acetoin have different values of a^C_{CO} and a^C_α; $a^C_{CO}(trans)$ and a^C_α are smaller (0.058 mT and 0.45 mT) than $a^C_{CO}(cis)$ and $a^C_\alpha(cis)$ (0.114 mT and 0.52 mT). Secondly, a^C_{CO} for a *trans*-semidione increases as the alkyl substituent goes through the series methyl to t-butyl, and a^C_α decreases as the alkyl substituent changes from methyl to t-butyl. The low values of a^C_{CO} (0.06—0.15 mT) indicate that the semidiones are strictly planar and have sp^2-hybridized carbon atoms. This is because a few per cent of *s* character in the original utilized by carbon would lead to values of a^C_{CO} much greater than the experimental value. The a^C_{CO} values are reasonably well accounted for by the Karplus–Fraenkel equation[29] because the authors have chosen to interpret the changes in a^C_{CO} as changes in $Q^C_{CC'}$. The a^C_α and a^C_β values are assumed to vary with the dihedral angle θ between the p^π-orbital of the carbonyl-carbon atom and the appropriate C_α- or C_β-atom orbital, and estimates of the average configuration are made. The second group of compounds described in the same paper is a series of ketyls of cyclic ketones of the type (30). Because of the different amounts of strain

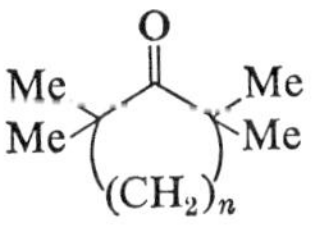

(30)

present at the carbonyl group on varying the value of *n*, these ketyls should provide a good test of whether their high values of a^C_{CO} are a manifestation of non-planarity. Experimentally there is only a small spread of the values of a^C_{CO} and furthermore they are in reasonable agreement with values predicted by the Karplus–Fraenkel equation. The conclusion is that the non-planarity of the ketyl is, rather surprisingly, not of major importance. However, some additional theoretical evidence upon the non-planarity of

[27] G. A. Russell, D. F. Lawson, H. L. Malkus, and P. R. Whittle, *J. Chem. Phys.*, 1971, **54**, 2164.
[28] E. T. Strom and G. A. Russell, *J. Chem. Phys.*, 1964, **41**, 1514; G. A. Russell and G. R. Underwood, *J. Phys. Chem.*, 1968, **72**, 1074.
[29] M. Karplus and G. K. Fraenkel, *J. Chem. Phys.*, 1961, **35**, 1312.

ketyls has also been presented recently. Following a suggestion[30] that the ketyl of hexamethylacetone is non-planar and that this non-planarity is a result of the bulkiness of the t-butyl group, INDO–SCF MO calculations have been performed on hexamethylacetone ketyl and a number of other ketyls and semidiones to determine the most stable geometries.[31] The calculations reported in this paper suggest that not only hexamethylacetone ketyl but even formaldehyde ketyl is non-planar. In the latter case it is calculated that the C—O bond is 32° out of the plane defined by the C—H bonds. The ketyls of 2,2,5,5-tetramethylcyclopentanone and 2,2,4,4-tetramethylcyclobutanone, in which steric repulsions are eliminated by the constraints of the four- and five-membered rings, are also calculated to have non-coplanar carbonyl-carbons. These two compounds are of the type (30) studied by Russell,[27] and it appears that Russell and the author of the present paper[31] place a somewhat different emphasis on the effect of non-planarity on the spin densities. The author of reference 30 attributes the non-planarity of ketyls to the presence of the odd electron in the anti-bonding orbital of the carbonyl group rather than to steric effects, especially in the case of hexamethylacetone.

There is also considerable interest in the effect of fluorine atoms on spin-density distributions, and a recent paper combines this interest with work on the structure of ketyls in a study of ketyls with perfluorinated alkyl side-chains.[32] The ketyl radical anions of hexafluoroacetone (HFA), perfluorodiethyl ketone (PFDK), and perfluorocyclobutanone (PFCB) have been generated electrochemically *in vacuo*, with a view to investigating the delocalization of the unpaired spin into the side-chain fluorine atoms. The primary mechanism is again thought to be direct, across-space interaction of the non-bonding $2p$ orbitals of fluorine with the $2p_z$ orbitals of the π-system, *i.e.* by analogy with protonated side-chains:

$$a_{\beta}^{\mathrm{F}} = B_{\mathrm{X}} \cos^{2}\theta \rho_{\mathrm{X}}^{\pi} \quad (3)$$

where a_{β}^{F} is the isotropic β-fluorine hfsc, ρ_{X}^{π} is the π-spin density on X adjacent to the perfluorinated side-chain, B_{X} is a constant, and θ is the dihedral angle between the axes of the $2p^{\mathrm{F}}$ and the $2p_{\mathrm{X}}^{\pi}$ orbitals. The object is to see if equation (3) has general application to perfluorinated side-chains for ketyls where the main spin density is on the carbonyl group. Thus, any changes in fluorine hfsc should reflect changes in the side-chain only. The experimental β-fluorine hfsc values range between 3.49 mT for HFA and 8.44 mT for PFCB, and the γ-fluorine hfsc are 3.80 mT for PFCB and 0.39 mT for PFDK. Equation (3) is not completely applicable because it does not correctly predict the fluorine splitting constants for PFCB and PFDK. The β-fluorine hfsc are not in proportion to the π-spin density on the adjacent π-centre, even when the large perfluoroalkyl

[30] E. T. Strom, *J. Phys. Chem.*, 1968, **72**, 4715.
[31] G. R. Underwood, *Mol. Phys.*, 1971, **22**, 729.
[32] W. R. Knolle and J. R. Bolton, *J. Amer. Chem. Soc.*, 1971, **93**, 3337.

groups are allowed to rotate freely. In general, INDO–SCF MO calculations give much better spin densities but only when all the valence electrons are taken into account. It is still thought, though, that the primary mechanism is a direct overlap between fluorine *p*-orbitals with the $2p_z$ orbitals of the π-system.

Several other new carbonyl-containing radical anion species have been prepared and their e.s.r. spectra recorded. Alkali-metal treatment of solutions of 7-norbornenone (31) and 9-benzonorbornenone (32) in DME or THF results in the formation of the semidiones (33) and (34).[33] The semidiones are identified as such because similar e.s.r. spectra are obtained on reduction of the corresponding diketones. Apparently the spectra of the semidiones are not split by the alkali metals, the ketyls of (31) and (32) are not formed, and no explanation of the mechanism of formation of (33) and (34) has yet been found.

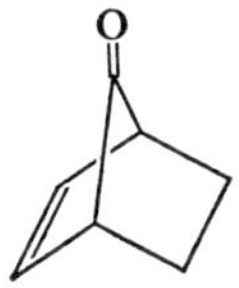

(31)

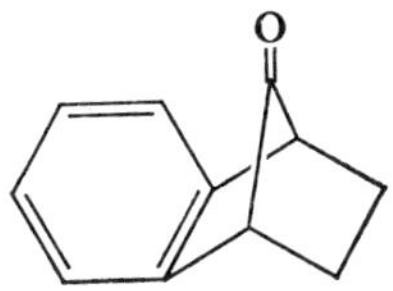

(32)

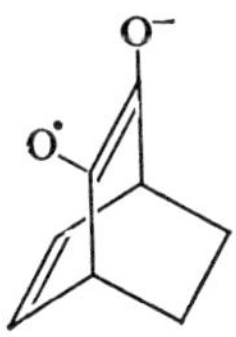

(33)

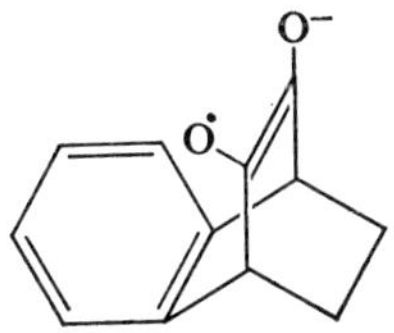

(34)

The non-cyclic enones α-ionone (35), β-ionone (36), and 5,6-dihydro-β-ionone (37) provide an interesting basis for the study of the ketyls of αβ-unsaturated ketones.[34,35] The anion radicals are generated electrochemically in THF. Compounds (35) and (37) give identical e.s.r. spectra, but the spectrum of the anion of (36) has extra splitting from the ring methyl group in conjugation with the carbonyl bond, and from the methylene ring protons adjacent to the methyl group. The splitting constants for the C-4 protons are large in each case (1.16 mT for α-ionone and 1.59 mT for β-ionone) and reflect the complete conjugation with the carbonyl group. Hückel and McLachlan calculations on β-ionone radical anion, in which the π-system extends into the cyclohexane ring, indicate that there is steric interaction between the methyl groups at positions 6 and 10 and the

[33] J. P. Dirlam and S. Winstein, *J. Org. Chem.*, 1971, **36**, 1559.
[34] J. Harbour and A. V. Guzzo, *Mol. Phys.*, 1971, **20**, 565.
[35] J. Harbour and A. V. Guzzo, *Mol. Phys.*, 1972, **23**, 151.

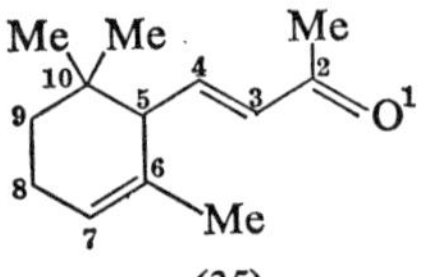

(35)

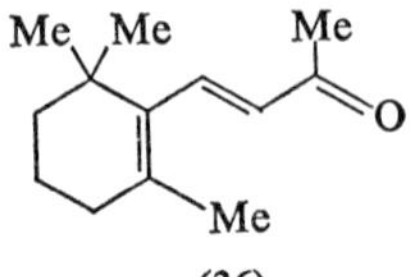

(36)

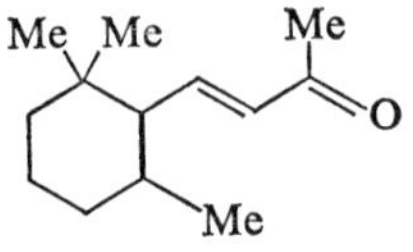

(37)

hydrogens at 3 and 4. This forces the system into a configuration where the π-system is twisted about the C(4)—C(5) bond, at a dihedral angle of approximately 30°.

Dialkyl-semidiones of compounds of the type R^1—O—CO—CO—O—R^2 have been prepared by a process of internal electrolysis and studied by e.s.r. methods. R^1 and R^2 include primary, secondary, and tertiary alkyl groups.[36] Once again the author's main interest is in the splitting constants arising from the α-protons – in this case α with respect to the oxalate linkage. Over a dozen different dialkoxy-semidiones have been prepared, and from the work on these it appears that primary alkyl groups exhibit time-averaged dihedral angles θ of *ca.* 45° between the direction of the α-C—H bond and the z-axis of the π-electron system. The actual value of the angle θ depends upon the size of the β-substituent. However, the spread in values of the dihedral angle over all the primary alkyl substituents is only 5°, which probably has as much to do with the simplifications involved in the assumed spin-density mechanism as it has physical meaning. The α-C—H bond of secondary groups is rotated out of the π-orbital z-axis to an angle of *ca.* 60°. Finally, the α-proton splitting constants are between 0.134 and 0.148 mT for primary substituents and 0.060 and 0.090 mT for secondary substituents.

An e.s.r. study of the polarographic reduction products of dimethyl fumarate and maleate, *trans*-dibenzoylethylene, and diphenyl fumarate shows that the initial species formed are the radical anions of the above compounds.[37] The initially formed *cis* radical anions then very rapidly isomerize to the *trans*-isomers. It is found that, on further electrolytic reduction of the radical anions, more radicals are formed. The structures of these radicals are discussed in detail; radical anions of dimers and trimers are said to be formed. Diphenyl maleate and fumarate radical anion are too unstable to obtain satisfactory e.s.r. spectra.

[36] J. Voss, *Tetrahedron*, 1971, **27**, 3753.
[37] A. V. Il'yasov, Yu. M. Kargin, N. N. Sotnika, V. Z. Kondranina, B. V. Mel'nikov, and A. A. Vafina, *Izvest. Akad. Nauk S.S.S.R., Ser. khim.*, 1971, 932.

Amongst aromatic ketyl radical anions, the perfluorobenzophenone anion has been generated in alcoholic solution by a photochemical method.[38] The anion is prepared in alkaline solution by flow methods to prevent nucleophilic degradation of the perfluoroketone prior to photolysis. A comparison is made of the fluorine hfsc with those of partially fluorinated radical anions. A report on the radical anions derived from *m*-dibenzoylbenzene and a t-butyl derivative illustrates, once again, the value of e.s.r. studies which are carried out in conjunction with optical spectrophotometric measurements.[39] Electron transfer from sodium biphenylide to *m*-dibenzoylbenzene in hexamethylphosphoramide (HMPA) leads to the formation of the radical anion which has a very well resolved e.s.r. spectrum. The analysis of the spectrum and assignments of the splitting constants have been carried out with the aid of ENDOR spectra, the e.s.r. spectrum of a t-butyl derivative, and MO calculations. The main feature of the spectrum is the large (0.626 mT) splitting from the two equivalent protons on the central ring. This value of the splitting constant is consistent with proton splitting constants of other radical anions with protons in positions *ortho* and *para* to electron-withdrawing substituents which are *meta* to one another on a benzene ring. The optical studies reveal that the radical anion is in equilibrium with another diamagnetic, dimeric species even in a strong solvating agent such as HMPA. This fact is not obvious from the e.s.r. spectrum because the dimer accounts for only about 5% of the species present in a 10^{-3} mol l^{-1} solution. The equilibrium constant for the monomer $\rightleftharpoons$ dimer equilibrium is 15 l mol^{-1} at room temperature. It is suggested that the dimer arises through π–π-bond association to form a kind of sandwich structure. In THF, which is a much weaker solvating agent, the visible spectra show that the dimeric species predominates. The e.s.r. spectrum of the radical anion in THF is a single line, indicating that it too is associated with other anions in such a way that there is strong relaxation due to interaction between unpaired spins. The basic difference between the behaviour of *m*-dibenzoylbenzene radical anions in HMPA and THF is due to the formation of ion pairs in THF, whilst in HMPA only free ions are present. Hückel and McLachlan calculations, when compared with the observed splitting constants in solutions in HMPA, point to a non-planarity of the π-system, *i.e.* the outer rings are not coplanar with the central ring and the two carbonyl groups. The position of the benzene rings with respect to one another and to the carbonyl groups gives rise to three possible conformations of the anion. In the light of the experimental hfsc, it is possible to say that the actual conformation is symmetrical with respect to the central ring, but there is insufficient evidence to distinguish between the two possible symmetrical configurations.

Finally in this section, a report on the radical anions of several dialkylmaleic anhydrides (semifuraquinones) should be mentioned. Several

[38] F. P. Sargent and M. G. Bailey, *Canad., J. Chem.*, 1971, **49**, 2350.
[39] J. Chaudhuri, R. F. Adams, and M. Szwarc, *J. Amer. Chem. Soc.*, 1971, **93**, 5617.

compounds with basic structures of the type (38) and (39) have been prepared in reduced form.[40] The compounds (38) are Diels–Alder adducts of dichloromaleic anhydride and cyclic dienes. The main interest of the authors is in using the semifuraquinone entity as a spin label for studying

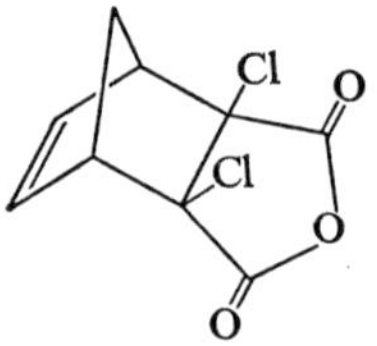

(38)

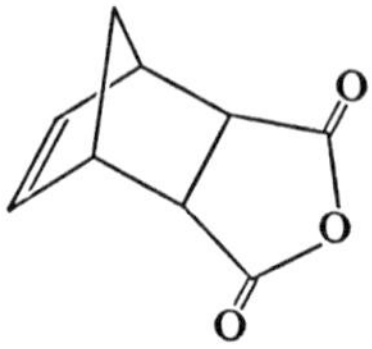

(39)

long-range proton hyperfine coupling. Extensive and precise splitting constants are given for the bridge protons, and the importance of hyperconjugation and spin-polarization mechanisms in the long-range coupling is assessed for each individual radical anion.

Other Radical Anions.—It is often the case that e.s.r. methods are used to supplement research in which the main emphasis is on other aspects of radical anion chemistry. A polarographic reduction study of several benzonitrile *N*-oxides in non-aqueous solvents (mainly DMF) has utilized e.s.r. techniques to identify the radical anions formed.[41] It is found that for most 4-substituted 2,6-dimethylbenzonitrile *N*-oxides (the substituents are H, Me, Br, OMe, or NO_2), the potential of the first polarographic wave is due to the deoxygenation of the $-C{\equiv}N{\rightarrow}O$ group. Similar behaviour is observed in aqueous solutions. The second wave potential is a result of the reduction of the corresponding benzonitrile anions. This has been confirmed by comparison of the e.s.r. spectra with those of the benzonitrile anions which have been prepared directly. The only exceptional compound is 2,6-dimethyl-4-nitrobenzonitrile *N*-oxide, whose radical anion is stable enough for its e.s.r. spectrum to be recorded. This *N*-oxide radical anion has very similar splitting constants to the nitrile anion except, as one would expect, for the nitrile-nitrogen splitting. In the *N*-oxide it is 0.152 mT, whereas in the nitrile it is 0.082 mT.

The radical anions of cyclo-octatetraene and its derivatives have been the subject of a number of interesting papers over the past few years. The interest has centred on the configuration of the monoanion and dianion as compared with the neutral molecule. The radical anion of cyclo-octatetraene itself has a doubly degenerate ground state and, like the dianion, it is planar. All the proton hyperfine splittings in the e.s.r. spectrum are equivalent. The latest radical anion of a cyclo-octatetraene derivative to be prepared is that of 1,3,5,7-tetraphenylcyclo-octatetraene (TPCOT).[42]

[40] S. F. Nelsen, E. F. Travecedo, and E. D. Seppanen, *J. Amer. Chem. Soc.*, 1971, **93**, 2913.
[41] H. Mujazaki, K. Nishikada, and T. Kubota, *Bull. Chem. Soc. Japan*, 1971, **44**, 277.
[42] R. D. Rieke and R. A. Copenhafer, *Tetrahedron Letters*, 1971, 4097.

In contrast to cyclo-octatetraene, electrochemical reduction of TPCOT occurs by a single, two-electron, reversible step. Alkali-metal reduction in ethereal solution only results in the formation of the yellow dianion, and even the addition of a large amount of neutral TPCOT has no discernible effect on the disproportionation equilibrium. However, as is so often the case, addition of a strong solvating agent such as HMPA ensures that there are enough free ions in solution such that the disproportionation equilibrium shifts sufficiently for the radical anion to be observed. The splitting constants of the octagonal ring protons are all the same and very close to the value for cyclo-octatetraene itself, *i.e.* 0.330 mT as compared with 0.323 mT for cyclo-octatetraene.[43] Clearly, the ring is planar, and probably ring flattening occurs during both one- and two-electron reduction.

Radical anions, which are interesting from the point of view of e.s.r. spectroscopists as well as being biologically important, include those derived from vitamin A and related Schiff bases.[44] It is possible that the mechanism of visual excitation may include radical anion and/or charge-transfer intermediates, and so the products of the reduction of retinal and related compounds are important. *In vacuo* electrolysis of a THF solution of retinal at − 70 °C results in the formation of a radical with a partially resolved e.s.r. spectrum, which cannot be fully analysed. Fortunately, electrolysis of the related *N*-retinylidene-n-butylamine (40) results in the

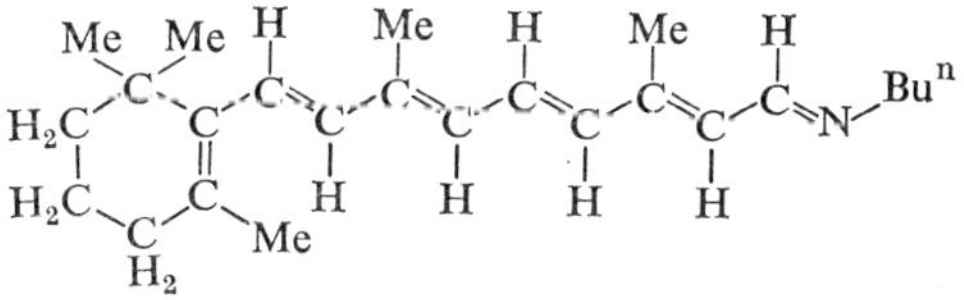

(40)

formation of its radical anion. From a tentative analysis of the e.s.r. spectrum it seems that the unpaired electron is distributed over the whole conjugated system. Most of the splitting constants have been identified, although the assignments are not certain. The radical anion is not observed at low temperatures, possibly because it either disproportionates or dimerizes. Lastly, Livingston *et al.*[45] have prepared the radical anions (or their protonated forms) of alloxan, parabanic acid, and *N*-methylparabanic acid by photolysis of aqueous solutions of these compounds with acetone and isopropyl alcohol. The radical anions are produced by electron transfer from the initially formed $(CH_3)_2\dot{C}OH$ radical.

3 Radical Cations

The number and variety of radical cations which have been prepared and studied by e.s.r. spectroscopy continues to grow, so that it rivals the

[43] R. D. Allendoerfer and R. H. Rieger, *J. Amer. Chem. Soc.*, 1965, **87**, 2336.
[44] C. M. Lang, J. Harbour, and A. V. Guzzo, *J. Phys. Chem.*, 1971, **75**, 2861.
[45] J. K. Dohrmann, R. Livingston, and H. Zeldes, *J. Amer. Chem. Soc.*, 1971, **93**, 3343.

field of radical anions as an area of intense interest. The focus of attention is usually upon the relationship between the splitting constants and *g*-values of corresponding radical anions and cations. Other important subjects are the effects of oscillations and restricted rotations on splitting constants and linewidths, the effects of near-orbital degeneracies, and new methods of preparation of novel radical cations. Though many methods of radical cation preparation have been used, the most popular reagents are still concentrated sulphuric acid and the nitromethane–aluminium chloride mixture. All other things being equal, the second method is in many ways more useful, especially where measurements of temperature-dependent effects are involved.

As far as crystalline salts of hydrocarbon radical cations are concerned, there has been further work on pyrene, perylene, and azulene perchlorates (py_2ClO_4, $pery_2ClO_4$, and az_2ClO_4, respectively). Single crystals of these compounds have been produced on electrode surfaces and their e.s.r. spectra measured.[46] Single crystals of py_2ClO_4 have $g_{\parallel}$ (with respect to the main crystal axis) equal to 2.00226 and $g_{\perp} = 2.00250$. The single line of the spectrum is very narrow (0.001 mT), which indicates a strong-exchange narrowing effect on the hyperfine structure. The *g*-values cannot be interpreted in any way quantitatively because they probably represent orientations spread over a number of different molecules and unfortunately the crystal structure is unknown. In order to obtain a sufficiently large signal for susceptibility measurements, polycrystalline samples have been prepared. Apparently the main source of the signal is a small amount of 'impurities', *i.e.* defects in the lattice structure which segment the linear chains of aromatic radicals. It is shown that the signal from py_2ClO_4 and $pery_2ClO_4$ consists of two components. One follows a Curie law and dominates at low temperatures and is attributed to the lattice 'impurities'. The other component, which is more important at high temperatures, arises from thermally generated Wannier triplet excitons.

Owen and Vincow[47] have looked at the spectra of naphthalene, anthracene, and perylene radical cations in boric acid glass. Using the theory of polycrystalline spectra propounded by Lefebvre and Maruani,[48] they have been able to simulate the spectra and determine the hyperfine tensor components.

Probably of more interest to the chemically biased e.s.r. spectroscopist is the *p*-quaterphenyl cation radical, which has been prepared, under vacuum conditions, in fuming sulphuric acid.[49] The radical cation is stable at − 10 °C but decomposes at room temperature. Its e.s.r. spectrum consists of seventeen partially resolved lines which have been computer simulated using splitting constants of 0.127, 0.010, and 0.194 mT for, respectively,

[46] T. C. Chiang, A. H. Reddoch, and D. F. Williams, *J. Chem. Phys.*, 1971, **54**, 2051.
[47] G. S. Owen and G. Vincow, *J. Chem. Phys.*, 1971, **54**, 368.
[48] R. Lefebvre and J. Maruani, *J. Chem. Phys.*, 1965, **42**, 1480.
[49] M. K. Carter, *J. Phys. Chem.*, 1971, **75**, 902.

the *ortho*-, *meta*-, and *para*-protons of the end rings, and 0.127 and 0.183 mT for the other ring protons. Comparison of these values with those of the anion shows that the *ortho* and *meta* splitting constants are *smaller* in the cation than in the anion, and this is attributed to the non-planarity of the radical cation. Accepting that steric interaction between ring hydrogens enhances the probability that two connected phenyl rings will not be coplanar, and that resonance effects will tend to oppose the steric effect, it seems reasonable that the phenyl rings of the radical cation will be more out of plane than those of the anion because resonance effects will be less important, there being two fewer electrons.

Among the more interesting results of e.s.r. work on radical cations has been the number of dimeric species that have been discovered. The radical cation of 1,4-dimethylnaphthalene (1,4-DMN) is prepared by oxidation of the parent compound by antimony pentachloride in dichloroethane–sulphur dioxide solution.[50] It is observed that there is an equilibrium in solution between the monomer cation and the dimer cation, with the dimer cation being favoured at lower temperatures. However, the isomeric 1,5- and 1,8-DMN do not form dimer cations and the 1,8-DMN cation is a short-lived species. The most noteworthy features of the splitting constants for 1,4-DMN radical cations are firstly that the methyl hfsc of the cation is almost three times that of the anion (0.963 and 0.326 mT, respectively), and secondly the splitting constants of the dimer cation do not correspond to one half the monomer cation values, thus showing that the unpaired electron distribution in the dimer cation is more delocalized over the two phenyl rings than in the monomer. Simple MO calculations fail to explain the discrepancies in the monomer cation splittings; however, more success is obtained with the dimer cation. Of the two possible sandwich arrangements [see (41) and (42)], (42), which has C_{2h} symmetry, is favoured on the

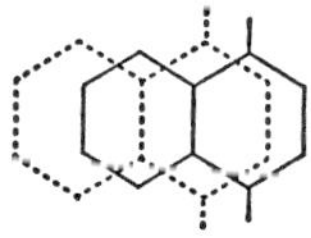

(41)

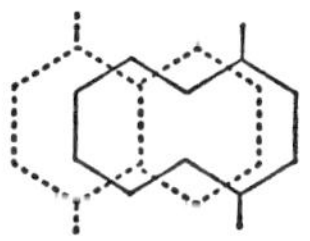

(42)

basis that the change in splitting constants is more correctly predicted and that the energy of the dimer is lower. The calculations have been performed assuming resonance integrals between the upper and lower rings to be proportional to the overlap integral, and assuming an interplanar separation of 0.3 nm.

A similar type of dimer cation structure is postulated in order to explain the removal of the orbital degeneracy in the mesitylene dimer cation.[51]

[50] H. Yoshini and K. Kuwata, *Mol. Phys.*, 1972, **23**, 297.
[51] S. Shih and R. M. Dessau, *J. Chem. Phys.*, 1971, **55**, 3757.

The splitting constants of the cation dimer quite clearly show that the two-fold degeneracy has been lifted by dimer formation and that the B_{3g} symmetric form of the original degenerate pair of molecular orbitals has been stabilized. Because of steric hindrance the dimer cation probably prefers a sandwich structure of D_{3d} symmetry in which the methyl groups are not directly over one another. The reason for the lifting of the degeneracy is a static Jahn–Teller effect brought about by a co-operative vibronic coupling of the two molecules. This co-operative vibronic coupling is probably a sliding motion of the two rings along the x- and y-axes (the z-axis being the three-fold symmetry axis). On the assumption that the force constant for the motion is an order of magnitude lower than the intramolecular stretching constant, the vibronic distortion energy required to produce a Jahn–Teller effect is of the order of 10^5 m^{-1}.

Triptycene (43) radical cation is prepared by cobaltic ion oxidation in trifluoroacetic acid using rapid-flow techniques.[52] The e.s.r. spectrum at

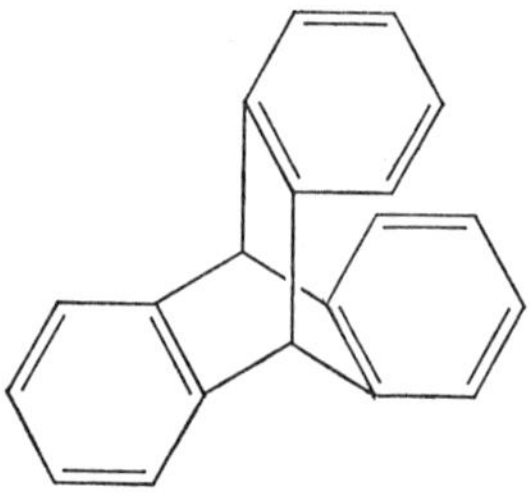

(43)

room temperature consists of seven equally spaced lines separated by 0.23 mT, the line intensities being those expected for six equivalent protons. This is taken to mean that there is a rapid intramolecular electron transfer between the rings due either to complete delocalization or rapid exchange between isolated π-systems. If triptycene is considered as three *o*-xylene units arranged at 120°, then the remote hydrogens are expected to be the ones producing the splitting because the unpaired electron in *o*-xylene cation occupies an asymmetric orbital of this type. No hyperfine interaction is observed for the bridgehead protons, which supports the conclusion that a hyperconjugative method of spin transmission between the rings is in operation. The unpaired electron in the radical cation of 5-nitrotriptycene only interacts with the nitro-substituted ring because of the stabilizing effect of the nitro-group.

The formation of radical cations from aromatic sulphides and sulphoxides has become well established in recent years. In most cases the same cation radical may be prepared by one-electron oxidation of the sulphide and by the homolytic deoxygenation of the sulphoxide. The radical cations of

[52] R. M. Dessau, *J. Chem. Phys.*, 1971, **54**, 5430.

bis-(4-hydroxyphenyl) sulphide, bis-(4-methoxyphenyl) sulphide, and the corresponding ethers have been prepared in order to assess the comparative effects of sulphur and oxygen bridging atoms on spin-density distributions.[53] All the compounds have well-resolved e.s.r. spectra and the splittings are easily analysed in terms of unpaired spin density distributed over both phenyl rings, *i.e.* rapid electron transfer takes place between the two rings. As in the case of the triptycene radical cation above, nitro-substitution on one or both rings affects the electron-transfer rate so that in bis-(2-nitrophenyl) ether transfer is slow. The spin density on the bridging sulphur atom in the cations of the diphenyl sulphides is much greater than that on the bridging oxygen in the diphenyl ether cations. A final useful point about this particular paper is that the authors bring together all the accepted MO parameters for sulphur and oxygen atoms and for C—S and C—O bonds.

A formidable array of radical cations prepared from halogenated tetrahydroxybenzene diethylene ethers is the subject of e.s.r. and visible spectrophotometric investigation.[54] Radical cations are obtained in sulphuric acid and nitromethane–$AlCl_3$ from (44), (45), (48), and (49) but not

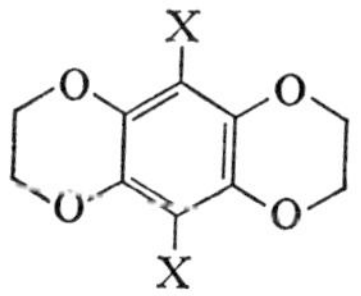

(44) X = Cl
(45) X = Br

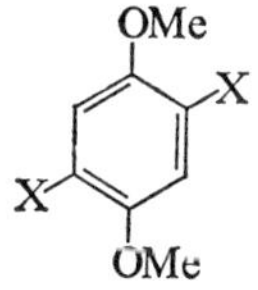

(46) X = Cl
(47) X = Br

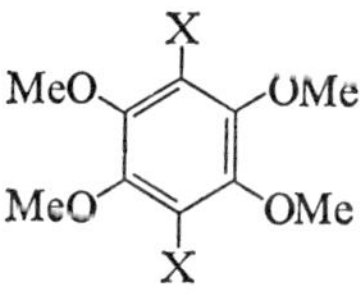

(48) X = Cl
(49) X = Br

from (46) and (47). The e.s.r. spectra are noteworthy for their small number of resolved lines, with the exception of (49) which has a much more complicated spectrum, due perhaps to restricted rotation of the methoxy-groups. Unfortunately the authors have not made a complete analysis of the spectra, and consequently only meagre conclusions concerning the spin-density distribution can be made.

[53] P. D. Sullivan and H. J. Shine, *J. Phys. Chem.*, 1971, **75**, 411.
[54] T. Asahara, M. Senō, and T. Teshirogi, *Bull. Chem. Soc. Japan*, 1971, **44**, 1125.

Evidence for hyperfine interaction between halogen anions and the unpaired electron of the metalloporphyrin cation radical is presented in a paper on π-radicals of zinc tetraphenylporphyrin [Zn(tpp)].[55] The π-radical $\overset{+}{\text{Zn(tpp)}}^{\bullet},ClO_4^-$ has an e.s.r. spectrum in solution consisting of nine lines which are attributed to four equivalent nitrogens. In the presence of 0.5 mol l^{-1} bromine, a new radical with the same optical spectrum but with a four-line e.s.r. spectrum is produced. Treatment with silver perchlorate regenerates the original e.s.r. spectrum. The following sequence of reactions seems to be in operation:

$$\overset{+}{\text{Zn(tpp)}}^{\bullet},ClO_4^- \xrightarrow{0.5\ \text{mol}\ l^{-1}\ Br_2} \overset{+}{\text{Zn(tpp)}}^{\bullet},Br^- \xrightarrow{AgClO_4} \overset{+}{\text{Zn(tpp)}}^{\bullet},ClO_4^- + AgBr$$

The $\overset{+}{\text{Zn(tpp)}}^{\bullet},X^-$ ion pairs (X may be chlorine or bromine) are said to be complexes rather than weak ion pairs, though it is not quite clear how the distinction is made. In chloroform, $a_{^{79}Br} = 0.818$ mT, $a_{^{81}Br} = 0.882$ mT, $a_{^{35}Cl} = 0.166$ mT, and $a_{^{37}Cl} = 0.138$ mT.

As a conclusion to this section on radical cations, mention of one paper on simple MO calculations is appropriate. Although theoretical calculations on radicals are considered in detail elsewhere in this volume, many workers, whose interest is mainly in radical ions and whose use of MO calculation methods is confined to simple Hückel and McLachlan methods, will be interested in any further enquiries into the theory behind such methods.

It has been pointed out[56] that the equations given by McLachlan[57] for the calculation of spin densities are only strictly applicable to the anions of even alternants. The spin densities predicted for the corresponding cations are different from those for the anions, although this should not be the case because deviations from the pairing theorem are only the result of a charge effect which modifies the McConnell equation. The discrepancy arises because McLachlan treats the anion as the addition of an unpaired electron to a closed shell and considers exchange perturbations to the α-spins only. Applied to the cation the perturbation is considered as the addition of an unpaired electron to the closed-shell dication. Its exchange field perturbs the α-electrons in the bonding MOs with the exception of the highest MO, which contains the unpaired electron. To make the results for the cation consistent with the anion, the cation should be treated as a removal of the β-electron from the neutral molecule. Apparently, incorrect McLachlan spin densities have been published in the past because of the failure to realize this problem. The solution is, of course, very simple. McLachlan spin densities calculated for the anion are equally applicable to the cation.

[55] A. Forman, D. C. Borg, R. H. Felton, and J. Fajer, *J. Amer. Chem. Soc.*, 1971, **93**, 2790.
[56] J. C. Bickerton and R. E. Moss, *Mol. Phys.*, 1971, **21**, 341.
[57] A. D. McLachlan, *Mol. Phys.*, 1960, **3**, 233.

4 Ion Pairing and Dynamic Processes

It is virtually impossible to separate problems concerned with the structure of ion pairs involving interaction between a radical anion and an alkali-metal counterion from a consideration of dynamic processes which affect the e.s.r. linewidths. Such dynamic processes as intramolecular and intermolecular cation exchange, electron and atom transfer, exchange between different types of ion pairs, *etc.*, are all a function of the nature of the ion pairs involved. Therefore, this section will deal with the whole range of e.s.r. studies which are principally concerned with finding out about the nature of ion pairs in solution. The previous sections have been concerned with novel radical ions; in the work described in this section, the radical ion is the tool for probing into a solution system as well as being the model system for describing the binding of certain types of anions to their counterions.

There have been many attempts in recent years to find a suitable MO model which will describe satisfactorily the mechanism of alkali-metal splitting in the e.s.r. spectra of many aromatic radical anions and the perturbation of proton hyperfine splittings due to the proximity of a cation. Any satisfactory model must not only explain these effects but also why the metal hfsc values tend to become negative with increasing atomic number and why the metal splitting constants often decrease with temperature.

Most MO models of an aromatic ion pair have treated the cation as a point positive charge and the perturbation as a straightforward coulombic interaction. In a series of recent McLachlan calculations,[58] the McClelland model[59] for the electrostatic interaction is used to determine the modification of the radical-anion coulomb- and resonance-integrals required by the presence of the counterion. A series of different radical anions have been chosen as representative of different types of ion pairs, *i.e.* anthracene, fluorenone, anthraquinone, and 2,2′-bipyridyl. The Fraenkel–Karplus formula and Stone's[60] theory are used to calculate the ion-pairing effect on ^{13}C splittings and *g*-factors, respectively. From the results obtained, it is quite clear that the effect of ion pairing on aromatic proton hfsc can be predicted qualitatively, but quantitative agreement, as one would expect from such a crude model, is not very good. The authors show that an electrostatic model of this type may be used to predict the most likely ion pair structures, and they can account for changes of the proton hfsc with alkali-metal ion and solvent. Apparently, change of solvent may be treated merely as a modification of the screening factor employed in the McClelland formula, at least for the radical ion systems used in the calculations. Rather more sophisticated MO calculations on ion-pair systems require a consideration of the nature of the mechanism for transfer of unpaired spin

[58] T. Takeshita and N. Hirota, *J. Amer. Chem. Soc.*, 1971, **93**, 6421.
[59] B. J. McClelland, *Trans. Faraday Soc.*, 1961, **57**, 1458.
[60] A. J. Stone, *Mol. Phys.*, 1963, **6**, 509; 1964, **7**, 311.

on to the metal nucleus. It is necessary not only to set up molecular orbitals for the ground-state ion-pair system, but also to consider admixtures of excited configurations which might contribute to the unpaired spin density on the metal. This is the approach taken by de Boer and co-workers.[61] In their calculations, the ground-state ion-pair MOs are considered to be basically anion orbitals with admixtures of the metal valence *s*-orbital. The results for the sodium naphthalenide system show that this zeroth-order treatment is not sufficient to explain the metal spin-density and that it is necessary to modify the metal spin-density by a configuration interaction procedure. The most important contributions arise from aromatic to metal excitations, *i.e.* $\bar{A}^{\cdot},Me^{+} \rightarrow A^{*},Me$ and A^{*},Me^{*}, because the spin density at the metal nucleus is large in these charge-transfer states and their contributions to the zeroth-order spin density will be negative. The authors report a systematic investigation of the effect of these contributions and also of the effect of changing the interionic distance, and they make some definite conclusions about ion-pair MO interactions – at least in the case of sodium naphthalenide. It is not sufficient to include only the zeroth-order contributions to the metal spin-density. Contributions from the excited charge-transfer states described above are very often of the same order of magnitude and opposite in sign. Even more important is the conclusion that the metal hfsc may have different signs in different regions of space. A region of negative spin density will often occur, particularly when there is an antisymmetry plane in the first antibonding MO of the aromatic ion. Further, electrostatic polarization of the π-MO's of the ion pair enhances the zeroth-order spin density compared with the first-order contribution, so that the spin density on the larger cations is increasingly determined by the negative first-order contribution. Thus, as the atomic number of the cation increases, in many ion pairs the metal hfsc decreases and eventually becomes negative.

E.s.r. investigations have enabled the complexities of ion pair formation to be more clearly seen, and it has become possible to progress from a mere distinguishing of 'free' ions from ion pairs, to identification of such species as 'contact', 'solvent-separated', and 'solvent-shared' ion pairs. Many investigators are now turning their attention to the arrangement and number of molecules which are closely associated with these different types of ion pair. However, first of all an ion pair must be identified as such, and this may not be very easy if the metal splitting is so small as to be unresolved. Jones *et al.*[62] have described the relationship between pure Lorentzian lines and lines which are non-Lorentzian owing to unresolved metal splitting. The non-Lorentzian lines fall off more sharply at the wings than the Lorentzian lines and this difference becomes more marked as the ratio between the metal hfsc and the peak-to-peak linewidth increases.

[61] G. W. Canters, C. Corvaja, and E. de Boer, *J. Chem. Phys.*, 1971, **54**, 3026.
[62] M. T. Jones, M. Komarynsky, and R. D. Rataiczak, *J. Phys. Chem.*, 1971, **75**, 2769.

By means of a complete analysis of the lineshape plus a comparison of the experimental and computed curves, the unresolved metal hf.s.c. may be measured, and this is in fact carried out for the benzene radical anion pair. The same group of investigators present details of a CW saturation study [63] on the same anion because it is representative of orbitally degenerate radical anions. As a further study of the spin relaxation properties (see also refs. 6 and 7) of orbitally degenerate ions this paper is extremely useful. It is also important because a full description of techniques for the measurement of H_1 and the cavity Q are given. Its relevance to this particular section of this Report lies in the detailed description of the temperature-dependence of T_1 and T_2 for the benzene ion pair in solution. This work and a subsequent paper [64] will be considered in detail later in the Report.

Returning to the more detailed investigations of the solvent arrangement around ion pairs, it is noteworthy that those old favourites, the nitrobenzene anions, still provide plenty of material for worthwhile investigation, particularly because the ^{14}N hfsc of the nitro-group is a sensitive indicator of its environment. It has been shown [65] that the effect of adding sodium or lithium salts to nitrobenzene and *meta*- and *para*-dinitrobenzene radical anions in methanol or ethanol is to increase the nitrogen splitting (no metal splitting is observed). The ion pairs are considered to be solvent-shared and the change in nitrogen splitting is thought to be due to a small increase in the bonding between the alcohol and the anion, although it is not immediately apparent what the precise nature of this is. The alkali-metal cation is thought not to displace any solvent molecules from around the anion, in contrast to t-butylammonium cations which are thought to displace solvent molecules from around the anion, and thus cause the ^{14}N hfsc to decrease quite markedly. In water, *m*-dinitrobenzene radical anion is asymmetrically solvated because one nitrogen coupling constant is observed to be much larger than the other,[66] and this therefore means that the mean lifetime of one solvent configuration is approximately 10^{-6} s. Addition of t-butyl alcohol (up to 0.02 mol l^{-1}) causes this lifetime to increase slightly, but above this concentration of the alcohol the lifetime decreases. An interesting model is used to explain these observations, *i.e.* one nitro-group obviously carries most of the negative charge when the ion is in water and this is strongly solvated; the rest of the molecule is surrounded by a cage of hydrogen-bonded water molecules. The t-butyl alcohol molecules may be accommodated in this cage without undue disruption until at a critical concentration alcohol molecules have to share two or more cages. It is the cage sharing that disorganizes the water molecules such that changes of hydration of the nitro-groups occur frequently enough to significantly shorten the lifetime of any one configuration. Postulated models of this kind are very useful as a basis for building

[63] R. D. Rataiczak and M. T. Jones, *J. Chem. Phys.*, 1972, **56**, 3898.
[64] M. T. Jones and M. Komarynsky, *J. Chem. Phys.*, 1972, **56**, 4404.
[65] J. Oakes, J. Slater, and M. C. R. Symons, *Trans. Faraday Soc.*, 1970, **66**, 546.
[66] D. Jones and M. C. R. Symons, *Trans. Faraday Soc.*, 1971, **67**, 961.

up a picture of what is going on in solution, but it requires a lot of careful measurements of the effect of solvent additions on radical ion lifetimes, and as yet no such comprehensive results are available.

In several investigations already referred to (see Section 2, p. 90), it has become apparent that the alkali-metal ketyls are extremely interesting species as far as their structures in solution are concerned. For example, the alkali-metal splittings in the spectra of xanthone and fluorenone ion pairs in DME and THF are dependent upon temperature and upon the addition of a polar solvent such as DMF.[67] The temperature-dependence is most marked for the caesium ketyls. In addition, the caesium lines exhibit broadening that is consistent with a dynamic process involving exchange between ion pairs with different metal hfsc. If it is assumed that the metal splitting constants for each individual type of ion pair are temperature-independent, then the ion-pair equilibrium may be quantified from measurements of the observed average metal splitting. In the caesium–xanthone system, $\Delta H = 5.5$ kJ for the equilibrium between the two different ion pairs in DME, and a similar value is obtained in THF. This value, taken together with the small value for ΔS of 24 J deg^{-1}, suggests that there is very little reorganization of the solvent molecules around the ion pair, much less in fact than in a solvent-separated $\rightleftharpoons$ contact ion-pair interconversion. The ion pairs in the caesium–xanthone system are probably always contact ion pairs and the changes in metal splitting arise because in one ion pair the average position of the cation is in the nodal plane of the $2p^{\pi}$ oxygen orbital and in the other ion pair the average position is out of the nodal plane. Addition of small amounts of DMF to the ketyl ion pair solutions results in the displacement of one solvent molecule by DMF and a reduction of the alkali-metal splitting. At higher DMF concentrations only contact ion pairs and free ions are present in solution because a superimposition of the spectra of these two species is observed. Apparently no intermediate solvated species such as solvent-separated ion pairs are formed.

Biacetyl semidione has been known for some time to exist as a mixture of *cis*- and *trans*-forms in DMSO.[68] In the presence of the large alkali-metal cations (caesium and rubidium) the *trans*-form is favoured, but with lithium, sodium, and potassium the *cis*-isomer predominates.[69] The radical anion of perfluorobiacetyl behaves in a similar way but is more convenient to study as it may be prepared by alkali-metal reduction in THF. In this solvent the e.s.r. spectrum of the *cis*-isomer is split by the lithium and sodium nuclei.[69] The *trans*-isomers exhibit pronounced linewidth alternation consistent with the cation jumping between the two oxygens. The rate of jumping is low for lithium cations, but in the intermediate region

[67] K. S. Chen, S. W. Mao, K. Nakamura, and N. Hirota, *J. Amer. Chem. Soc.*, 1971, **93**, 6004.
[68] G. A. Russell and R. D. Stephens, *J. Phys. Chem.*, 1966, **70**, 1320.
[69] G. A. Russell, J. L. Gerlock, and D. F. Lawson, *J. Amer. Chem. Soc.*, 1971, **93**, 4088.

for sodium and potassium. The e.s.r. spectra of the lithium salts of perfluorobiacetyl radical anion rather unusually vary with time following reduction with lithium metal. The suggested reaction scheme which would explain the changes involves the reduction of the diketone all the way to the dianion, followed by reaction with a second diketone molecule to yield a dimeric form of the ion pair, which is then slowly converted into a mixture of the *trans* ion pair and a new contact ion pair with a large lithium splitting. The observed spectral changes are due to kinetic effects associated with the rate of dissolution of lithium metal in THF.

Two investigations of alkali-metal naphthalenides warrant close attention. An n.m.r. study of the proton, deuterium, and metal hfsc of ion pairs of deuteriated and non-deuteriated naphthalene radical anions in DME and THF [70] presents a complete picture of the ion-pair situation in these pure solvents. The splitting constants determined from n.m.r. spectra are in close agreement with the e.s.r. results, and the alkali-metal n.m.r. spectrum shows that the spin density is positive at the lithium and sodium nuclei and negative for rubidium and caesium. In the case of potassium, there is a reversal of the sign of the spin density at the nucleus on changing the temperature. The plot of sodium splitting *versus* temperature in DME is strongly curved, suggesting that solvent-separated ion pairs exist in equilibrium with contact ion pairs in this region. The sign reversal of the potassium spin density may best be rationalized by assuming the existence of contact ion pairs over the whole temperature range, because for solvent-separated ion pairs a sign reversal would not be expected. Rubidium, like sodium, forms contact ion pairs at temperatures greater than $-20\,^{\circ}C$ and solvent-separated ion pairs at lower temperatures. Lastly, it is thought that the almost constant value of the caesium splitting suggests the formation of contact ion pairs throughout the whole temperature range. For naphthalene ion pairs in general it is possible to conclude that the tendency to form contact ion pairs is greater the larger the radius of the metal cation, because the cation is less solvated, and that the metal hfsc tend to become negative for the larger cations. The latter observation has been accounted for theoretically, as has been mentioned above (ref. 61). In mixed solvent systems the situation is obviously more complicated, and a combination of techniques is required to elucidate the equilibria set up in individual systems. An e.s.r. and optical investigation of sodium and potassium naphthalenide ion pairs in tetrahydropyran (THP) and THP + DME mixtures as well as in diethyl ether (DEE) and DEE + THF mixtures has revealed the existence of various ion-pair species.[71] For example, on addition of small amounts of DME to a solution of naphthalene in THP which is in contact with a sodium mirror, the following equilibria

[70] B. M. P. Hendricks, G. W. Canters, C. Corvaja, J. W. M. de Boer, and E. de Boer, *Mol. Phys.*, 1971, **20**, 193.

[71] L. Lee, R. Adams, J. Jagur-Grodzinski, and M. Szwarc, *J. Amer. Chem. Soc.*, 1971, **93**, 4149.

are set up:

$$Na(metal) + C_{10}H_8 \rightleftharpoons C_{10}H_8^{\overline{\cdot}},Na^+$$

$$C_{10}H_8^{\overline{\cdot}},Na^+ + DME \rightleftharpoons C_{10}H_8^{\overline{\cdot}},Na^+,DME$$

$$C_{10}H_8^{\overline{\cdot}},Na^+,DME + DME \rightleftharpoons C_{10}H_8^{\overline{\cdot}},Na^+,DME_2$$

These equilibria may be determined quite straightforwardly from measurements of the visible spectrum of sodium naphthalenide. The additional information gained from the e.s.r. spectrum of this system is that the contact ion pair $C_{10}H_8^{\overline{\cdot}},Na^+,DME_2$, is in dynamic equilibrium with its solvent-separated isomer, *i.e.*

$$C_{10}H_8^{\overline{\cdot}},Na^+,DME_2 \rightleftharpoons C_{10}H_8^{\overline{\cdot}},DME,Na^+,DME$$

and this is revealed by the temperature- and solvent-dependence of the sodium lines. In the DEE + THF system, the dominant species is $C_{10}H_8^{\overline{\cdot}},Na^+,THF_2$ at the concentrations of sodium naphthalenide which were investigated (up to 1.1 mol l^{-1}). The mono-solvated ion pair apparently does not exist in a significant concentration as far as optical measurements are concerned, although its concentration is large enough to contribute to the temperature-dependence of the metal splitting.

It is clear that in recent published work an increasingly detailed picture is being obtained of the structures of ion pairs and the arrangement of the associated solvent molecules. Many questions still remain, however, not the least of which are: what constitutes an associated solvent molecule and what is the significance of measurements of these numbers of solvent molecules in the ion-pair entity? Do the numbers, in fact, serve to rationalize experimental data or do they have physical significance? In what way are the solvent molecules most closely associated with the ion pair different from those in the next solvent shell?

The simple aromatic semiquinones have been arguably the class of radical anions which have been most widely investigated by e.s.r. methods. It is not hard to see the reasons for this great interest in their properties. The radical anions are easily generated and quite stable when formed, the spectra are, on the whole, reasonably simple, and the presence of two electronegative centres (*i.e.* the oxygen atoms) creates the possibility of a wide range of ion-pairing effects. One of the most interesting of these is the phenomenon of linewidth alternation due to intramolecular cation exchange between equivalent sites, and this exchange process has been most thoroughly investigated for semiquinones.

Until recently, intramolecular cation-exchange parameters have not been systematically measured with a view to comparison of rates in different solvents and for different cations. However, Warhurst and Wilde [72] have

[72] E. Warhurst and A. M. Wilde, *Trans. Faraday Soc.*, 1971, **67**, 605.

now presented their measurements of the effect of solvent and cation on the Arrhenius parameters in the 2,5-di-t-butyl-*p*-benzosemiquinone (2,5-TBPBZQ) system. All the alkali metals are used as cations by them and most of the systems are studied in DME, THF, and 2-MeTHF. It is apparent from the results that even a relatively simple ion pair such as this is not easily described quantitatively in terms of a simple model. The activation energy varies between 22 and 30 kJ mol^{-1} and $\log_{10} A$ between 8.3 and 11.1. The trends in E_a and A are interpreted qualitatively on the basis of plausible configurations for the initial and transition states of the reaction. Only two types of interaction energy are considered, (i) the electrostatic interaction energy between the cation and the anion, and (ii) the solvation energies of the two ions when they are associated as an ion pair. A simple model is set up in which the initial state is a contact ion pair. The electrostatic energy is then $\delta e^2/r$, where δ depends on the partial charge on the oxygen at the site occupied by the cation, and the solvation energy of anion is S_0^{I}. S_0^{I} represents the solvation energy of the unoccupied site. The solvation of the cation is not taken into account because it is assumed to be the same in both the initial and the transition states. In the transition state the cation is located an equal distance between the two oxygens, and the solvation energy $[2S_0 = 2(S_0^{\mathrm{I}} + \sigma)]$ of the anion increases because the charge density on the two oxygen atoms in the transition state is greater than that on the vacant oxygen in the initial state. The activation energy E_a for the process is therefore given by equation (4). Then the change, ΔE_a, in activation energy in going from cation A to cation B is given by equation (5).

$$E_a = E_{\mathrm{T.S.}} - E_{\mathrm{I.S.}} = (\delta e^2/r) - S_0 - \sigma \tag{4}$$

$$\Delta E_a = (\delta e^2/r)_{\mathrm{A}} - (\delta e^2/r)_{\mathrm{B}} - (\sigma_{\mathrm{A}} - \sigma_{\mathrm{B}}) \tag{5}$$

If A has a larger radius than B, $(\delta e^2/r)_{\mathrm{A}} < (\delta e^2/r)_{\mathrm{B}}$ and ΔE is decreased. However, since the larger cation also attracts less charge away from the oxygen of the vacant site, $\sigma_{\mathrm{A}} < \sigma_{\mathrm{B}}$ and ΔE is increased. The dominance of one or other of these two opposing effects will decide the direction of the trend in E. A 'good' solvent will favour the dominance of the solvation factor whereas in a 'poor' solvent the electrostatic factor will be the more important. Since the transition state is more solvated than the initial state, the A factors should be lower than normal for a unimolecular reaction. Warhurst and Wilde show that this simple model is capable of explaining most of the trends in their experimental results. They do not pretend that it is anything other than a very great simplification of the real situation and that it is limited in its application to other systems. It is a basis upon which more sophisticated models may be built.

A comparison of the activation energies and A factors obtained for 2,5-TBPBZQ with those recently measured for the durosemiquinone system[73] is instructive because it shows that in comparable systems the activation

[73] T. E. Gough and P. R. Hindle, *Canad. J. Chem.*, 1971, **49**, 2412.

energies are usually similar but the *A* factors are consistently higher for durosemiquinone. In ref. 73, Δa_{Me}, the difference between the splitting constants of the methyl groups in the unsymmetrical locked configuration, is regarded as a measure of the strength of the interionic interactions in the initial state, an approach which is not dissimilar to that of Warhurst and Wilde. A correlation is found between the free energy of activation, $\Delta G^{\ddagger}$ and Δa_{Me}, and it is also possible to correlate the observed trends in $\Delta H^{\ddagger}$ and $\Delta S^{\ddagger}$ with the extent of solvation and the interionic electrostatic interaction energy. The results of Pasimeni *et al.*[74] of e.s.r. experiments on the same durosemiquinone systems appear to differ somewhat from those mentioned above, although the largest differences are in the *A* factors where the greatest experimental error is expected. Any comparison, however, is of limited value because the number of systems actually studied by Pasimeni is much smaller than the number investigated by Gough and Hindle.[73] Taking the results quoted in refs. 73 and 74 together, and applying the criteria of Warhurst and Wilde,[72] it is apparent that their simple model copes reasonably well with the observed trends if one allows for the disagreement in the experimental values of the activation energies. Pasimeni *et al.* also attempt to calculate the position of the counterion relative to the durosemiquinone anion with the help of McClelland's model for modifying the Hückel parameters, and the expression of Goldberg and Bolton[75] for the energy of ion association. The calculated contour map of the associated energy has minima at *ca.* 0.015 nm from the oxygen atoms along the C=O bonds. If the interionic distance is set at 0.35 nm, then the energy barrier to jumping is reasonably close to the experimental activation energy. It is further shown that the main contribution to the spin density at the metal nucleus is from the zeroth-order term (see above and ref. 61) at all positions of the counterion, except when it is located close to the centre of the benzene ring. The spin density on the metal is therefore predicted to be positive and confirmed to be such from analysis of the anisotropic contributions to the linewidths. The mechanism of spin density distribution on to the cation must then be direct overlap between the valence *ns*-orbital of the cation and the first antibonding anion orbital, which in this case has a region of high electron density around the C=O bond.

Pyracene ion pairs have also long been known to exhibit linewidth alternation because of the intramolecular cation exchange between two equivalent positions. The question is, which two equivalent positions? It has been supposed until recently that the cation does not jump between the two methylene bridges but rather between equivalent positions over the two rings of the central naphthalene system. This is because, (i) positions over the methylene bridge have unfavourable electrostatic energy, and (ii) there is no evidence of inequivalence among the aromatic protons as

[74] L. Pasimeni, M. Brustolon, and C. Corvaja, *J. C. S. Faraday II*, 1972, **68**, 223.
[75] I. B. Goldberg and J. R. Bolton, *J. Phys. Chem.*, 1970, **74**, 1965.

might be expected if the cation is asymmetrically placed with respect to the anion. However, it is pointed out [76] that there is a further possibility: the counterion may exchange between positions above and below the plane of the aromatic ring (see also ref. 16). This possibility is ingeniously investigated [76] by inspection of the spectra of [$^2H_4$]pyracene radical anion (one ethylene bridge deuteriated) and [$^2H_2$]pyracene radical anion (one methylene group deuteriated). For potassium at least, it is shown that the counterion does in fact exchange between positions above and below the plane of the aromatic system.

It is now well established that intermolecular cation-exchange between radical ion pairs and salts with a common cation in ethereal solvents involves exchange between ion pairs of both species,[77] and that in many systems the intermediate triple ion may be identified by e.s.r. Many examples of triple ions, *i.e.* species with two cations associated with one radical anion, have been prepared, and in this connection the work of Gough and co-workers is especially important.[78] Triple ions have been prepared by addition of sodium tetraphenylboride to sodium–pyrazine and sodium–tetramethylpyrazine ion pairs;[79] the splitting from two equivalent alkali-metal counterions has been observed in each case. This investigation has now been extended to the formation of triple ions between pyrazine and tetramethylpyrazine and the cations of lithium, sodium, and potassium in DME, THF, and 2-MeTHF.[80] All the triple ions are symmetrical, one cation being associated with each electronegative site of the anion, *i.e.* close to the nitrogen atoms. For all the azine triple ion systems prepared, the nitrogen and proton hfsc are, within experimental error, independent of temperature. However, the alkali-metal splittings are dependent on temperature and solvent.

In order to obtain more information about the energy of association in the lithium–pyrazine system, the widths of the lithium lines have been measured and fitted to the Barton and Fraenkel [81] equation for anisotropic linewidths. The constants of the equation, which relates the linewidth to the metal magnetic spin quantum states associated with each line, provide a measure of the strength of the interionic interaction. The sodium–pyrazine triple ion spectral lineshapes are more fully explored by computer techniques in another paper from the same group.[82] The problem is to decide whether the sodium ions lie along the N $\cdots$ N axis of the pyrazine ring or alternatively whether they are situated above the aromatic ring over

[76] A. H. Reddoch, *Chem. Phys. Letters*, 1971, **10**, 108.
[77] A. W. Rutter and E. Warhurst, *Trans. Faraday Soc.*, 1970, **66**, 1866.
[78] T. E. Gough and P. R. Hindle, *Trans. Faraday Soc.*, 1970, **66**, 2420; S. A. Al-Baldawi and T. E. Gough; *Canad. J. Chem.*, 1970, **48**, 2799; T. E. Gough and P. R. Hindle, *ibid.*, 1969, **47**, 1968, 3393.
[79] S. A. Al-Baldawi and T. E. Gough, *Canad. J. Chem.*, 1970, **48**, 2799.
[80] S. A. Al-Baldawi and T. E. Gough, *Canad. J. Chem.*, 1971, **49**, 2059.
[81] B. L. Barton and G. K. Fraenkel, *J. Chem. Phys.*, 1964, **41**, 695.
[82] T. E. Gough and R. G. Hacker, *J. Magn. Resonance*, 1972, **6**, 129.

each nitrogen. An analysis of the anisotropic linewidths shows that the linewidth variations are consistent with a negative value of the metal splitting constant. The cations must therefore be situated along the N ··· N axis because in these positions the principal interaction is between the 3*s* orbital of each sodium and the lone-pair electrons on the nitrogen and will give a negative constant in the following manner: the lone-pair electrons are formally spin-paired, but they will be partially polarized by the unpaired spin density in the $2p^{\pi}$ orbital on the nitrogen. The electrons will be polarized in a manner analogous to that existing in an sp^2 C—H fragment and thus the sodium hyperfine interaction will be negative.

Other problems concerning the structure of triple ions still remain, especially that of the role of the second diamagnetic ion. For example, addition of sodium tetraphenylboride or sodium perchlorate produces the triple ion species, but addition of sodium iodide generates a species in which the sodium splitting is unresolved.[83] Such a situation may be understood if the following mechanism is in operation:

$$\bar{A}^{\bullet},Me^+ + X^-,Me^+ \rightleftharpoons Me^+,\bar{A}^{\bullet},Me^+ + X^-$$

Salts of the type Me^+,X^-, which do not dissociate to any large extent in solution (*e.g.* NaI), may possibly cause the dissociation–association equilibrium to be rapid enough to average out the metal splitting. This is one possible explanation of what the diamagnetic ion is doing; perhaps there may be others, such as the formation of quadruple ions, *e.g.* $Me^+,\bar{A}^{\bullet},Me^+,X^-$, where the electrostatic effect of the diamagnetic ion governs the position of the cations and thus the magnitude of the metal splitting.

A quantitative estimation has been made of the kinetic parameters of the above equilibrium for the system sodium 2,5-di-t-butylbenzosemiquinone–sodium iodide–tetrahydrofuran.[84] In this case the intramolecular cation-exchange rate is low. A similar kinetic analysis of a system where the intramolecular exchange rate is both slow and fast has recently been carried out.[85] As the spectrum of the system potassium durosemiquinone in THF shows no metal splitting because of the intermolecular cation exchange, identifying the presence of the triple ion is not straightforward, and it is of interest to note the three criteria that the authors use to identify the triple ion. These are: (i) lines to the low-field side of the centre of the spectrum are broader than their high-field counterparts (because g and a_{Me} for the triple ion are slightly less than g and $\bar{a}_{Me}$ for the ion pair); (ii) the lineshapes are asymmetric, and (iii) the observed hfsc (that is of the methyl protons) varies by as much as 10% across the spectrum in a systematic large–small–large–small pattern. These criteria identify a triple-ion spectrum superimposed on the ion-pair spectrum and may be useful to those looking

[83] T. E. Gough and P. R. Hindle, *Canad. J. Chem.*, 1970, **48**, 3959.
[84] A. W. Rutter and E. Warhurst, *Trans. Faraday Soc.*, 1970, **66**, 1866.
[85] T. E. Gough and P. R. Hindle, *Canad. J. Chem.*, 1971, **49**, 1529.

for the participation of triple ions in their systems. The equilibria proposed to describe the durosemiquinone situation completely are as follows:

$$K^+, D^{\bar{\cdot}}, K^+$$

$$K^+ + D^{\bar{\cdot}}, K^+ \rightleftharpoons K^+, D^{\bar{\cdot}}, K^+ \rightleftharpoons K^+ + D^{\bar{\cdot}}, K^+$$

The modified Bloch equations have been used to analyse the lineshapes in accordance with the above equilibria, and from this analysis the Arrhenius parameters are extracted in the usual manner. For the intramolecular cation exchange, $\log_{10} A = 12.1$, while for the dissociation of the triple ion, $\log_{10} A = 11.7$; for both reactions the activation energy is 27.7 kJ mol^{-1}. This may possibly be a reflection of the similarity in energy requirements between moving a cation between one site and another on an anion and actually detaching the cation from the anion. No accurate parameters have been obtained for the association reaction, though it is estimated that the value of $\log_{10} A$ is *ca.* 14. For the present, it seems that this reaction scheme is the most satisfactory put forward to explain intramolecular cation-exchange processes.

The [*m*,*n*]paracyclophanes and the $\alpha\omega$-diarylalkanes are interesting from the point of view of the effect of inter-ring interactions on the properties of the radical anions and because there exists the possibility of intramolecular electron transfer between the rings. The investigations of Weissman [86] and Gerson and Martin [87] have shown that the unpaired spin on the radical anions of some paracyclophanes and diphenylalkanes is located on one ring and that in others intramolecular electron transfer takes place between the rings. The theory of the factors controlling intramolecular electron transfer was proposed some years ago by McConnell.[88] He considered that the electron is trapped on one of the rings by the electron vibrational interaction and by polarization of the surrounding solvent molecules by the charge. When the molecule receives sufficient energy from random thermal processes, the electron may jump to an adjacent ring. The properties of the radical anions of three [2,2]paracyclophanes, *i.e.* benzeneparacylophane, 1,4-naphthylparacyclophane, and 9,10-anthrylparacyclophane, and the $\alpha\omega$-diarylalkanes, $Ar(CH_2)_nAr$ (Ar = benzene, 1-naphthyl, 9-anthryl, or 1-pyrenyl) have been studied with a view to determining the extent of electron transfer.[89] The anion radicals are much less stable than their simple aromatic counterparts. The authors of ref. 87 speculate on the mechanism of decomposition but do not give any

[86] S. I. Weissman, *J. Amer. Chem. Soc.*, 1958, **80**, 6462.
[87] F. Gerson and W. B. Martin, jun., *J. Amer. Chem. Soc.*, 1969, **91**, 1883.
[88] H. M. McConnell, *J. Chem. Phys.*, 1961, **35**, 508.
[89] D. J. Williams, J. M. Pearson, and M. Levy, *J. Amer. Chem. Soc.*, 1971, **93**, 5478, 5483.

positive proof for their mechanisms, although, at least for the 1,2-di-(1-naphthyl)ethanes ($NpCH_2CH_2Np$), it has recently been shown[90] that they decompose *via* a rapidly established disproportionation equilibrium, *i.e.*

$$2[(NpCH_2CH_2Np)^{\overline{\cdot}},Me^+] \rightleftharpoons (NpCH_2CH_2Np)^{2-},2Me^+ + NpCH_2CH_2Np$$

followed by the rate-determining decomposition of the dianion to the carbanion $NpCH_2^-$, *i.e.*

$$(NpCH_2CH_2Np)^{2-},2Me^+ \longrightarrow 2(NpCH_2^-,Me^+)$$

The e.s.r. results for the paracyclophanes may be summarized as showing that delocalization of the unpaired spin over both rings occurs in all cases, or alternatively it may be said that the intramolecular exchange rate is greater than $10^8\ s^{-1}$. The spectrum of $NpCH_2CH_2Np^{\overline{\cdot}}$, which is also described in ref. 90, is only partially resolved and very temperature-dependent.[89] This is probably the result of selective line-broadening of the type associated with intramolecular electron transfer at intermediate rates. The evidence presented is not conclusive, although it is a reasonable assumption that this is the case in the light of the fact that in $C_6H_5CH_2CH_2C_6H_5^{\overline{\cdot}}$ the unpaired electron is located on one ring and that in 1,2-di-(9-anthryl)ethane the exchange is fast. Thus, the rates of intramolecular exchange in the diarylalkanes are in conformity with the trends predicted by McConnell. However, for $Np(CH_2)_nNp$, where $n = 3$ or 4, the dependence of exchange rate on chain length does not follow the prediction of McConnell's theory in that both exhibit exchange in the intermediate range. This may be due to the neglect of conformations where direct spatial overlap is appreciable. However, it must be said that a considerable amount of extra data is required from compounds where n is greater than 4 and where the spectra are fully resolved before definite conclusions can be made on this point.

The unpaired electron in the diphenylmethane radical anion ($DPM^{\overline{\cdot}}$) occupies the antisymmetric benzene-type orbital when it is localized on one of the rings[87] because of the destabilization of the symmetric orbital by the inductive effect of the benzyl substituent. In a mixture of MTHF and dicyclohexyl-18-crown-6-ether, the electron is in fact localized on one ring[91] and this is attributed to either an inter-ring non-classical interaction of the 1,3- or the 1,5-type or a dipole–induced-dipole interaction between the two rings. Another, perhaps more straightforward, explanation may be that the strong co-ordination of the cation and possibly the anion by the crown ether should stabilize the electron on one ring. Experiments with

[90] A. Lagendijk and M. Szwarc, *J. Amer. Chem. Soc.*, 1971, **93**, 5359.
[91] J. D. Young and N. L. Bauld, *Tetrahedron Letters*, 1971, 2251.

strongly polar solvents, in which it is certain that only free ions exist, might provide the answer.

Although there seems to be a great need for extensive correlation of e.s.r. measurements of homogeneous intermolecular electron transfer processes in order to gain more insight into the predominant interactions which determine the rates, there are no new experimental results to report at this time. This seems rather a pity because electron-transfer mechanisms are of prime importance in chemical and biological processes and e.s.r. offers a straightforward method of obtaining information on such processes.

The effect of certain dynamic processes on the measurement of the unresolved metal splitting in the benzene anion radical has been investigated by Jones and Komarynsky.[64] For example, as the frequency of any intermolecular cation-exchange process increases, for a given metal hfsc : ΔH_c (component linewidth) ratio the lineshape becomes more nearly Lorentzian and the overall linewidth, ΔH_{obs}, approaches ΔH_c. At sufficiently fast exchange rates, the lineshape is Lorentzian because the metal splitting is completely averaged out. The net effect of any undetected intermolecular cation exchange on the measurement of unresolved metal splittings is to make the observed value (obtained by comparing computed and experimental lineshapes) smaller than that which would be observed in the absence of exchange. The other dynamic processes which may contribute to the observed linewidth and shape, and therefore to the measurement of unresolved splitting, are Heisenberg exchange, electron and atom transfer, and intramolecular cation-exchange. Each of these processes will make the measured splitting lower than the true value, but fortunately each has such a sufficiently striking effect on the spectrum that they are easily detected and taken into account.

At least one new example[9] of a dynamic process which involves a change in configuration has already been mentioned, as well as several examples of the effect of restricted rotation on the e.s.r. spectra. The last part of this section will be concerned with more examples of such processes.

Linewidth alternation is observed in the spectra of the radical anions of 2-cyclohexenone, and its 5,5-dimethyl and its 3-methyl derivatives when they are prepared in a liquid ammonia flow system.[92] The conformational change which is modulating the hyperfine splittings is the movement of the C-5 atom above and below the plane in which the ring carbons and the oxygen atom are located. For 3,5-dimethylcyclohexenone the anion is locked in one conformation in which the 5-methyl group is axial.

The lines of the e.s.r. spectra of the positive and negative ions of 1,2,3,6,7,8-hexahydropyrene (4) show an alternating linewidth effect which is attributed to the non-synchronized motion of the two aliphatic bridges in the molecule.[9, 93, 94] Curiously enough, the activation energy for the

[92] I. H. Elson, T. J. Kemp, and T. J. Stone, *J. Amer. Chem. Soc.*, 1971, **93**, 7091.
[93] E. de Boer and A. P. Praat, *Mol. Phys.*, 1964, **8**, 291.
[94] F. W. Pijpers, M. R. Arick, B. M. P. Hendricks, and E. de Boer, *Mol. Phys.*, 1971, **22**, 781.

interconversion of the anion obtained by de Boer *et al.*[94] is almost twice as large as that given in ref. 9, though the value given by de Boer (42 kJ mol^{-1}) is much closer to that determined by n.m.r. methods for the boat–chair interconversion in cyclohexane.[95] The activation energy for the interconversion of the cation[94] is only 14.7 kJ mol^{-1} and the rate constant is twice as large for the cation at 0 °C. What is most striking is the quite large and negative $\Delta S^{\ddagger}$ of − 18.0 e.u., which reflects the greater solvation of the activated complex of the cation compared with that of the anion. As the authors remark,[94] it would be interesting to see whether the positive and negative ions of other similar molecules reveal the same behaviour of the activation energy.

The hydroxyl-proton splitting constant is very much temperature-dependent in many cation radicals derived from hydroquinone and its derivatives, hydroxy-naphthalenes and -anthracenes, and dihydroxy-biphenyls.[96] The cause is the restricted rotation about the carbon–oxygen bond and it is possible to measure the barrier to rotation. Hydroquinone, duroquinol, 1,4-dihydroxynaphthalene, and 9,10-dihydroxyanthracene cations have been prepared in either the aluminium chloride–nitromethane system or in the sulphuric acid–nitromethane system.[94] Their hydroxyl-proton splitting constants are temperature-dependent through torsional oscillations in a way which may be described by equation (6). This expres-

$$a^{\mathrm{H}}_{\mathrm{OH}} = Q^{\mathrm{H}}_{\mathrm{OH}}\, \rho^{\pi}_{\mathrm{O}} \langle \cos^2 \theta \rangle + Q^{\mathrm{H}}_{\mathrm{COH}}\, \rho^{\pi}_{\mathrm{C}} \langle \sin^2 \theta \rangle \qquad (6)$$

sion takes into account the two major mechanisms contributing to the splitting constant of the hydroxyl proton, that is (i) the polarization of the electrons in the O—H bond arising from the unpaired spin density in the oxygen $2p^{\pi}$ orbital, and (ii) hyperconjugative interactions between the hydroxyl proton and the unpaired spin density in the carbon $2p^{\pi}$ orbital. It must be noted that even the spin polarization term is dependent on the dihedral angle θ, because on twisting the hydroxy-group out of the plane by an angle θ the overlap between the carbon and oxygen $2p^{\pi}$ orbitals will be decreased, thus decreasing the π-spin density on the oxygen atom. To a first approximation this decrease in spin density will be proportional to $\sin^2 \theta$. The results indicate that an equation of the form of equation (7)

$$a^{\mathrm{O}}_{\mathrm{OH}} = -\,(35 \pm 5)\, \rho^{\pi}_{\mathrm{O}} \langle \cos^2 \theta \rangle + (50 \pm 10)\, \rho^{\pi}_{\mathrm{C}} \langle \sin^2 \theta \rangle \qquad (7)$$

describes the hyperfine interaction of a hydroxyl proton. The potential energy barriers to rotation are highest for hydroquinone (42 kJ mol^{-1}), and this is the value obtained from the temperature-dependence of the proton splittings as well as from analysis of the linewidth alternation. For all the other cations prepared the energy barrier is much lower and the agreement

[95] F. R. Jensen, D. S. Noyce, C. H. Sederholm, and A. J. Berlin, *J. Amer. Chem. Soc.*, 1962, **84**, 386; T. M. Ivanova and G. P. Kutagova-Shenyakina, *Russ. Chem. Rev.*, 1970, **39**, 510.

[96] P. D. Sullivan, *J. Phys. Chem.*, 1971, **75**, 2195 and references cited therein.

between temperature-dependence measurements and linewidth measurements is less satisfactory. This is not unexpected because the hydroquinone cation will correspond most closely to the model chosen to describe the spin-density mechanism. When the barrier is less than 13 kJ mol^{-1}, then the calculations are in poor agreement with experiment and a different approach may be needed.

Evidence of a barrier to internal rotation in substituted benzaldehyde anions was first reported by Maki and Geske.[97] Restricted rotation about the Ar—C(X)=O bond has subsequently been noted in a number of aromatic aldehyde ions.[98] Kaminski and Mobius[99] have measured the potential energy barriers in a series of eight acetophenone and benzaldehyde anion radicals of the type X—C_6H_4—C(Y)=O, where X is a *para* substituent. X, Y are H, Me; NO_2, Me; OMe, Me; Me, Me; H, H; NO_2, H; Me_2N, H; and CN, H, respectively. The measured energy barriers correlate very well with the calculated resonance energies as shown in equation (8),

$$\Delta E_\pi = E_\pi(\theta = 90°) - E_\pi(\theta = 0°) \qquad (8)$$

that is, there is evidence that the main contribution to ΔE comes from resonance interaction of the molecular fragments. Unfortunately, the resonance energies E_π have been calculated using Hückel parameters which have been adjusted to fit the experimental spin densities, and this presumably means that a proportional relationship between the spin density and the coupling constant may be assumed. In the light of the results described in ref. 96, this does not seem to be the case.

The e.s.r. spectra of the free anions of *m*-dinitrobenzene show linewidth alternation, owing to the out-of-phase correlations of the rotation of the nitro-groups out of the plane of the benzene ring.[100] The electrogenerated radical anions of 2,2′-dinitrobiphenyl and 1,8-dinitronaphthalene display e.s.r. linewidth effects which may be explained in the same way.[101] A complete analysis of the linewidths, by relaxation theory, accounts for the observed changes in the spectra.[101] An examination of the details of the modulating motions by a comparison of the experimental spin densities with those calculated by McLachlan and UHF methods shows that, in the 2,2′-dinitrobiphenyl anion, satisfactory agreement is obtained between calculated and observed spin densities for the geometry in which the two planes of the phenyl rings make an angle of 40°.

[97] A. H. Maki and D. H. Geske, *J. Amer. Chem. Soc.*, 1961, **83**, 1852.
[98] E. W. Stone and A. H. Maki, *J. Chem. Phys.*, 1963, **38**, 1999; P. H. Rieger and G. K. Fraenkel, *J. Chem. Phys.*, 1962, **37**, 2811; N. Steinberger and G. K. Fraenkel, *J. Chem. Phys.*, 1964, **40**, 723.
[99] W. Kaminski and K. Mobius, *J. Magn. Resonance*, 1971, **5**, 182.
[100] R. J. Faber and G. K. Fraenkel, *J. Chem. Phys.*, 1967, **47**, 2462.
[101] J. Subramanian and P. T. Narasimhan, *J. Chem. Phys.*, 1971, **56**, 2572.

6

Free Radicals in the Solid State

BY J. E. BENNETT AND K. D. J. ROOT

Introduction.—Interest in free radicals in the solid state has been maintained during the period covered by this Report. In the main, the free radicals investigated have been formed by high-energy *X*- or γ-rays, and the interest has centred both on obtaining the electron spin resonance (e.s.r.) spectra of new radicals and on elucidating the reactions which occur in the solid phase during and after irradiation. In the latter studies the effects of temperature and of photolysis with light of a selected wavelength have often been examined. Other methods for preparing the radicals have been used relatively infrequently, but nevertheless some important results have been obtained by these methods, which include photolysis at low temperatures, direct chemical reaction at low temperatures, and trapping out from the gas phase on to a cold substrate.

The present Report will inevitably overlap to some extent with others in this volume, but as far as possible the discussion is limited to the consideration of (i) free radicals formed in the solid state by high-energy radiation (but excluding polymers) and (ii) free radicals formed by other methods, *i.e.* chemical reaction, photolysis, and trapping from the gas phase (but excluding the solid-state spectra of free radicals which are relatively stable in solution at room temperature).

1 Radiation Damage

In the study of radiation damage it is important to remember that not all the products will be paramagnetic, and hence not all will be observed by e.s.r. spectroscopy; furthermore, the initial species may be so unstable that they cannot be observed by any spectroscopic technique. For certain systems where the initial species are very unstable, it is sometimes possible to observe the radical by continuous irradiation in the e.s.r. spectrometer; however, the most general technique is to irradiate the specimen outside the spectrometer, at a low temperature, and to record the spectrum without warming the sample. For a full study of the system the sample should be irradiated at a low temperature, immersed in liquid helium if possible. The spectra should then be recorded at gradually increasing temperatures, starting at the irradiation temperature. The whole process should then be repeated at a number of other irradiation temperatures. If there is any change in the spectra on warming the sample, the temperature should be reduced again to see whether the change is reversible. The matrix has in some cases quite an important effect on the spectra, and phase changes in the matrix should be taken into account.

In this way it is possible, in favourable cases, to identify both the products of irradiation and the products observed at room temperature, which, in some cases, can be identified chemically.

Since only some of the species in the reaction chain are paramagnetic, complementary studies with other spectroscopic techniques should be considered. Unfortunately, most other techniques are less sensitive than e.s.r., but certain i.r. absorption bands are narrow and strong enough to be used. An example is the CN ligand in certain transition-metal complexes. Radiation-induced species are usually highly susceptible to further damage, even from visible light, and so all studies should be carried out with the sample in a light-proof container. This sensitivity often precludes the use of electronic spectroscopy for the study of these systems, but the effect of visible light is often a fruitful sideline in itself, and has been used in the past to gain structural information on the radiation products. Bleaching with polarized light, for example, has been used to study the structure of colour centres in alkali-metal halide crystals.

Apart from the identification of paramagnetic species, e.s.r. spectroscopy can yield information on the concentration and environment of the paramagnetic species. Quantitative studies of the absolute concentration can be performed using suitable standard samples, and relative concentrations within a system can often be found to within 10%. The linewidth of the e.s.r. spectrum can give information on the microenvironment and concentration of the species. Usually, the products are not distributed evenly throughout the sample, but occur in clusters. This is often characterized by an unusually broad signal. E.s.r. spectra from irradiated crystals yield information about the orientation of the species relative to the crystal axes and the number of distinct species in the unit cell; both pieces of information can be of use in determining reaction chains.

Extra hyperfine structure, other than from the nuclei that constitute the species, is sometimes observed, especially with simple species such as atoms. This again gives information on the microenvironment.

Impurities in the system can have an important effect on the radiation products. In fact the presence of even a minute quantity of an electron trap or an electron-hole trap may completely invalidate a study of a model system. Impurities are, of course, present in every real system, and cannot be completely eliminated even in model systems.

In this section we have attempted to review some of the e.s.r. work involving radiation damage whether or not this was the motive for the study. Many of the systems have been studied before, and although in some cases very little extra information on radiation-damage aspects of the work is included, we have mentioned the new work since the papers refer to earlier studies.

Simple Inorganic Crystals.—The study of irradiated inorganic salts, especially those of oxyacids, has been carried out for many years. The effect of ionizing radiation can often be described by the following simple mechanism:[1]

$$R{-}A \longrightarrow R{-}A^+ + e^-$$
(ionization)

$$R{-}A + e^- \longrightarrow R{-}A^* \longrightarrow R^\bullet + A^-$$
(cage back-reaction followed by bond fission of excited intermediate)

$$R{-}A^+ + R{-}A \longrightarrow R{-}A_2 + R^\bullet$$
(reaction with neighbouring species)

A recent example of this can be found[2] in the effect of γ-radiation on KPO_2F_2 at 77 K, where both the PO_2F^- and PO_3F radicals can be identified from the e.s.r. spectrum. All the reaction products can be explained by the above mechanistic scheme.

X-Irradiation of $KClO_3$ at 77 K,[3] and γ-irradiation of $NaClO_3$ at a similar temperature,[4] yielded both O_3^- and ClO_2. The suggested mechanism is:[4]

$$ClO_3^- \xrightarrow{\gamma} ClO_3 + e^-$$
$$ClO_3 + ClO_3^- \longrightarrow ClO_2 + ClO_4^-$$
$$ClO_3^- + e^- \longrightarrow ClO_3^{2-}$$
$$ClO_3^{2-} \longrightarrow Cl^- + O_3^-$$

A possible alternative for the last two reactions is that the final products are obtained *via* a ClO_4 intermediate:

$$ClO_3 + ClO_3^- \longrightarrow ClO_4 + ClO_2^-$$
$$ClO_4 + ClO_3^- \longrightarrow 2ClO_2 + O_3^-$$

This latter mechanism is supported by a study[5] of the effect of γ-radiation on $KBrO_3$ at 10 and 26 K. At the lower temperature, species assigned to BrO_4 and BrO_4^{2-} were observed, and these decomposed into BrO_2 and O_3^-, respectively, at 20 K.

Symons and co-workers[6] have studied the effect of radiation on a number of crystalline bromates at 77 K. Sodium, potassium, and barium bromates gave the BrO_3 radical, in the first- and last-mentioned lattices being trapped in two different sites. In the case of potassium bromate a small quantity of BrO_2 was observed, and this was the only radical observed

[1] M. C. R. Symons, 'Advances in Chemistry Series', 1968, No. 82, p. 1.
[2] A. Begum, S. Subramanian, and M. C. R. Symons, *J. Chem. Soc.* (*A*), 1971, 700.
[3] S. Schlick, *J. Chem. Phys.*, 1972, **56**, 654.
[4] M. C. R. Symons, *J. Chem. Soc.* (*A*), 1971, 698.
[5] J. R. Byberg, *J. Chem. Phys.*, 1971, **55**, 4867.
[6] S. Subramanian and M. T. Rogers, *J. Phys. Chem.*, 1971, **75**, 3479.

in irradiated zinc bromate. In $BaSO_4$ crystals doped with bromate ions the BrO_3 radical was again observed, together with a further species that was tentatively assigned as the BrO_3^{2-} radical. Single crystals of KIO_2F_2, on γ-irradiation at 77 K, yielded the $IO_2F_2^{2-}$ ion by electron capture.[6] The e.s.r. spectrum obtained on warming to 140 K could be assigned to the IO_2F^- species.

Gamma irradiation of barium selenate[7] at 77 K yielded radicals assigned as SeO_2^-, SeO_3^-, and SeO_4^{3-}. The results were compared with those for irradiated aqueous solutions containing selenate ions, where SeO_4^{3-} was the only radical observed. Potassium antimonate,[8] on irradiation, yielded further oxy-radicals, and sodium hexafluoroantimonate gives a radical which is more likely to be an oxy-radical than a purely fluorinated species.

Holmberg has studied further the products obtained on γ-irradiating $KHCO_3$ at both 77 and 300 K.[9] At the lower temperature, both $(HCO_3)_2^-$ and CO_3^{3-} radicals were observed, with small amounts of HCO_3. At room temperature the only species observed was the well-characterized CO_2^- radical. The irradiation of nitrate and nitrite ions in a number of lattices[10, 11] has again shown the products to be NO_2 and NO_2^{2-}.

BH_3 and CH_3 radicals were observed in γ-irradiated tetramethylammonium borohydride[12] at 77 K. On raising the temperature, only the BH_3 radical could be observed. The NH_3^+ radical has been observed in irradiated ammonium chlorate.[13, 14] In the latter paper, it is noted that both NH_2 and H_2CN were also observed at 77 K. Irradiation of ammonium dihydrogen phosphate including 5% potassium dihydrogen arsenate[15] yielded NH_3^+, while γ-irradiation of ammonium sulphate yielded the $N(SO_3)_2^{2-}$ radical[16] and the SO_3^- and SO_2^- radicals.[17]

The effect of ionizing radiation on alkali-metal halides has been studied by a wide range of techniques for many years. A number of paramagnetic centres have been studied by e.s.r. including the V_k-centre, which is a Hal_2^- ion. In crystals with mixed halide ions, XY^- centres have been detected. Delberg, Schoemaker, and Yaster[18] have studied the effect of *X*- and γ-irradiation on KCl crystals doped with bromide ions and electron traps such as Ag^+, Tl^+, NO_2^-, and Pb^{2+} ions. At 77 K they observed mainly Cl_2^- with a small concentration of the mixed species $BrCl^-$. On

7 K. V. S. Rao and M. C. R. Symons, *J.C.S. Dalton*, 1972, 147.
8 M. C. R. Symons, *J. Chem. Soc.* (*A*), 1971, 2393.
9 R. W. Holmberg, *J. Chem. Phys.*, 1971, **55**, 1730.
10 B. L. Silver, J. Koresh, S. Schlick, and Z. Luz, *Mol. Phys.*, 1971, **22**, 225.
11 A. M. Fuller and C. E. Tarr, *J. Chem. Phys.*, 1972, **56**, 438.
12 E. D. Sprague and F. Williams, *Mol. Phys.*, 1971, **20**, 375.
13 J. Junecka and M. Fujimoto, *J. Magn. Resonance*, 1971, **4**, 47.
14 K. V. S. Rao and M. C. R. Symons, *J. Chem. Soc.* (*A*), 1971, 2163.
15 I. Suzuki and R. Abe, *J. Phys. Soc. Japan*, 1971, **31**, 951.
16 I. Barbar, *Physica*, 1972, **58**, 324.
17 I. Suzuki and R. Abe, *J. Phys. Soc. Japan*, 1971, **30**, 586.
18 C. J. Delberg, D. Schoemaker, and P. H. Yaster, *Phys. Rev.* (*B*), 1971, **3**, 473.

warming the sample to 210 K the concentration of $BrCl^-$ increased at the expense of the homonuclear species. They also studied the reorientation kinetics.

Irradiation of azide crystals and other lattices doped with the azide ion normally yields nitrogen atoms and defects such as the *F*-centre, which is an electron trapped at a cation vacancy. One recent study involved photolysis of azide ions in alkali-metal halide lattices under controlled conditions.[19] The initial products observed at 77 K were nitrogen atoms, which on warming disappeared, to be replaced completely by the N_3 radical and an *F*-centre. *F*-Centres were also observed at 77 K in γ-irradiated sodium azide.[20] On annealing, the centres were destroyed and the nitrogen atoms observed. Potassium azide on similar treatment yielded only nitrogen atoms.

Eachus and Herring[21, 22] have studied the results of γ-irradiation of crystalline mercuric and cadmium acetates. Irradiation at 77 K yielded Hg^{I} (ref. 21) or Cd^{I} ions (ref. 22). They postulated that the reaction proceeds *via* the electron-excess $CH_3CO_2^{2-}$ species.

Ionic crystals on irradiation often yield paramagnetic species that cannot be discussed in terms of a simple chemical species. One very common species is the *F*-centre. These centres have been studied for many years in alkali-metal halides, but more recently they have been observed in other lattices. Rose and Schneider[23] irradiated β-PbF_2 at 80 K and obtained an e.s.r. spectrum which they attributed to an *F*-centre-like species, and Unruh and co-workers observed an *F*-centre in MgF_2.[24]

Garlick and co-workers[25] observed two paramagnetic species in hexagonal GeO_2. These they attribute to an oxygen vacancy and an oxygen vacancy interstitial pair. Purcell and Weeks have studied the effect of γ-radiation on TiO_2 crystals.[26] At 140 K they irradiated with neutrons and found that the main effect was the displacement of ions. On further irradiation with γ-rays at 77 K, electrons were displaced and they observed Ti^{3+} ions.

Frozen Solutions.—The paramagnetic species observed in irradiated frozen solutions are normally the products of reactions with solvent fragments, which are the primary reaction species.

Symons and Marov have studied the results of γ-irradiation of frozen aqueous solutions of alkali-metal bromides and iodides at 77 K.[27] They observed hydroxyl radicals and V_k-centres, which are normally found in irradiated alkali-metal halides, *viz.* Br_2^- or I_2^- and the mixed species

[19] J. F. Testa and D. W. Robinson, *J. Chem. Phys.*, 1971, **55**, 3056.
[20] A. Begum and M. C. R. Symons, *J. Chem. Soc.* (*A*), 1971, 2062.
[21] R. S. Eachus and F. G. Herring, *Canad. J. Chem.*, 1971, **49**, 2868.
[22] R. S. Eachus and F. G. Herring, *Canad. J. Chem.*, 1971, **49**, 562.
[23] B. F. Rose and E. Schneider, *Phys. Letters* (*A*), 1971, **34**, 27.
[24] W. P. Unruh, L. G. Nelson, T. J. Lewis, and J. L. Kolopus, *J. Phys.* (*C*), 1971, **4**, 2992.
[25] G. F. J. Garlick, J. E. Nicholls, and A. M. Ozer, *J. Phys.* (*C*), 1971, **4**, 2230.
[26] T. Purcell and R. A. Weeks, *J. Chem. Phys.*, 1971, **54**, 2801.
[27] I. Marov and M. C. R. Symons, *J. Chem. Soc.* (*A*), 1971, 201.

$BrOH^-$ or IOH^-. These doped frozen solutions yielded a much lower concentration of hydroxyl radicals, suggesting that the reaction:

$$^{\bullet}OH + hal^- \longrightarrow HOhal^-$$

competes with direct ionization of the halide ion followed by the reactions:

$$Br^{\bullet} + H_2O \longrightarrow BrOH_2 \cdots BrOH$$

$$Br^{\bullet} + Br \longrightarrow Br_2^-$$

Trapped electrons were not observed in the e.s.r. spectrum, but the irradiated samples were violet, which is often indicative of trapped electrons. A further investigation of γ-irradiation of alkali-metal halides in frozen solutions revealed signals due to trapped hydrogen atoms and the MH^+ species.[28] Frozen solutions of alkali-metal azides and cyanides yielded radicals on γ-irradiation that were identified as $(CN)_2^-$, H_2CN, and HCN^-.[29, 30] Again, these species had been observed earlier in cyanide-doped alkali-metal halide lattices.

Selenite ions in frozen solutions are attacked by hydroxyl radicals, forming the SeO_4^{3-} species,[7] which is perhaps weakly protonated. In the same study, bromate ions were irradiated and, unlike pure alkali-metal bromates, only the BrO_2 species was observed. This was attributed to the reaction of the BrO_3^{2-} species, which is formed by electron capture, with water molecules, even at 77 K.

Frozen solutions of gallium(III), thallium(III), and mercury(II) salts, when γ-irradiated at 77 K, gave e.s.r. spectra which could be assigned to Ga^{2+}, Tl^{2+}, and Hg^+, respectively.[31] Thallium(I) salts also yielded the Tl^{2+} species, and annealing experiments confirmed that the ion was formed by electron loss, and not by electron capture of impurity Tl^{III} ions. A more detailed study of γ-irradiated aqueous and alcoholic solutions of mercurous and mercuric salts[32] showed the presence of the Hg^+ species, together with HgH^{2+}, $HgOH^{2+}$, $HgOEt^{2+}$, and a dimeric mercury species. In the case of mercuric salts in frozen acidic solution, $^{199}Hg^+$ and $^{201}Hg^+$ species were observed. The $HgOH^{2+}$ species could be observed most strongly in systems of low acidity. On annealing, a further species could be observed which was assigned to HgH^{2+}. This was formed in highest concentrations in strongly acidic solutions. A possible reaction mechanism was suggested:

$$H_2O \longrightarrow H_2O^+ + e^-$$

$$H_2O^+ \longrightarrow (H^+) + {^{\bullet}OH}$$

$$Hg^{2+} + e^- \longrightarrow Hg^+$$

28 I. S. Ginns and M. C. R. Symons, *J.C.S. Dalton*, 1972, 143.
29 I. S. Ginns and M. C. R. Symons, *J.C.S. Dalton*, 1972, 185.
30 I. S. Ginns and M. C. R. Symons, *Chem. Comm.*, 1971, 893.
31 M. C. R. Symons and J. K. Yandell, *J. Chem. Soc.* (*A*), 1971, 760.
32 R. J. Booth, H. C. Starkie, and M. C. R. Symons, *J. Chem. Soc.* (*A*), 1971, 3198.

$$Hg^{2+} + {}^{\bullet}OH \longrightarrow HgOH^{2+}$$
$$(H^+) + e^- \longrightarrow H^{\bullet}$$
$$Hg^2 + H^{\bullet} \longrightarrow HgH^{2+}$$
$$Hg^+ + (H^+) \longrightarrow HgH^{2+}$$

The mercuric salts were dissolved in dilute perchloric or sulphuric acid solutions with similar results. Mercurous salts, on the other hand, only yielded the Hg^+ species in perchloric acid solution, whereas in sulphuric acid solution the main feature was the dimeric species, either Hg_2^+ or Hg_2^{3+}. On annealing, the Hg^+ species was observed, suggesting that the dimeric species is thermally unstable.

More fundamental species, such as trapped electrons, have been studied by a number of different authors, but little has been added to our knowledge of the reaction mechanisms. They have been studied in alkaline ices [33, 34] and in aqueous sugar ices.[35]

Glasses and Amorphous Solids.—A number of radicals have been observed [36] in γ-irradiated $POCl_3$, including $POCl_3^-$, $POCl_2$, and PO_2Cl_2. The primary reaction for this system is:

$$POCl_3 \longrightarrow POCl_3^+ + e^-$$
$$POCl_3 + e^- \longrightarrow POCl_3^-$$

$POCl_3^+$ then reacts, possibly with the hydrated species:

$$POCl_3^+ + HPO_2Cl_2 \longrightarrow HPO_2Cl_2^+ + POCl_3$$
$$HPO_2Cl_2^+ \longrightarrow PO_2Cl_2 + H^+$$

and $POCl_3^-$ dissociates:

$$POCl_3^- \rightleftharpoons POCl_2 + Cl^-$$

Similar radiation treatment of PCl_3 yielded PCl_2 and PCl_4, and treatment of PCl_5 gave only PCl_4. γ-Irradiation of SO_2Cl_2 yielded the anion $SO_2Cl_2^-$.[37]

Banerman [38] has reported the e.s.r. spectra of electron-irradiated ammonia glasses. At temperatures between 20 and 100 K, H_2^- and NH_2^- were observed. γ-Irradiation of NH_3 adsorbed on silica gel yielded signals that were attributed to NH_2 and to a radical associated with the surface.[39]

33 K. Ohne, I. Takemura, and J. Sohma, *J. Chem. Phys.*, 1972, **56**, 1202.
34 P. Hambet and L. Keyan, *J. Amer. Chem. Soc.*, 1972, **93**, 1102.
35 J. D. Zimberick and L. S. Myers, *J. Chem. Phys.*, 1971, **54**, 2899.
36 A. Begum and M. C. R. Symons, *J. Chem. Soc.* (*A*), 1971, 2065.
37 G. Lind and R. Kewley, *Canad. J. Chem.*, 1971, **49**, 2514.
38 G. Banerman, *Z. Physik.*, 1971, **247**, 25.
39 O. I. Broakovski, G. M. Zhidomirov, V. B. Kozanski, A. I. Mashchenko, and B. N. Shelimov, *Kinetika i Kataliz*, 1971, **12**, 700.

γ-Radiolysis of H_2O on silica at 77 K leads to trapped hydrogen atoms. Other hydrogenous compounds were studied, and for all except ethane the yield of hydrogen atoms depends upon surface coverage and the stability depends on the nature of the surface.

A number of e.s.r. studies have been of irradiated borate glasses.[40–43] Usually the centre is assigned to a boron–oxygen hole centre together with a trapped-electron centre associated with alkali-metal ions. A zinc oxide borate glass on *X*-irradiation yielded a signal attributed to Zn^+.[44]

Transition Metals.—Interaction between neighbouring ions usually broadens the e.s.r. spectrum of undiluted paramagnetic transition-metal ions so that a signal can only be observed at low temperatures, if at all. A common way of preparing microscopically dilute samples of paramagnetic ions is to irradiate diamagnetic transition-metal complexes in the hope of producing the required paramagnetic ion. Both electron-deficient and electron-excess centres can be formed under favourable conditions. By deliberately incorporating specific impurities, which act as electron sources or traps, it is possible to influence the type of centre formed. Because of the great complexity of the systems, care has to be exercised in the interpretation of the results, but, just as with simple inorganic systems, useful data have been obtained both on the structure of the complexes and on the mechanism of radiation-induced reactions in these systems.

The effect of electron irradiation on hexacyanocobaltate(III) has been studied by Danon and co-workers.[43] They showed that the end-product was the $Co^{II}(CN)_5{}^{3-}$ ion formed by loss of one CN ligand.

Transition-metal complexes dissolved in alkali-metal halide lattices often yield useful information since the halide lattice magnetically insulates the complexes. Danon has studied $Rh(CN)_6{}^{3-}$ ions dissolved in a KCl lattice [44] and found that the complex ion exchanged two of the cyanide ions with the halide ions in the lattice in addition to trapping an electron, so that the final species could be considered to be $Rh(CN)_4Cl^{4-}$. A number of other species were observed, but not identified.

γ-Irradiation of the related pentacyanonitrosyl complexes in alkali-metal halides has been reported by Symons and co-workers.[45] The e.s.r. results were complemented by a study of the effect on the i.r. spectra of these complexes. The success of the i.r. study was due to the fact that the cyanide and nitrosyl ligands both have strong i.r. bands, particularly when the complex is dilutely distributed in an alkali-metal halide lattice. For $Cr(CN)_5NO^{3-}$ in KCl, a new i.r. absorption

40 P. K. Wong, *J. Phys. Chem.*, 1971, **75**, 201.
41a D. L. Griscom, *J. Chem. Phys.*, 1971, **55**, 1113.
41b D. L. Griscom, *J. Non-cryst. Solids*, 1971, **6**, 275.
42 R. Berger, *Compt. rend.*, 1972, **274**, *B*, 262.
43 F. D. Tsay, H. B. Gray, and J. Danon, *J. Chem. Phys.*, 1971, **54**, 3760.
44 R. P. A. Muniz, N. V. Vagman, and J. Danon, *J. Chem. Phys.*, 1971, **54**, 1284.
45 M. B. D. Bloom, J. B. Raynor, K. D. J. Root, and M. C. R. Symons, *J. Chem. Soc.* (*A*), 1971, 3213.

band was observed on γ-irradiation which corresponds with the published data for $K_4[Cr(CN)_5NO],H_2O$. For $Mn(CN)_5NO^{3-}$ and $Fe(CN)_5NO^{2-}$ the i.r. spectra were similar, but less easily interpreted. The e.s.r. spectra of the irradiated manganese complex in KCl were attributed to the $[Mn(CN)_5NO]^{2-}$ ion. At 77 K only the Cl_2^- ion was observed, but on warming to 300 K this species disappeared and was replaced by the manganese complex. $Fe(CN)_5NO^{2-}$ ions in KCl on γ-irradiation yielded a weak e.s.r. spectrum which was attributed to the $Fe(CN)_5NO^{2-}$ species, but there is still some conjecture over the identity of a similar spectrum in irradiated sodium nitroprusside. $[Mn(CN)_5(NO)]^{3-}$ dissolved in KCl, on γ-irradiation, shows evidence for the migration of the Mn ion to a site surrounded by six chloride ions.[46] γ-Irradiation of anhydrous sodium nitroprusside has been studied further by Symons and co-workers [47] who have reported the e.s.r. spectra of the NO radical in this system.

Fujiwara and co-workers [48] have studied the reduction of Co^{III} complexes to Co^{II} on γ-irradiation and have used e.s.r. spectroscopy to study the electronic and spin states of these ions.

Changes in ligands and co-ordination number have been reported in a number of halide complexes. Fujiwara and Nakamura,[49] and also Krigas and Rogers,[50] have studied the effect of γ-irradiation on the $PdCl_4^{2-}$ ion in both K_2PdCl_4 and $(NH_4)_2PdCl_4$ crystals. Evidence was found for both the $PdCl_4^{3-}$ and $PdCl_5^{2-}$ ions. The latter authors also studied the effect of γ- and electron-irradiation at 77 K on K_2PtCl_4 and observed the spectra of $(PtCl_4)_2^{3-}$ and $PtCl_3^{2-}$.[51] The dimeric form may thus be an intermediate in both the platinum and palladium systems.

Silver atoms have been studied by e.s.r. and optical spectroscopy in *X*-irradiated KBr and KI crystals containing Ag^+ ions,[52] and in Ag^+-doped K_2SO_4 crystals.[53] Gee and Wan have studied the positive ion formed by γ-irradiation of the silver complexes with cyclohexadiene and methylcyclopentane.[54]

E.s.r. spectra were used to examine the multiple Ag^0 trapping sites in γ-irradiated frozen $AgNO_3$ ices [55] and also the sites of the dipositive ions V^{II} and Mn^{II} in LiF and NaF crystals [56] after *X*-irradiation.

Ionic crystals containing rare-earth ions often form centres which, like *F*-centres, cannot be classed as a pure chemical species. Calcium fluoride

46 M. B. D. Bloom, J. B. Raynor, and M. C. R. Symons, *J. Chem. Soc.* (*A*), 1971, 3209.
47 M. B. D. Bloom, J. B. Raynor, and M. C. R. Symons, *Chem. Comm.*, 1971, 1069.
48 S. Fujiwara, T. Watanabe, and H. Tadano, *J. Co-ordination Chem.*, 1972, **1**, 195.
49 S. Fujiwara and N. Nakamura, *J. Chem. Phys.*, 1971, **54**, 3378.
50 T. Krigas and M. T. Rogers, *J. Chem. Phys.*, 1971, **54**, 4769.
51 T. Krigas and M. T. Rogers, *J. Chem. Phys.*, 1971, **55**, 3035.
52 N. I. Mel'nikov, R. A. Zhitnikov, and P. G. Baranov, *Fiz. tverd. Tela*, 1971, **13**, 1337.
53 V. F. Krutikov, N. I. Silkin, and L. A. Trofanchuk, *Fiz. tverd. Tela*, 1971, **13**, 617.
54 D. R. Gee and J. K. S. Wan, *Canad. J. Chem.*, 1971, **49**, 160.
55 B. L. Bates and L. Kevan, *J. Chem. Phys.*, 1971, **55**, 1327.
56 K. K. Chan, L. Shields, and G. J. Weston, *J. Chem. Soc.* (*A*), 1971, 2769.

crystals containing rare-earth ions such as lanthanum, cerium, and gadolinium are photochromic. Anderson and Sabisky[57] have studied these systems using e.s.r. spectroscopy. Photo-ionization led to a reversible colour change, and the e.s.r. results confirmed earlier optical results which had suggested that the centre was a combination of a rare-earth ion and a halide vacancy. The e.s.r. spectra of the ground state showed that the centre was a tervalent rare-earth ion next to two electrons trapped in the halide vacancy (an F'-centre). The stable state is ionized on irradiation, leaving the tervalent rare-earth ion next to a perturbed F-centre, the electron being trapped by a rare-earth ion in a cubic site to form a bivalent rare-earth ion.

Tzalmona and Persham studied the radiation damage of Eu^{3+}-doped CdF_2 crystals.[58] The main species observed in this case was Eu^{2+}, the tripositive ion being a deep electron trap.

Organic Radicals.—At present there is a welcome trend for organic systems to be studied in depth and for a careful analysis of the e.s.r. spectra to be made, isotopic substitution often being used to assist in the positive identification of a radical species. As a consequence of this approach, much of the work has been carried out with single crystals, which usually permits a fuller analysis of the spectra to be made. However, there is still a tendency to make rather speculative assignments of the spectra, and thus the radiation mechanism deduced may be incorrect. On the other hand, the radical species may be assigned according to the radiation mechanism expected and then the observed spectrum interpreted in terms of that radical. Both approaches have their dangers, and it should be emphasized that relatively minor changes in the environment can alter completely the nature of the radical which is trapped.

For clarity, the subject has been divided into several sections, though it has not been possible to keep to clear-cut boundaries in every case.

Alkyl and Alkenyl Radicals. Single crystals[59] of the n-alkanes with 4—16 carbon atoms have been γ-irradiated at 77 K, and the spectra have been observed between 77 and 195 K. It is assumed that two basic radicals are formed, namely $CH_3\dot{C}HCH_2{-}R^1$ and $R^2{-}CH_2{-}\dot{C}H{-}CH_2{-}R^3$. The observed spectra were analysed into the two spectra of these radicals by computer, and the anisotropic hyperfine coupling constants were obtained. The relative yields of the two types of radicals as determined from the e.s.r. spectra are not consistent with those obtained by product analysis. This discrepancy immediately highlights the problems associated with the elucidation of the reaction mechanism by an e.s.r. study alone.

Cyclopentyl radicals have been formed in pure cyclopentane or mixtures of cyclopentane and neohexane[60] by u.v.- or γ-irradiation at 77 K. The

57 C. H. Anderson and E. S. Sabisky, *Phys. Rev. (B)*, 1971, **3**, 527.
58 A. Tzalmona and P. S. Persham, *J. Chem. Phys.*, 1971, **55**, 4804.
59 A. Lund, *J. Phys. Chem.*, 1972, **76**, 1411.
60 C. Fauquenoit, J. P. Dodelet, and P. Claes, *Bull. Soc. chim. belges*, 1971, **80**, 315.

spectra in the mixed glass were slightly better resolved than those in pure cyclopentane. The cyclopentyl radical is considered to exist in two very non-planar configurations and the assignment of the hyperfine couplings is most unusual. It should be pointed out that the usual assignment accounts satisfactorily for the additional structure observed in the mixed matrix because of the slightly non-integral relationship of the coupling constants.[61, 62] The dependence of the yield of cyclopentyl radicals on the concentration of cyclopentane in neohexane has also been investigated.[63]

Vinyl and 1-methylvinyl radicals have been stabilized [64] on silica gel by the γ-irradiation of acetylene or methylacetylene adsorbed on the surface at 77 K. At 77 K the complete eight-line spectrum of the vinyl radical is observed, though the lines exhibit considerable anisotropy.[65] As the temperature is increased the central four lines decrease in intensity and are virtually absent at 150 K, while the outer lines become more symmetrical, as is observed in the liquid phase.[66] When the temperature is raised above 200 K the central two lines appear of the sextet (1 : 1 : 2 : 2 : 1 : 1) that is expected for rapid inversion at the α-carbon atom. The effect of temperature is completely reversible and the radicals are relatively stable at 300 K. This appears to be the first report of the observation of the spectrum in the rapid exchange region. It is only possible to reach the intermediate exchange region for the 1-methylvinyl radical at 300 K, because inversion at the α-carbon atom is much slower for the methyl substituent.

In similar experiments, attempts have been made [67] to identify the spectra of the α-acrylic radicals $CH_2{=}\dot{C}{-}CO_2H$ and $CH_2{=}\dot{C}{-}CO_2Me$, stabilized on molecular sieves at 77 K. However, because of lines from other species in the centre of the spectra, it was not possible to determine whether the two proton-coupling constants were equal or unequal (corresponding to a π-radical or a σ-radical, respectively). On the basis of INDO calculations, both radicals are expected to have a π-configuration.

The γ-irradiation at 77 K of 1,3-butadiene adsorbed on silica gel gives three different radicals.[68] A quintet spectrum [A(H) = 11 G] is attributed to the radical cation $(CH_2CHCHCH_2)^+$. At high concentrations of butadiene an allylic radical $RCH_2\dot{C}H{-}CH{=}CH_2$ is formed by polymerization, whereas at low concentrations the 1-methylallyl radical $CH_3\dot{C}H{-}CH{=}CH_2{-}$ is formed by addition of a hydrogen atom to a terminal carbon atom. A spectrum attributed to $-CH_2{-}\dot{C}H{-}CH{=}CH{-}CH_2$ is observed [69] in a γ-irradiated single crystal of 1,3-butadiene

[61] P. B. Ayscough and C. Thomson, *Trans. Faraday Soc.*, 1962, **58**, 1477.
[62] J. E. Bennett and A. Thomas, *Proc. Roy. Soc.*, 1964, **A280**, 123.
[63] C. Fauquenoit and P. Claes, *Bull. Soc. chim. belges*, 1971, **80**, 323.
[64] S. Nagai, S. Ohnishi, and I. Nitta, *Chem. Phys. Letters*, 1972, **13**, 379.
[65] E. L. Cochran, F. J. Adrian, and V. A. Bowers, *J. Chem. Phys.*, 1964, **40**, 213.
[66] R. W. Fessenden and R. Schuler, *J. Chem. Phys.*, 1963, **39**, 2147.
[67] L. Bonazzola, S. Fenistein, and R. Marx, *Mol. Phys.*, 1971, **22**, 689.
[68] T. Shiga, A. Lund, and P. O. Kinell, *Internat. J. Radiation Phys. Chem.*, 1971, **3**, 131.
[69] T. Shiga, A. Lund, and P. O. Kinell, *Internat. J. Radiation Phys. Chem.*, 1971, **3**, 145.

at 77 K. It is suggested that the radical is produced in an oligomer formed by 1,4-*trans*-addition, followed by isomerization.

There have been several attempts to observe the bridgehead radical (1-adamantyl) in adamantane for, because of the ring structure, it should be constrained to a non-planar configuration. The earlier results [70] have been shown to be due to impurities [71] or to a radical formed by ring opening.[72] A more recent study [73] has claimed to observe 1-adamantyl in zone-refined adamantane after irradiation. However, once again the observed radical appears to be due to an impurity or a scission product. Experiments [74] in the liquid phase have finally answered the problem and show that the 1-adamantyl radical is pyramidal, though it may be slightly flattened from the tetrahedral configuration. The six β-protons are equivalent, with a hyperfine coupling constant of 6.58 G, and there is also considerable coupling to the three γ-protons (4.66 G) and to three of the δ-protons (3.08 G).

The 2-adamantyl radical has been observed in *X*-irradiated adamantane.[71] The spectrum consists of a broad doublet [$A(H_\alpha) = 20.5$ G] with signs of an additional triplet splitting of *ca.* 5 G.

In this context it should be mentioned that adamantane is a very good matrix for trapping other radicals. The guest radicals are trapped in relatively large cavities in the adamantane and are able to rotate much more freely than in most other matrices. In consequence, well-resolved isotropic e.s.r. spectra are observed, with linewidths in the range 0.5—2.0 G.

The spectra from radicals formed [75] by the loss of the tertiary hydrogen atom from 2-methylbutane, 2-methylpentane, or 3-methylpentane have been observed in adamantane. The hyperfine couplings to the β-CH_2— and β-CH_3 groups were readily obtained for each radical [$A(CH_3) = 23.0$ G; $A(CH_2) = 15.0$ G]. A series of cycloalkenes [76] included in adamantane has also been studied after γ-irradiation at room temperature. The corresponding substituted cyclic allyl radicals were formed and the hyperfine coupling constants were obtained. The terminal α-protons had couplings of *ca.* 14 G, β-protons in the range 20—26 G, and the central CH-proton of *ca.* 3 G, all of which are typical of allylic radicals.

Halogenated Alkyl Radicals. The spectrum of the $\dot{C}H_2Cl$ radical has been observed [77] in a single crystal of CH_3Cl after γ-irradiation at 77 K. The chlorine quadrupole splittings were observed, in addition to the normal chlorine and proton hyperfine couplings. The distribution of the unpaired spin density shows that the radical is only slightly non-planar.

[70] D. R. Gee, L. Fabes, and J. K. S. Wan, *Chem. Phys. Letters*, 1970, **7**, 311.
[71] J. R. Ferrell, G. R. Holdren, R. V. Lloyd, and D. E. Wood, *Chem. Phys. Letters*, 1971, **9**, 343.
[72] L. Bonazzola and R. Marx, *Chem. Phys. Letters*, 1971, **8**, 413.
[73] W. G. Filby and K. Gunther, *Chem. Phys. Letters*, 1972, **14**, 440.
[74] P. J. Krusic, T. A. Rettig, and P. von R. Schleyer, *J. Amer. Chem. Soc.*, 1972, **94**, 995.
[75] S. W. Kanick, R. E. Linder, and A. C. Ling, *J. Chem. Soc.* (*A*), 1971, 2971.
[76] D. R. Gee and J. K. S. Wan, *Canad. J. Chem.*, 1971, **49**, 20.
[77] J. P. Michaut and J. Roncin, *Chem. Phys. Letters*, 1971, **12**, 95.

The $\dot{C}Cl_3$ radical has been studied[78] in polycrystalline carbon tetrachloride at 77 K. The isotropic ^{13}C hyperfine coupling (= 115 G) shows that the radical has a non-planar structure with a $\widehat{ClCCl}$ angle of *ca.* 116°. On admission of oxygen the spectrum changes to that of the peroxy-radical ($Cl_3COO^{\bullet}$), which has $A(^{13}C) = 6.2$ G. Thus, there is negligible spin density on the carbon atom, as the observed splitting probably arises mainly from spin polarization by the unpaired electron density on the adjacent oxygen atom.

γ-Irradiation of aligned crystals[79] of bromobutane yields Br_2^- and n-butyl radicals. Interestingly, equal hyperfine interaction with the two β-protons [$A(H_\beta) = 33.3$ G] occurs in the crystalline material, whereas in glassy bromobutane the two β-protons have unequal hyperfine coupling constants (22 and 44 G, respectively).

Experiments[80] have been continued in an effort to identify the paramagnetic species that gives rise to the multi-line spectrum covering about 1000 G in polycrystalline n-alkyl iodides, after γ-irradiation at 77 K. Studies with six isotopically substituted ethyl iodides containing ^{127}I, ^{129}I, ^1H, or ^2H atoms show that the main splittings arise from interaction with iodine nuclei and the methyl protons. It is tentatively suggested that the spectrum results from physically trapped electrons similar to those originally postulated[81] to exist in γ-irradiated CD_3CN.

The growth rates of the radicals formed in glassy and polycrystalline tritiated ethyl iodide by self-radiation have also been studied.[82] The multi-line spectrum is again observed in the polycrystalline material. Detailed measurements[83] of the decay of ethyl radicals in glassy ethyl iodide show that the decay involves the recombination of geminate pairs ($C_2H_5^{\bullet}\ I^-$). Experiments with $CH_3{}^{13}CH_2I$ show that the proton-exchange reaction $CH_3{}^{13}\dot{C}H_2 \leftrightharpoons \dot{C}H_2{}^{13}CH_3$ does not occur. The effects of the matrix and of the structure and size of the alkyl radical on the decay rates have been examined,[84] and a model has been developed[85] for the growth of radicals in solids during irradiation.

Of particular interest are the β-substituted halogenated radicals[86] $R^1R^2\dot{C}{-}CH_2X$, where X = Cl, Br, or I, which have been prepared by γ-irradiation of the t-butyl halides at 77 K. The initial radical formed is presumably $\dot{C}H_2{-}C(X)Me_2$, which then undergoes a 1,2-halogen migration to give the observed radical $Me_2\dot{C}{-}CH_2X$. The isotropic halogen hyperfine

[78] C. Hesse, N. Leray, and J. Roncin, *Mol. Phys.*, 1971, **22**, 137.
[79] M. L. Bonin, M. A. Bonin, and F. Williams, *J. Chem. Phys.*, 1971, **54**, 2641.
[80] R. J. Egland, P. J. Ogren, and J. E. Willard, *J. Phys. Chem.*, 1971, **75**, 467.
[81] M. A. Bonin, K. Takeda, K. Tsuji, and F. Williams, *Chem. Phys. Letters*, 1968, **2**, 263.
[82] P. J. Ogren and J. E. Willard, *J. Phys. Chem.*, 1971, **75**, 3359.
[83] H. W. Fenrick, N. B. Nazhat, P. J. Ogren, and J. E. Willard, *J. Phys. Chem.*, 1971, **75**, 472.
[84] C. R. Roy and J. E. Willard, *J. Phys. Chem.*, 1972, **76**, 1405.
[85] P. J. Ogren, *J. Phys. Chem.*, 1972, **76**, 1324.
[86] A. R. Lyons and M. C. R. Symons, *J. Amer. Chem. Soc.*, 1971, **93**, 7330.

couplings are very large [$A(^{81}Br)$ = 280 G; $A(^{127}I)$ = 400 G] and show that the radicals are 'locked' in a configuration in which the dihedral angle between the axis of the *p*-orbital containing the unpaired electron and the C—X bond direction is close to zero. Similar splittings have been observed in the radicals $\dot{C}H_2—CH_2X$ prepared by the photolysis of bromine or iodine in ethylene at 77 K.

Very similar effects [87, 88] are found for the radicals $\dot{C}H_2—CH_2MR_n$, where M = Sn, P, As, Si, or Ge. The metal hyperfine-coupling constants are again very large, corresponding to spin densities of *ca.* 6 and 20% in the metal *s*- and *p*-orbitals, respectively. In contrast, for the series $R^1_2\dot{C}MR^2_n$ the metal hyperfine couplings [89] are less than 1%. The α-proton splittings (R^1 = H) and the β-proton splittings (R^1 = Me) are both close to those observed for the ethyl radical and also indicate that there is virtually no delocalization on to the metal atom. This conclusion is in conflict with that [90] previously advanced to explain the trend in *g*-factors for these radicals, and tentative mechanisms are given which can explain these trends without invoking $p(\pi)$ delocalization.

In general, fluorinated alkyl radicals tend to be more nearly comparable to the hydrogenated species. The radical anion $(CF_3CO_2H)^-$ has been observed [91] in γ-irradiated polycrystalline trifluoroacetic acid at 77 K. The complete fluorine hyperfine tensors were not obtained but two fluorines have about equal coupling [$A_\parallel(F)$ = 150 G] whereas the third coupling is different. This observation suggests that the CF_3 group is not freely rotating, and thus the situation is comparable to that of $(CH_3CO_2H)^-$ at 77 K. The coupling to the OH-proton is *ca.* 16 G, which seems a very high value, but it appears to be substantiated by deuterium substitution. The radical anion is considered to be the precursor of $\dot{C}F_2CO_2H$, while the cation is not trapped but breaks down spontaneously to give $\dot{C}F_3$. Thus

$$(CF_3CO_2H)^- \xrightarrow{\text{slow}} \dot{C}F_2CO_2H + F$$

$$(CF_3CO_2H)^+ + CF_3CO_2H \xrightarrow{\text{fast}} \dot{C}F_3 + CO_2 + CF_3CO_2H_2^+$$

Similar results are found for other fluorinated carboxylic acids, and hindered rotation of CF_3 groups has been observed in $CF_3\dot{C}HOH$ and $CF_3(CH_3)\dot{C}OH$, which were formed by γ-irradiation of the corresponding alcohols.

More-detailed measurements of the fluorine hyperfine couplings have been made for $^-O_2C—CF_2\dot{C}F—CO_2^-$ trapped in a single crystal of sodium perfluorosuccinate [92] and for $\dot{C}F_2COO^-$ in a single crystal of sodium

87 A. R. Lyons and M. C. R. Symons, *Chem. Comm.*, 1971, 1068.
88 A. R. Lyons and M. C. R. Symons, *J. C. S. Faraday II*, 1972, **68**, 622.
89 A. R. Lyons, G. W. Neilson, and M. C. R. Symons, *J. C. S. Faraday II*, 1972, **68**, 807.
90 J. H. Mackey and D. E. Wood, *Mol. Phys.*, 1970, **18**, 783.
91 C. Chacaty and M. Shiotani, *J. Chim. phys.*, 1971, **68**, 300.
92 L. D. Kispert and M. T. Rogers, *J. Chem. Phys.*, 1971, **54**, 3326.

chlorodifluoroacetate [93] at room temperature. The complete fluorine hyperfine coupling tensors were obtained and their structural implications discussed. Partial analysis of the ^{13}C couplings for the latter radical shows that it is probably non-planar. A similar study [94] of the radicals $\dot{C}ClFCONH_2$ and $\dot{C}ClFCOND_2$ in dichlorofluoroacetamide has been made and complete or partial hyperfine tensors have been obtained for ^{35}Cl, ^{19}F, ^{13}C, and ^{14}N nuclei together with the g-tensors and chlorine quadrupole couplings. The results have been compared with INDO and CNDO/2 calculations, and it is concluded that $\dot{C}ClFCONH_2$ is a planar π-radical.

The isotropic spectra of $\dot{S}iH_2F$, $\dot{S}iHF_2$, and $\dot{S}iF_3$ have been observed [95] from the electron irradiation of the corresponding silanes in SF_6 at 77 K. The fluorine hyperfine couplings are very similar to those of the corresponding fluorinated methyl radicals, but the proton couplings differ considerably and the absolute values are positive. The ^{29}Si hyperfine coupling is $-$ 498 G in $\dot{S}iF_3$ and corresponds to 30% 3*s*-character on the silicon atom. Thus $\dot{S}iF_3$ has a very pyramidal structure.

Aromatic Radicals. *X*-Irradiation of toluene and α-, 2-, 3-, or 4-azatoluene in adamantane [96] at 300 K gave the corresponding benzyl and N-substituted benzyl radicals. Linewidths were *ca.* 0.5 G and all the hyperfine coupling constants were assigned. INDO calculations severely underestimate the effect of the nitrogen atom on the spin distribution and predict that all the radicals should be very similar to benzyl, whereas the experimental results show marked differences between the different radicals.

The α-styryl radical [67] has been stabilized on a molecular sieve and the observed spectrum was very similar to that reported earlier,[97] which was interpreted in terms of the π-radical α-methylenebenzyl. INDO calculations also indicated that the α-styryl radical should have a π-configuration, and the calculated hyperfine-coupling constants are in fair accord with the experimental values.

The toluene cation is formed by γ-irradiation at 77 K of toluene adsorbed on silica gel or porous Vycor glass.[98] No evidence for the negative ion is found. The hyperfine coupling constants for $(C_6H_5CH_3)^+$ are: $A(ortho\text{-H}) = 2.1$ G; $A(meta\text{-H}) = 0.75$ G; $A(para\text{-H}) = 11.8$ G; $A(CH_3) = 18.0$ G, and for $(C_6H_5CD_3)^+$, $A(CD_3) = 2.75$ G. These values are in accord with those expected from simple theoretical considerations which predict that the unpaired electron will be in the symmetrical bonding molecular orbital of the benzene ring.

Further attempts have been made to characterize the duryl radical (2,4,5-trimethylbenzyl radical) in irradiated durene. Single crystals of

93 M. T. Rogers and L. D. Kispert, *J. Chem. Phys.*, 1971, **55**, 2360.
94 L. D. Kispert and F. Myers, *J. Chem. Phys.*, 1972, **56**, 2623.
95 M. V. Merritt and R. W. Fessenden, *J. Chem. Phys.*, 1972, **56**, 2353.
96 R. V. Lloyd and D. E. Wood, *Mol. Phys.*, 1971, **20**, 735.
97 J. E. Bennett and J. A. Howard, *Chem. Phys. Letters*, 1971, **9**, 460.
98 T. Komatsu, A. Lund, and P. O. Kinell, *J. Phys. Chem.*, 1972, **76**, 1721.

durene and [$^2H_2$]durene (ring positions deuteriated) were irradiated but no conclusive results were obtained.[99] However, the 13-line spectrum observed[100] in γ-irradiated durene at 300 K has been attributed to the tetramethylcyclohexadienyl radical (1). Observation of the dispersion

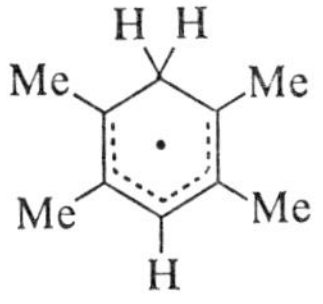

(1)

e.s.r. signal instead of the absorption signal, which overcomes problems caused by saturation, showed an intense single-line spectrum, about 20 G wide. On the basis of INDO calculations of the hyperfine coupling constants for the duryl radical and spectral simulation, it is shown that this line is consistent with the predicted spectrum.

The 2-naphthyl radical has been prepared[101] by the self-radiolysis at 77 K of [2-^{3}H]naphthalene incorporated in a single crystal of [$^2H_8$]-naphthalene. The hyperfine coupling constants are very close to those observed for 2-naphthyl trapped in argon[102] at 4 K. However, the principal values of the g-tensor are all larger by about 0.0011 in [$^2H_8$]naphthalene and may indicate that there is some interaction with the argon molecules. INDO calculations using a standard hexagonal geometry give the wrong order for the relative magnitude of the coupling constants. The molecular geometry determined by X-ray structural measurements gives the right relative order, but the r.m.s. fit to the experimental results is slightly poorer than that using the standard geometry!

The $(C_6H_5)_2\dot{P}O$ radical has been observed[103] in a single crystal of diphenylphosphine oxide. Analysis of the ^{13}P-hyperfine couplings shows that the unpaired electron has 11% s-character and 60% p-character on the phosphorus atom, and that, unlike diphenyl nitroxide [$(C_6H_5)_2NO^\bullet$], it is non-planar.

Alcohols and Ethers. The γ-irradiation[104] of glassy allyl alcohol at 77 K gives approximately twice the yield of allyl radicals $CH_2{=}CH{-}\dot{C}H_2$ as that of 1-hydroxyallyl radicals $CH_2{=}CH{-}\dot{C}HOH$. Subsequent u.v. irradiation converts the hydroxyallyl radical into the allyl radical.

The free radical[105] formed in a single crystal of deoxyadenosine monohydrate by X-irradiation at 300 K has been identified as an alcohol

99 J. Heurtaux, *J. Chim. phys.*, 1972, **69**, 315.
100 J.-L. Monge and M. Schott, *J. Chim. phys.*, 1972, **69**, 324.
101 R. V. Lloyd and D. E. Wood, *J. Chem. Phys.*, 1972, **56**, 916.
102 P. H. Kasai, P. A. Clark, and E. B. Whipple, *J. Amer. Chem. Soc.*, 1970, **92**, 2640.
103 M. Geoffroy and E. A. C. Lucken, *Mol. Phys.*, 1971, **22**, 257.
104 S. Noda, K. Torimoto, K. Fueki, and Z. Kuri, *Bull. Chem. Soc. Japan*, 1971, **44**, 273.
105 C. Alexander and C. E. Franklin, *J. Chem. Phys.*, 1971, **54**, 1909.

radical $\rangle CH\dot{C}HOH$, by e.s.r. spectroscopy. The spin density (0.64) on the free-valence carbon atom is typical of that found for the α-radicals in alcohols.

Trapped electrons are formed[106] by γ-irradiation at 77 K of methyl vinyl ether in 3-methylhexane (3-Mehex) or 2-methyltetrahydrofuran (2-MeTHF). Subsequent exposure to visible light causes dissociation of the methyl vinyl ether to give $\dot{C}H_3$ radicals:

$$CH_3OCH{=}CH_2 + e^- \longrightarrow \dot{C}H_3 + CH_2{=}CHO^-$$

In contrast, alkyl halides included in 3-Mehex or 2-MeTHF dissociate directly on γ-irradiation.

γ-Irradiation[107] of tetrahydrofuran hydrate gives a mixture of the 2- and 3-furanyl radicals by loss of a ring proton. 1,3-Dioxolan also gives a mixture of the two possible radicals at 77 K.

Carboxylic Acids. Extensive results have been reported for the free radicals formed by *X*- or γ-irradiation of single crystals of carboxylic acids and their salts. A very careful and intriguing investigation has been made on γ-irradiated single crystals of maleic acid and its Na and K salts. The interpretation of the e.s.r. results has been made by relating the principal axes of *g*- and hyperfine-tensors to the crystallographic axes, and thence to the actual configuration of the molecules in the undamaged crystals. For potassium hydrogen maleate[108] (KHM) at 77 K, a symmetrical σ-radical (2) is formed in which the unpaired electron is located in a σ-antibonding orbital between the in-plane *p*-orbitals of the ring oxygen atoms. At 300 K a symmetrical π-radical (3) is formed in KHM, with the unpaired electron in a delocalized π-orbital extending over the whole ring. In

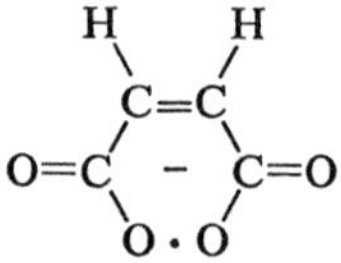

77 K (2) σ

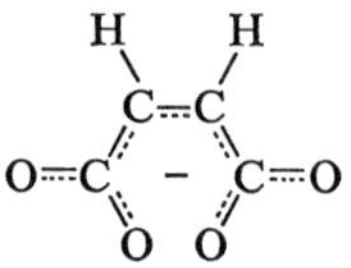

300 K (3) π

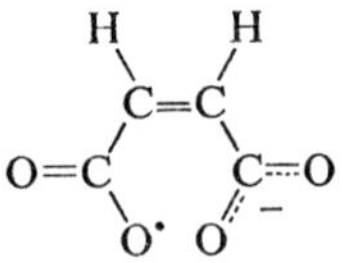

77 K (4) σ

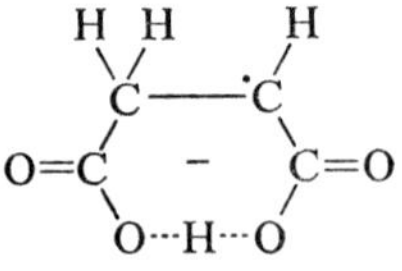

300 K (5) *p*

[106] M. Irie, K. Hayashi, S. Okamura, and H. Yoshida, *J. Phys. Chem.*, 1971, **75**, 476.
[107] V. I. Trofimov and I. I. Chkheidze, *Doklady Akad. Nauk S.S.S.R.*, 1971, **198**, 354.
[108] K. Toriyama and M. Iwasaki, *J. Chem. Phys.*, 1971, **55**, 2181.

contrast, for sodium hydrogen maleate[109] (NaHM) an unsymmetrical σ-radical (4) is formed at 77 K, with the unpaired electron located on one oxygen atom. At 300 K hydrogen addition occurs selectively to one end of the double bond in NaHM, giving a normal type of alkyl radical (5) with the unpaired electron located in a carbon *p*-orbital.

X-Irradiation[110] of deuteriated maleic acid ($DO_2C-CH=CH-CO_2D$) at 77 K in the dark gives the σ-radical (6) with the unpaired electron located in the in-plane *p*-orbitals of a non-ring oxygen atom. It is suggested that the σ-radical is formed from the primary positive ion by transfer of the acid deuteron to a neighbouring molecule. The radical fragments to give a substituted vinyl radical (7).

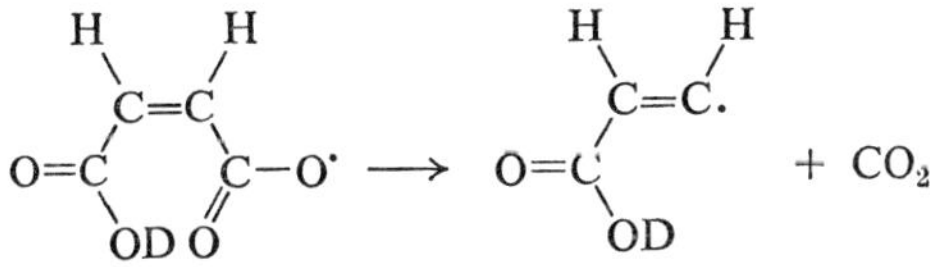

(6) (7)

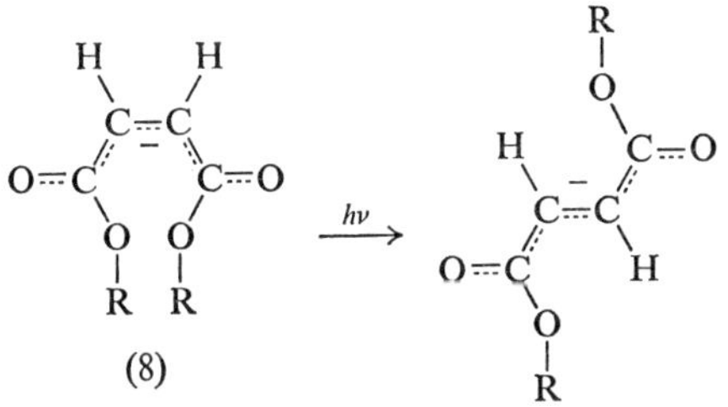

(8) (9)

X-Irradiation[111] of KHM in the dark at 77 K also yields the asymmetrical σ-radical (2), which on exposure to visible light is transformed into the unsymmetrical radical (4).

Maleate radical anions[112] (8), formed by γ-irradiation of the corresponding maleates in 2-methyltetrahydrofuran at 77 K, isomerize to the corresponding fumarate anions (9) on exposure to visible light. The reverse reaction does not occur, and irradiation of the parent compounds with visible light does not cause any isomerization. Analysis of the e.s.r. spectrum of the maleate anion shows that there is about 40% of the unpaired spin density on each of the olefinic carbon atoms. Thus, the C=C bond can be considered to have considerable single-bond character in the anion, which will facilitate isomerization.

[109] K. Toriyama, H. Muto, and M. Iwasaki, *J. Chem. Phys.*, 1971, **55**, 1885.
[110] B. Eda and M. Iwasaki, *J. Chem. Phys.*, 1971, **55**, 3442.
[111] K. Toriyama, M. Iwasaki, S. Noda, and B. Eda, *J. Amer. Chem. Soc.*, 1971, **93**, 6415.
[112] A. Torikai, T. Suzuki, T. Miyazaki, K. Fueki, and Z. Kuri, *J. Phys. Chem.*, 1971, **75**, 482.

γ-Irradiation in the dark at 77 K of single crystals of potassium hydrogen fumarate ($K^+ \, ^-O_2CCH{=}CHCO_2H$) yields two distinct σ-radicals.[113] One is attributed to the loss of the acidic hydrogen from the CO_2H group whereas the other is formed by the ejection of the electron from the CO_2^- group. The latter radical can be regarded as a positive hole trapped on the oxygen non-bonding orbital.

Hydroxyl radicals[114] have been observed in NaHM trihydrate and are trapped in two different sites. The proton hyperfine coupling tensors are almost identical but the *g*-tensors differ considerably in the two sites. The *g*-shifts from the free-spin value in one site are about twice those in the other site. The larger *g*-anisotropy is attributed to a weaker interaction of the $\dot{O}H$ radical with the environment causing a smaller energy separation of the two oxygen $p\pi$ orbitals, and *vice versa*.

Three distinct paramagnetic species have been observed[115] in a single crystal of potassium hydrogen malonate ($K^+ \, ^-O_2CH_2CO_2H$) after *X*-irradiation at 4 K. First there is the radical anion, giving the typical doublet splitting [$A(H_{iso})$ = 25.2 G] from one of the β-hydrogen atoms, secondly, a delocalized π-radical (11) formed by the loss of a hydrogen atom shared between two adjacent molecules (10), and thirdly, a radical pair

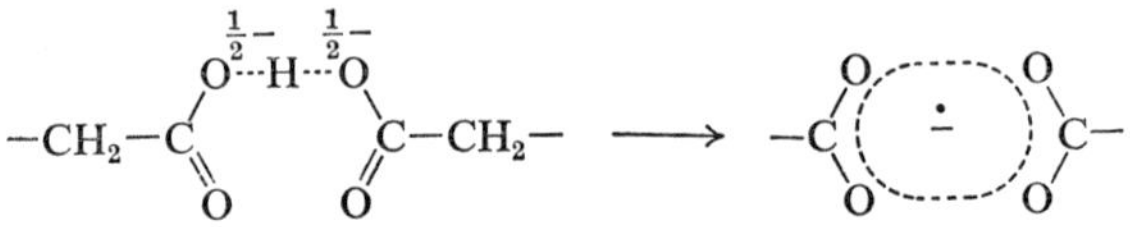

(10) (11)

formed by loss of H_2 from molecules in adjacent chains in the crystal. Analysis of the observed dipole–dipole couplings gives a separation of 0.56 nm for the radical pair.

On the other hand, irradiation of potassium hydrogen malonate[116] at 77 K leads to $HO_2C\dot{C}HCO_2^- K^+$. The proton hyperfine tensors have been analysed in detail, and hyperfine coupling occurs with four extra-molecular protons. The hyperfine tensors for the naturally abundant carbon-13 atoms in the α- and β-positions have also been obtained.

A single crystal of succinic acid[117] which has been doped with 0.005 mole fraction of fumaric acid gives the molecular anion of fumaric acid when γ-irradiated at 77 K. On irradiation at 300 K, hydrogen-atom addition occurs to the double bond in fumaric acid. In both cases the free radicals usually observed from succinic acid are suppressed, showing the marked effect that impurities can have.

A similar effect has been observed[118] in a single crystal of strontium acetate hemihydrate containing a trace of propionate. γ-Irradiation of the

[113] M. Iwasaki, K. Minakata, and K. Toriyama, *J. Amer. Chem. Soc.*, 1971, **93**, 3533.
[114] K. Toriyama and M. Iwasaki, *J. Chem. Phys.*, 1971, **55**, 1890.
[115] H. C. Box, E. E. Budzinski, and W. Potter, *J. Chem. Phys.*, 1971, **55**, 315.
[116] R. C. Nicklin, H. A. Farach, and C. P. Poole, *J. Chem. Phys.*, 1972, **56**, 1279.
[117] M. Iwasaki, H. Muto, and K. Toriyama, *J. Chem. Phys.*, 1971, **55**, 1894.
[118] W. M. Tolles, R. A. Sanders, and R. G. Gisch, *J. Chem. Phys.*, 1971, **54**, 1532.

doped crystal at room temperature gives $CH_3\dot{C}HCO_2^-$, instead of $\dot{C}H_2CO_2^-$ observed in the pure crystal. At 77 K the acetate dianion (12) is formed in the pure crystal. The isotropic hyperfine coupling [A(Me) = 11.8 G] agrees well with that previously reported,[119] but is smaller than that observed[120, 121] for the acetic acid anion (13)

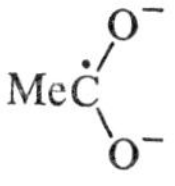

(12)

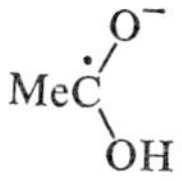

(13)

[A(Me) = 13.5 G]. The reduction in the methyl hyperfine coupling is consistent with the greater delocalization expected for $-O^-$ than for $-OH$.

The ^{13}C hyperfine-coupling tensors have been obtained[122] for $^{13}\dot{C}H_2CO_2^-$ and $\dot{C}H_2{}^{13}CO_2^-$ trapped in single crystals of sodium acetate. γ-Irradiation of sodium acetate,$3D_2O$ at 77 K gives[123] an e.s.r. spectrum which is attributed to methyl radicals interacting with neighbouring sodium nuclei. However, the hyperfine structure on each of the four methyl hyperfine lines appears to be closer to a 1 : 3 : 3 : 1 quartet than the 1 : 1 : 1 : 1 quartet that would be expected from interaction with a sodium nucleus.

The mechanism for the γ-radiolysis of acetic acid at 77 K has been studied in detail.[121] The proposed mechanism is compatible with the product yields obtained from the radiolysis in the liquid phase and relates the final products to the primary species formed by radiolysis, *i.e.* $CH_3\dot{C}O_2H^+$ and e^-. The study has been extended[124] to other carboxylic acids and it is shown that the main features of the reaction mechanisms are similar to those observed for acetic acid.

Relaxation measurements[125] have been made on the free radicals in an irradiated single crystal of malonic acid over the temperature range 1.1—300 K. It is concluded that the spin–lattice relaxation times are determined by phonon modulation of the spin–orbit coupling rather than by the hyperfine interactions.

Amino-acids. The evolution of the various radicals with increasing temperature has been followed[126] in single crystals of glycine (and deuteriated glycines) from 77 to 300 K. Starting with the radical anion $N^+H_3CH_2\dot{C}O_2H^-$ at 77 K, four other radicals have been observed as the temperature is increased ($\dot{C}H_2\overset{+}{N}H_3$ at 77 K; $\overset{+}{\dot{N}}H_2CH_2CO_2^-$ at 110 K;

[119] M. D. Sevilla, *J. Phys. Chem.*, 1970, **74**, 669.
[120] J. E. Bennett and L. H. Gale, *Trans. Faraday Soc.*, 1968, **64**, 1174.
[121] P. B. Ayscough, K. Mach, J. P. Oversby, and A. K. Roy, *Trans. Faraday Soc.*, 1971, **67**, 360.
[122] M. Fujimoto and J. Janecka, *J. Chem. Phys.*, 1971, **55**, 5.
[123] J. Janecka, H. M. Vyas, and M. Fujimoto, *J. Chem. Phys.*, 1971, **54**, 3229.
[124] P. B. Ayscough and J. P. Oversby, *Trans. Faraday Soc.*, 1971, **67**, 1365.
[125] L. R. Dalton, A. L. Kwiram, and J. A. Cowen, *Chem. Phys. Letters*, 1972, **14**, 27.
[126] J. Sinclair, *J. Chem. Phys.*, 1971, **55**, 245.

$\overset{+}{N}H_3\dot{C}HCO_2^-$ at 120 K; and $\dot{C}H_2CO_2H$ at 145 K). In contrast, in a similar study[127] the anion is the only paramagnetic species observed at 77 K, and neither $\dot{C}H_2\overset{+}{N}H_3$ nor $\overset{+\bullet}{N}H_2CH_2CO_2^-$ is detected as the temperature is raised. However, subsequent u.v. irradiation of the γ-irradiated crystals yielded $\dot{N}HCH_2CO_2H$.

Controversy[128, 129] has arisen over the earlier postulate[130] that deuterium–hydrogen exchange reactions can occur in partially deuteriated glycines at 77 K. Obviously, until this matter is clearly understood the conclusions obtained from deuterium-substitution experiments should be treated with caution.

The study[131] of single crystals of β-alanine or ε-aminocaproic acid at 4 K shows that the primary reduced species is the radical anion $NH_2(CH_2)_n\dot{C}O_2H^-$ [or $^+NH_3(CH_2)_nCO_2^{2-}$], which gives the typical doublet spectrum [A(H) = 25—28 G]. Alkyl radicals are also observed and are postulated to originate from the positive ion, which is unstable and undergoes decarboxylation. Broadly comparable results have been obtained[132] for single crystals of β-alanine and of γ-aminobutyric and ε-aminocaproic acids at 77 K. The main radicals observed are $\dot{C}H_2CH_2X$, where X = $(CH_2)_{n-2}\overset{+}{N}H_3$, arising from decomposition of the radical cation, and X = $(CH_2)_{n-2}CO_2^-$, arising from decomposition of the radical anion. At 300 K, loss of a hydrogen adjacent to the acid group occurs exclusively, to give $^+NH_3(CH_2)_n\dot{C}HCOO^-$.

In the light of these results, it is doubtful whether the multi-line spectra observed in single crystals of alanine[133] (X-irradiated at 80 K) or α-aminoisobutyric acid[134] (γ-irradiated at 77 K) should be attributed to the radical cations. In the latter the original assignment[135] is probably correct. It also seems probable that the single-line spectrum ($g_\parallel$ = 2.0160; $g_\perp$ = 2.0147) observed in polycrystalline alanine[136] after X-irradiation at 300 K arises from a peroxy-radical (g_{iso} = 2.0154), and thus the conclusion that the hyperfine spectra observed in other work arise from impurities is unjustified.

E.s.r. studies have also been made on single crystals of glycylglycine,[137] N-benzoylalanine,[138] sarcosine hydrochloride,[139] *dl*-serine,[140] homoserine,[141]

[127] P. B. Ayscough and K. Mach, *J. C. S. Faraday I*, 1972, **68**, 1139.
[128] I. Mayer, *Chem. Phys. Letters*, 1971, **8**, 117.
[129] M. J. A. de Bie and R. Braams, *Chem. Phys. Letters*, 1971, **8**, 208.
[130] J. R. Morton, *J. Amer. Chem. Soc.*, 1964, **86**, 2325.
[131] H. C. Box and E. E. Budzinski, *J. Chem. Phys.*, 1971, **55**, 2446.
[132] Yu. D. Tsvetkov and G. A. Almanov, *Khim. vysok. Energii*, 1971, **5**, 537.
[133] E. A. Friday and I. Miyagawa, *J. Chem. Phys.*, 1971, **55**, 3589.
[134] A. Minegishi, Y. Shinozaki, and G. Meshitsuka, *J. Chem. Phys.*, 1972, **56**, 2481.
[135] D. G. Cadena, R. E. Linder, and J. R. Rowlands, *Canad. J. Chem.*, 1969, **47**, 3429.
[136] A. M. Ozer, A. P. Kulshreshtha, and T. Mookherji, *Chem. Phys. Letters*, 1971, **10**, 204.
[137] T. B. Meloe, *Internat. J. Radiation Biol.*, 1971, **20**, 557.
[138] G. Lassman and W. Damerau, *Mol. Phys.*, 1971, **21**, 551.
[139] G. Lassman and W. Damerau, *Mol. Phys.*, 1971, **21**, 555.
[140] B. W. Castleman and G. C. Moulton, *J. Chem. Phys.*, 1971, **55**, 2598.
[141] Yu. D. Tsvetkov and G. A. Almanov, *Zhur. strukt. Khim.*, 1971, **12**, 1092.

and taurine.[142] Two sulphur-centred radicals were detected in taurine, $H_3N^+CH_2CH_2SO_3^-$.[142] The ^{33}S hyperfine-coupling tensors were obtained and the radicals identified as SO_3^- and $H_3N^+CH_2CH_2\dot{S}O_2$, respectively.

Amines, Amides, and Cyanates. Radicals of the type $R^1_2\dot{C}NR^2_2$ have been prepared[143] by the γ-irradiation of amines at 77 K. From an analysis of the nitrogen and proton hyperfine couplings, it is concluded that the radicals have a nearly planar π^3-structure with *ca.* 30% of the spin density on the nitrogen atom and 70% on the carbon. The carbon centre is considered to be planar, and the deviation from planarity at the nitrogen is not greater than 5°. These conclusions conflict with those arrived at from a study[144] of similar radicals trapped in adamantane and from associated INDO calculations which suggest that there are large deviations from planarity at both centres. It would appear desirable to obtain the complete hyperfine coupling tensors for the protons, nitrogen, and carbon-13 nuclei to resolve these structural discrepancies.

In contrast to the open-chain amines, the *X*-irradiation of pyrrolidine and 3-pyrroline in adamantane[145] yields the corresponding nitrogen-centred radicals (14) and (15). The isotropic nitrogen hyperfine couplings

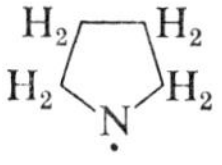

(14)

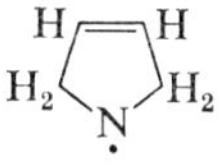

(15)

(14.4 G) are very close to those obtained for simple dialkylamino-radicals in the liquid phase.[146] In conjunction with INDO calculations, it is concluded that the radicals are π-type with very little σ-character, and that the $\widehat{CNC}$ angle is only slightly less than 120°.

A series of aliphatic amines[147] have been γ-irradiated at 77 K in a polycrystalline state, and deuterium- and ^{15}N-substitution has been used to help identify the radicals. The primary amines yield predominantly $R^1\dot{N}H$ radicals, but when warmed these either isomerize or abstract from a neighbouring amine molecule to give $R^2\dot{C}HNH_2$ radicals. The disubstituted amines give both $R^1_2\dot{N}$ and $R^2\dot{C}HNHR^1$ radicals on irradiation at 77 K. The ^{14}N hyperfine coupling for all the nitrogen-centred radicals is $A_{\parallel}(^{14}N) = 35$ G, and on the assumption that $A_{\perp} = 0$, the isotropic coupling is *ca.* 12 G, which is in reasonable agreement with those (= 14.5 G) reported above.

Similar results were observed[148] for less than monolayer amounts of the amines adsorbed on silica gel and irradiated at 77 K.

142 G. Lind and R. Kewley, *Canad. J. Chem.*, 1972, **50**, 43.
143 A. R. Lyons and M. C. R. Symons, *J. C. S. Faraday II*, 1972, **68**, 502.
144 D. E. Wood and R. V. Lloyd, *J. Chem. Phys.*, 1970, **53**, 3932.
145 D. W. Pratt, J. J. Dillon, R. V. Lloyd, and D. E. Wood, *J. Phys. Chem.*, 1971, **75**, 3486.
146 W. C. Danen and T. T. Kensler, *J. Amer. Chem. Soc.*, 1970, **92**, 5235.
147 P. Wardman and D. R. Smith, *Canad. J. Chem.*, 1971, **49**, 1869.
148 P. Wardman and D. R. Smith, *Canad. J. Chem.*, 1971, **49**, 1880.

The *X*-irradiation [149] of a single crystal of dicyandiamide N≡C—N-=C(NH_2)$_2$ at 77 K gives a radical which shows coupling to one proton (= 75 G, nearly isotropic) and two nitrogen nuclei [$A_1(N_1)$ = 31.7; $A_2(N_1)$ = 7.2; $A_3(N_1)$ = 2.4 and $A_1(N_2)$ = 8.5; $A_2(N_2)$ = 6.1; $A_3(N_2)$ = 4.9 G]. The larger nitrogen coupling is typical of those for the electron in a *p*-orbital on the nitrogen, and it is concluded that the radical is N≡C—N⋯C(NH_2)⋯NH with the unpaired electron in a π-molecular orbital located mainly on the terminal nitrogen atom. At 300 K three radicals are formed,[150] one of which exhibits a large proton coupling (70—80 G) and a maximum nitrogen coupling of 25 G. The second radical has *A*(H) = 30 G and *A*(N) = 15 G. Both spectra are attributed to the same radical as before, but that with the large proton and nitrogen hyperfine couplings is postulated to be a σ-radical, whereas that with the smaller couplings is taken to be the π-radical. These interpretations are open to the same criticisms as those put forward for RCOṄH radicals,[151, 152] and it appears that the radical Ṅ=CH—N=C(NH_2)$_2$ could well be responsible for the spectrum observed at 77 K.

The assignments [153, 154] of the spectra observed in γ-irradiated malonamide and cyanoacetyl urea to a σ-radical R—C(=O)—Ṅ—H have been questioned,[151, 152] because these radicals are expected to be π-radicals. It is pointed out that although the proton hyperfine coupling is comparable with that expected for a σ-radical (*e.g.* H_2CṄ),[155, 156] the nitrogen coupling is that observed for a normal π-radical. It is tentatively suggested either that the observed spectrum arises from a radical R(H)C=Ṅ or that not all the proton hyperfine lines were detected in the centre of the spectrum and that it is due to NH_3^+. A spectrum has been observed [157] in urea at 77 K which is attributed to H_2NCOṄH. The proton and nitrogen hyperfine couplings show that this RCOṄH radical has a π-structure, with the electron located mainly in the nitrogen 2*p*-orbital (75%).

X-Irradiation [158] of alkyl isocyanates and isothiocyanates in adamantane at 300 K yields the corresponding radicals R^1R^2ĊNCO and R^1R^2ĊNCS,

[149] P. W. Lau and W. C. Lin, *J. Chem. Phys.*, 1971, **54**, 823.
[150] G. Lind and R. Kewley, *Canad. J. Chem.*, 1971, **49**, 2514.
[151] M. C. R. Symons, *J. Chem. Soc.* (*A*), 1971, 3205.
[152] M. C. R. Symons, *J. Chem. Phys.*, 1971, **55**, 1493.
[153] N. Cyr and W. C. Lin, *J. Chem. Phys.*, 1969, **50**, 3701.
[154] P. W. Lau and W. C. Lin, *J. Chem. Phys.*, 1969, **51**, 5139.
[155] J. A. Brivati, K. D. J. Root, M. C. R. Symons, and D. J. A. Tinling, *J. Chem. Soc.* (*A*), 1969, 1942.
[156] F. J. Adrian, E. L. Cochran, and V. A. Bowers, *J. Chem. Phys.*, 1962, **36**, 1938.
[157] H. J. Bower, J. A. McRae, and M. C. R. Symons, *J. Chem. Soc.* (*A*), 1971, 2400.
[158] D. E. Wood, R. V. Lloyd, and W. A. Lathan, *J. Amer. Chem. Soc.*, 1971, **93**, 4145.

respectively. Comparison with INDO calculations shows that the radicals have a planar π-structure. The effects of sulphur substitution are in line with the lower electronegativity of sulphur compared with that of oxygen. Comparable results are obtained[159] for $\dot{C}H_2NCO$ and $CH_3\dot{C}HNCO$ trapped in polycrystalline isocyanates at 77 K. However, from the reported values of the α-proton coupling constants, it would appear that there is an anisotropic contribution included.

Acetonitrile. The paramagnetic species trapped[160] in the upper crystalline phase of acetonitrile metastabilized at 77 K is shown to be the dimer radical anion. Analysis of the ^{14}N and ^{13}C hyperfine coupling constants shows that the unpaired electron resides entirely in the *p*-orbitals of the CN groups of two molecules. Thus it is now clear that this species should be designated as a dimer radical anion rather than as a 'dipole-trapped electron'.[161, 162]

The reaction of methyl radicals in γ-irradiated acetonitrile[163] has been followed over a small temperature range above 77 K and found to have an activation energy of only 5.9 kJ mol^{-1}, in contrast to the gas-phase values ($>$ 20 kJ mol^{-1}) for hydrogen abstraction by methyl radicals.

$$\dot{C}H_3 + CH_3CN \longrightarrow CH_4 + \dot{C}H_2CN$$

It is suggested that the low value may be caused by the occurrence of quantum-mechanical tunnelling in the solid phase.

Further evidence[164] for quantum-mechanical tunnelling is provided by comparable studies in methyl isocyanide between 77 and 125 K. An anomalous Arrhenius plot is obtained, giving an activation energy of 5.9 kJ mol^{-1} at 90 K and 18.8 kJ mol^{-1} at 120 K.

From studies[165] of the reactions of trapped electrons with acetonitrile in 8M-NaOH at 77 K, it is found that the yield of CH_3 is only about 10% that of $CH_3CHN^{\bullet}$. This result is taken as evidence that dissociative electron capture is not very efficient, but this conclusion depends on the assumption that on visible bleaching all the trapped electrons form CH_3CN^-, which then reacts further to give either one radical or the other. However, it appears feasible that some of the electrons could react in the frozen aqueous matrix to form hydrogen atoms which could subsequently add to acetonitrile to give $CH_3CHN^{\bullet}$.

Evidence[166] for a methyl radical–halide ion pair has been obtained from the γ-irradiation of methyl bromide in [$^2H_3$]acetonitrile. The hyperfine coupling constants for the radical pair $\dot{C}H_3$–Br^- are $A_{\parallel}(^{81}Br) = 56.7$ G; $A_{\parallel}(^{79}Br) = 52.6$ G; A(Me) $= 20.6$ G. These values are consistent with an

159 Y. J. Chung and F. Williams, *J. Phys. Chem.*, 1971, **75**, 1893.
160 E. D. Sprague, K. Takeda, and F. Williams, *Chem. Phys. Letters*, 1971, **10**, 299.
161 M. A. Bonin, K. Takeda, K. Tsuji, and F. Williams, *Chem. Phys. Letters*, 1968, **2**, 363.
162 K. Takeda and F. Williams, *Mol. Phys.*, 1969, **17**, 677.
163 E. D. Sprague and F. Williams, *J. Amer. Chem. Soc.*, 1971, **93**, 787.
164 J.-T. Wang and F. Williams, *J. Amer. Chem. Soc.*, 1972, **94**, 2930.
165 F. P. Sargent and M. G. Bailey, *Canad. J. Chem.*, 1971, **49**, 3898.
166 E. D. Sprague and F. Williams, *J. Chem. Phys.*, 1971, **54**, 5425.

unpaired spin density of 90% on the methyl radical and 10% in the bromine 4*p*-orbital.

Reversible line-broadening effects observed for the t-butyl radical trapped in γ-irradiated t-butyl isothiocyanate[167] are also attributed to the existence of a radical pair $Me_3\dot{C}$–NCS^-.

Uracil Derivatives. Results have been reported for a number of uracil derivatives, but the interpretation of the spectra is by no means unambiguous. It should be borne in mind that it may be possible to form some of the postulated species in the liquid phase.[168] The compounds irradiated in the solid phase include uracil,[169] single crystals of 5-nitrouracil[170] (though the original assignment may be incorrect[171]), single crystals of 6-methyldihydrouracil,[172] thymine in alkaline, neutral, or acidic aqueous glass (photolysis),[173] single crystals of 5,5-diethylbarbituric acid (comprehensive e.s.r. results suggest an unambiguous assignment),[174] single crystals of 6-azathymine (fairly unambiguous assignment),[175] and thymidine in aqueous glasses (mechanistic study).[176]

Sulphur Compounds. Most of the sulphur compounds studied appear to be of biological interest and are not discussed here. The *X*-irradiation of single crystals of aromatic sulphones[177] has led to the identification of $Ph\dot{S}O_2$ and $PhCH_2\dot{S}O_2$ radicals. The ^{33}S hyperfine-coupling tensors have been obtained for the naturally abundant isotope and indicate that the unpaired spin density has about 8% 3*s*-character and 42% 3*p*-character on the sulphur atom.

Inclusion Compounds. An interesting series of experiments which must be mentioned in this Report is that using urea and perhydrotriphenylene inclusion compounds. Complexes of relatively long-chain molecules can be formed with these compounds, in which the guest molecules are oriented along the chain direction. The e.s.r. spectra, although not giving as much information as can be obtained from single crystals, are considerably better resolved than those in the polycrystalline or amorphous samples. The recent work has been concerned with the free radicals formed by *X*-irradiation of heptanal $[CH_3(CH_2)_4\dot{C}HCHO]$,[178] decanol oxime *O*-methyl ether $[CH_3(CH_2)_8CH_2\dot{N}OCH_3]$,[179] 6-undecanol, and 1-decanol [radical formed by loss of hydrogen α to OH group in both alcohols; large coupling constant

[167] Y. J. Chung and F. Williams, *J. Phys. Chem.*, 1972, **76**, 1793.
[168] J. K. Dohrmann and R. Livingston, *J. Amer. Chem. Soc.*, 1971, **93**, 5363.
[169] J. N. Herak and M. Hudomaly, *Chem. Phys. Letters*, 1972, **12**, 460.
[170] B. Benson and W. Snipes, *Mol. Phys.*, 1971, **20**, 357.
[171] M. C. R. Symons, *Mol. Phys.*, 1971, **22**, 551.
[172] R. Bergene and T. Henriksen, *Internat. J. Radiation Biol.*, 1971, **20**, 201.
[173] M. D. Sevilla, *J. Phys. Chem.*, 1971, **75**, 626.
[174] R. Haak and B. Benson, *J. Chem. Phys.*, 1971, **55**, 3693.
[175] J. N. Herak and G. Schoffa, *Mol. Phys.*, 1971, **22**, 379.
[176] A. D. Lenherr and M. G. Ormerod, *Proc. Roy. Soc.*, 1971, **A235**, 81.
[177] M. Geoffroy and E. A. C. Lucken, *J. Chem. Phys.*, 1971, **55**, 2719.
[178] Z. Ciecierska-Tworek, G. B. Birrell, and O. H. Griffith, *J. Phys. Chem.*, 1972, **76**, 1009.
[179] Z. Ciecierska-Tworek, G. B. Birrell, and O. H. Griffith, *J. Chem. Phys.*, 1972, **56**, 1001.

to OH proton in 1-decanol; A(OH) = 3.7 G],[180] and methyl 2-nonynoate [$CH_3(CH_2)_3CH_2-\dot{C}H-C{\equiv}C-C({=}O)-OCH_3$; hyperfine coupling with the methoxy-protons; A(Me) = 1.6 G].[181]

A series of radicals are identified from γ-irradiated hexadecanoic acid,[182] but the emphasis is on mechanistic aspects.

Radical Pairs. Basically, irradiation of a molecular substrate must form radicals in pairs whether the initial event is ejection of an electron from a molecule to form a radical cation, or the scission of a bond to form two neutral free radicals. In the first case the electron may travel some distance before being trapped or forming a molecular radical anion, and thus the ion pairs (or their scission products) will be formed at random and are unlikely to have any spatial constancy. For the scission of a C—H bond, the hydrogen atom may migrate at random before reacting, and so again the radical pairs will not have a constant geometry. However, in certain cases the radical pair may have a constant geometry, and then the spin–spin interaction between the two partners will be well defined. In many systems the spectra of the radical pairs are much weaker than those of the mono-radicals and so are not readily observable, but even so there have been several reports of radical-pair spectra.

X-Irradiation at 4 K of dibenzoyl peroxide incorporated in a single crystal of dibenzoyl sulphide[183] gives pairs of benzoyloxyl radicals by simple scission of the O—O bond. Analysis of the dipole–dipole coupling indicates that the radical pairs are separated by 0.36 nm (assuming a point-charge model). By contrast, u.v. photolysis yields pairs of phenyl radicals by elimination of two molecules of CO_2. As expected, the experimentally determined separation between the phenyl radicals (0.65 nm) is larger than that for the benzoyloxyl radical pairs. Radical pairs were also formed by the X-irradiation of single crystals of potassium hydrogen malonate,[183] hydroxyurea,[183], dimethylglyoxime,[183, 184] potassium deuterium fumarate,[185] and carbazide.[186]

A reinvestigation[187] of the radical pairs formed in polycrystalline diphenyl carbonate[188] has shown that, in addition to pairs of phenoxyl radicals formed by elimination of CO, asymmetrical phenoxyl–phenyl radical pairs are formed by elimination of CO_2. The calculated separations are 0.57 nm for phenoxyl–phenoxyl pairs, and 0.68 nm for phenoxyl–phenyl pairs.

[180] G. B. Birrell and O. H. Griffith, *J. Phys. Chem.*, 1971, **75**, 3489.
[181] G. B. Birrell, A. A. Lai, and O. H. Griffith, *J. Chem. Phys.*, 1971, **54**, 1630.
[182] A. Faucitano, A. Perotti, G. Allara, and F. Faucitano Martinotti, *J. Phys. Chem.*, 1972, **76**, 801.
[183] H. C. Box, *J. Phys. Chem.*, 1971, **75**, 3426.
[184] O. E. Yakimchenka and Ya. S. Lebedev, *Internat. J. Radiation Phys. Chem.*, 1971, **3**, 17.
[185] M. Iwasaki, K. Minakata, and K. Toriyama, *J. Chem. Phys.*, 1971, **55**, 1472.
[186] K. Reiss and W. Gordy, *J. Chem. Phys.*, 1971, **55**, 5329.
[187] A. R. Lyons, M. C. R. Symons, and J. K. Yandell, *J. C. S. Faraday II*, 1972, **68**, 495.
[188] J. A. McRae and M. C. R. Symons, *J. Chem. Soc. (B)*, 1968, 428.

2 Free Radicals formed by other Methods

In this section the studies of free radicals that have been formed by methods other than by high-energy irradiation are discussed. The formation of free radicals in the solid phase by photolysis, thermolysis, direct chemical reaction, and trapping out from the gas phase are considered. As far as possible the results are grouped under these headings.

Photolysis.—In this section, free radicals that have been formed directly by photolysis will be considered, but the effects of light on free radicals that have been formed by other methods will not be discussed.

In principle, photolysis should be a more specific method than high-energy irradiation for producing a given free radical. However, in practice a precursor with a strong absorption band in the readily accessible visible or u.v. range is not always available. Furthermore, the primary free radical may have a stronger absorption at longer wavelengths than the precursor has, and so can itself be photolysed to give secondary radicals. This situation occurs in some of the examples considered, and in extreme cases the primary radical may not be observed at all. Thus, caution should be exercised when attempts are made to elucidate the reaction mechanism by this method.

Methylene. One of the most important species studied is triplet-state methylene and its derivatives. The ^{13}C hyperfine-coupling constants for CD_2 trapped in xenon at 4 K have been reported,[189, 190] and from the values [$A_{iso}(^{13}C)$ = 88.3 G and $A_{dip}(^{13}C)$ = 12.5 G] the apex angle $\widehat{DCD}$ has been determined as 137°. This value is in good agreement with the earlier one[191] of 136° determined from the zero-field parameters of CH_2 and CD_2, and has been shown to be independent of the matrix, whereas the values of the zero-field parameters D and E are not. The values[192] of D_n for rigid methylene are very similar for CH_2 and CD_2 trapped in matrices of octafluorocyclobutane, SF_6, or krypton (average 0.76 cm^{-1}) whereas that in xenon is much higher (0.96 cm^{-1}). This difference is attributed to the spin–orbit interaction with the heavy xenon atoms, which causes about 50% increase in the observed value of D_n. The values in the other matrices are close to the most recent theoretical value[193] (0.71 cm^{-1}), and the average (0.76 cm^{-1}) is taken to be that of 'free' methylene, though some interaction in the case of krypton cannot be excluded.

In contrast to the other methylenes [CH_2, PhCH, PhCPh, (1- and 2-naphthyl)CH, and (9-anthryl)CH], which are all bent with an apex

189 E. Wasserman, V. J. Kuck, R. S. Hutton, E. D. Anderson, and W. A. Yager, *J. Chem. Phys.*, 1971, **54**, 4120.
191 R. A. Bernheim, H. W. Bernard, P. S. Wang, L. S. Wood, and P. S. Skell, *J. Chem. Phys.*, 1971, **54**, 3223.
192 E. Wasserman, V. J. Kuck, R. S. Hutton, and W. A. Yager, *J. Amer. Chem. Soc.*, 1970, **92**, 7491.
193 E. Wasserman, R. S. Hutton, V. J. Kuck, and W. A. Yager, *J. Chem. Phys.*, 1971, **55**, 2593.
194 J. F. Harrison, *J. Chem. Phys.*, 1971, **54**, 5413.

angle of 135—140°, 9,9′-dianthrylmethylene[194] appears to be linear, with the two anthryl planes perpendicular to each other, and may thus be considered as a substituted allene.

A few reactions of methylene in the solid state have also been reported.[195] CH_2 abstracted hydrogen from traces of water in xenon at 4 K to give CH_3, and when D_2O was added both CH_2D and CHD_2 were observed. The appearance of CHD_2 is explained by hydrogen–deuterium exchange occurring in the gas phase before the reactants are condensed. Of more interest is the formation of the formaldoxime radical CH_2NO by the addition of CH_2 to NO. The experimental value of the isotropic nitrogen coupling constant (33.3 G) is typical of those observed for oxime radicals in solution. The values of the isotropic coupling constants [$A(^{14}N) = 33.3$ G; $A(H_1) = 26.2$ G; $A(H_2) = 2.8$ G] are in good agreement with those calculated by SCF theory,[196] and confirm that the observed spectrum arises from the planar radical.

The reactions of diphenylmethylene (Ph_2C:), formed by photolysis of a single crystal of 1,1-diphenylethylene ($Ph_2C{=}CH_2$) containing diphenyldiazomethane ($Ph_2C{=}N_2$), have been followed by e.s.r., ENDOR, and optical techniques.[197] The rates of reaction of Ph_2C: with $Ph_2C{=}CH_2$ and $Ph_2C{=}N_2$ have been studied over the temperature range 90—120 K. The products are 1,1,2,2-tetraphenylcyclopropane and benzophenone azine ($Ph_2C{=}N{-}N{=}CPh_2$), respectively.

Alkyl Radicals. Methyl radicals[198] have been formed by the photolysis of methane adsorbed on porous Vycor glass at 77 K. Depending on the experimental conditions, a normal or an abnormal spectrum can be observed. The presence of a trace of a metal oxide, *e.g.* V_2O_5, enhances the formation of the abnormal species.[199] The normal species has a hyperfine splitting (23.0 G) and *g*-factor (2.003) close to the values normally observed for $\dot{C}H_3$, whereas the abnormal species has low values (19.3 G and 2.002, respectively). Similar results[200] are obtained for $\dot{C}D_3$ formed by the photolysis of CD_4. The ethyl radical has also been observed in both forms by the photolysis of ethylene.[200] The normal species has hyperfine splitting constants of $A(H_\alpha) = 21.5$ G and $A(H_\beta) = 26.7$ G, and a *g*-factor of 2.003, whereas for the abnormal radical $A(H_\alpha) = 15.1$ G, $A(H_\beta) = 27.3$ G, and $g = 2.002$. It is tentatively suggested that the drop in the α-proton coupling constants is caused by a slight change in the configuration of the

[194] E. Wasserman, V. J. Kuck, W. A. Yager, R. S. Hutton, F. D. Greene, V. P. Abegg, and N. M. Weinshenker, *J. Amer. Chem. Soc.*, 1971, **93**, 6335.
[195] J. B. Farmer, C. L. Gardner, M. C. L. Gerry, C. A. McDowell, and P. Raghunathan, *J. Phys. Chem.*, 1971, **75**, 2448.
[196] G. Berthier, H. Lemaire, A. Rassat, and A. Veillard, *Theor. Chim. Acta*, 1965, **3**, 213.
[197] D. C. Doctschman and C. A. Hutchinson, *J. Chem. Phys.*, 1972, **56**, 3964.
[198] N. Shimamoto, Y. Fujita, and T. Kwan, *Bull. Chem. Soc. Japan*, 1970, **43**, 580.
[199] Y. Fujita, K. Hatano, M. Yanagita, T. Katsu, M. Sato, and T. Kwan, *Bull. Chem. Soc. Japan*, 1971, **44**, 2884.
[200] T. Katsu, M. Yanagita, and Y. Fujita, *J. Phys. Chem.*, 1971, **75**, 4064.

α-carbon atom. The effect of such changes is very marked on the α-proton couplings but has very little effect on the β-coupling constants.

The earlier results[201] on the reaction of OH radicals, formed by photolysis at 77 K of H_2O_2 in methylenecyclobutane or methylenecyclopentane, have been reinterpreted.[202] For methylenecyclobutane (16) it is argued that the observed e.s.r. spectrum arises from the allenyl radical $CH_2{=}C{=}\dot{C}H$, and not from a substituted vinyl radical as originally

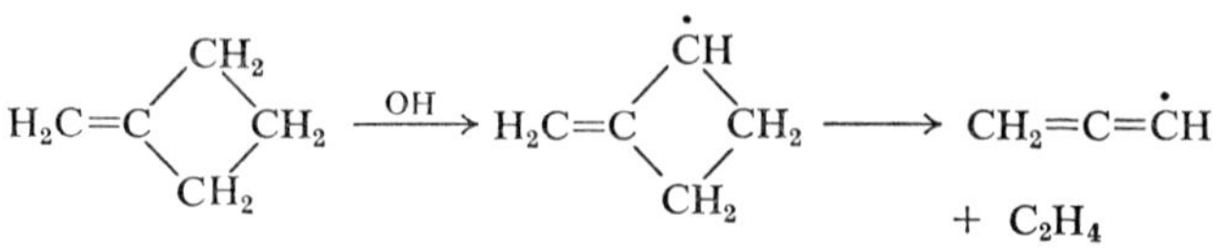

(16) (17)

supposed. The mechanism suggested for the formation of the allenyl radical involves abstraction of an allylic hydrogen followed by ring cleavage of (17). Similarly, it is suggested that the spectrum from methylenecyclopentane arises from the allyl or a substituted allyl radical, but not from a cyclic allyl radical.

The photolysis[203] of HI in polycrystalline isobutane at 77 K yields isobutyl radicals, whereas t-butyl radicals are formed predominantly in a glassy matrix of isobutane–methylcyclohexane. However, γ-irradiation gives only t-butyl radicals in both polycrystalline and glassy matrices. The change in the free radical formed is attributed to differences in energy of the hydrogen atoms formed by the dissociation of HI. However, the selectivity appears to be opposite to that expected from bond strengths and may indicate the participation of other matrix effects.

The photolysis[86] of iodine or bromine in ethylene at 77 K yields the corresponding halogenated ethyl radical $\dot{C}H_2CH_2X$. Both radicals display a very large hyperfine coupling to the halogen atom; $A(^{127}I) = 400$ G and $A(^{81}Br, ^{79}Br) = 250$ G. The large coupling shows that the radicals have a locked configuration in which the halogen atom lies above the nodal plane. In this configuration there is maximum overlap between the *p*-orbital of the unpaired electron and the orbitals of the halogen atom.

Aromatic Radicals. The e.s.r. spectrum of $PhC{\equiv}C^{\bullet}$ has been observed[204] in glassy $PhC{\equiv}CI$ after photolysis at 77 K. There is hyperfine coupling to the *ortho*-, *meta*-, and *para*-protons (2.22, 0.81, and 2.16 G, respectively), which shows that $PhC{\equiv}C^{\bullet}$ is not a σ-radical but a π-radical with a structure similar to that of benzyl.

[201] K. Takeda, H. Yoshida, K. Hayashi, and S. Okamira, *Bull. Inst. Chem. Res. Kyoto Univ.*, 1967, **45**, 55.
[202] D. H. Volman, *J. Phys. Chem.*, 1971, **75**, 714.
[203] T. Wakayama, T. Miyazaki, K. Fueki, and Z. Kuri, *Bull. Chem. Soc. Japan*, 1971, **44**, 2619.
[204] J. S. Coleman, A. Hudson, K. D. J. Root, and D. R. M. Walton, *Chem. Phys. Letters*, 1971, **11**, 300.

The u.v. irradiation at 4 K of dibenzoyl peroxide included in single crystals of dibenzoyl disulphide[183] gives pairs of trapped phenyl radicals. Analysis of the dipole–dipole coupling tensor (− 203, + 101, + 103 G) on the basis of point charges leads to a separation of 0.65 nm, which is in good agreement with that expected from the molecular structure. In contrast, *X*-irradiation of the crystals leads to pairs of benzoyloxyl radicals, with a calculated separation of 0.36 nm.

Radical pairs are also formed when tetraphenylhydrazine,[205] as a single crystal, polycrystalline, or in frozen solution in toluene, is photolysed at 77 K. The decay of the radical pairs in the single crystal has been studied over a range of temperature above 77 K.

Sodium triphenylacetate contained in a KBr pellet at room temperature gives the triphenylmethyl radical when photolysed in the absence of oxygen.[206] When oxygen is present, the triphenylmethyl peroxyl radical is formed.

Nitrogen and Sulphur Compounds. The photolysis of formaldazine $CH_2{=}N{-}N{=}CH_2$ at 77 K yields the methyleneimino-radical ($CH_2{=}N\cdot$), which was identified by its e.s.r. spectrum.[207]

Photolysis of azobisisobutyronitrile[208] at 77 K gives the 2-cyanopropyl radical $(CH_3)_2\dot{C}CN$. The anisotropy of the nitrogen hyperfine coupling is determined from the solid-state spectrum, and by comparison with the isotropic coupling constant obtained by liquid-phase measurements[209] the relative signs of the constants are determined: $A(^1H) = 21.3$ G; $A_\parallel(^{14}N) = + 13.4$ G; $A_\perp(^{14}N) = - 2.0$ G. The spin density on the nitrogen nucleus is estimated to be $\rho_N = 0.3$.

The photochemistry of a series of 1,2,3-thiadiazoles has been studied[210] by detecting the free radicals trapped at 77 K.

When aliphatic mercaptans RSH in a matrix of xenon at 77 K are photolysed with light of 254 nm, spectra of the corresponding thiyl radicals $RS^\bullet$ are observed.[211] Photolysis in a matrix of 3-methylpentane gives an intense even-line spectrum due to a radical from the 3-methylpentane. Thus:

$$R^1SH \xrightarrow{h\nu} R^1S^\bullet + H^*$$

$$H^* + R^2H \longrightarrow R^{2\bullet} + H_2$$

The photolysis at 254 nm of pyridine in methylcyclohexane or methanol yields a radical identified as the pyridinyl radical (18), formed by hydrogen

205 O. Ya. Grinberg, A. A. Dubinskii, and Ya. S. Lebedev, *Kinetika i Kataliz*, 1971, **13**, 660.
206 E. G. Janzen and M. Buckheit, *J. Phys. Chem.*, 1972, **76**, 937.
207 M. Kamachi, K. Kuwata, and S. Murahashi, *J. Phys. Chem.*, 1971, **75**, 164.
208 S. Moriuchi and J. Sohma, *Mol. Phys.*, 1971, **21**, 369.
209 R. Livingston and H. Zeldes, *J. Magn. Resonance*, 1969, **1**, 169.
210 P. Krauss, K.-P. Zeller, H. Meier, and E. Muller, *Tetrahedron*, 1971, **27**, 5953.
211 J. Shelton and F. C. Adam, *Canad. J. Chem.*, 1971, **49**, 3536.

abstraction from the matrix.[212] The spectrum is identical to that observed from γ-irradiated pyridine in ethanol [213] at 77 K, which is attributed to this radical.

(18)

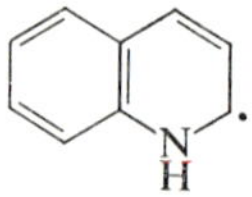

(19)

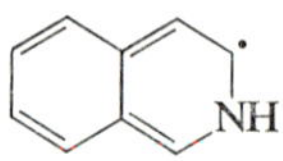

(20)

In similar experiments [214] the spectra observed at 77 K from a series of substituted quinolines and isoquinolines in MeOH, EtOH, or Et_2O are attributed to the coresponding quinolinyl (19) and isoquinolinyl (20) radicals formed by hydrogen abstraction from the solvent. Virtually no hyperfine structure is resolved and the identification depends on the comparison of the experimental and calculated second moments.

The photolysis of 2-bromopyridine at 77 K yields a three-line e.s.r. spectrum attributed to the 2-pyridyl radical.[215] Only single-line spectra are observed from 3- and 4-bromopyridine, thus confirming that the three-line spectrum previously observed [216] for β- or γ-irradiated pyridine arises from the 2-pyridyl radical.

The technique of trapping radicals in a matrix of adamantane has been used previously in the studies of radicals formed by chemical means [217] and by high-energy irradiation.[218] The greatly enhanced resolution of the e.s.r. spectra in adamantane makes the interpretation of the spectra much simpler than is often the case for the solid state. The technique has been extended to photolysis studies [219] by incorporating suitable precursors in adamantane at high pressure (10^4 bar). By this means, the radicals formed by the photolysis at 185 and 254 nm of some aliphatic amines have been observed at room temperature, but the results are complicated by the formation of secondary radicals. For instance, $CH_3\dot{C}HNH_2$, and

[212] S. Caplain, A. Castellano, J. P. Catteau, and A. Lablanche-Combier, *Tetrahedron*, 1971, **27**, 3541.
[213] C. Chacaty, *J. Chim. phys.*, 1967, **64**, 608.
[214] G. Allan, A. Castellano, J. P. Catteau, and A. Lablanche-Combier, *Tetrahedron*, 1971, **27**, 4687.
[215] K. Tsuji and T. Seiki, *Bull. Chem. Soc. Japan*, 1972, **45**, 923.
[216] H. J. Bower, J. A. McRae, and M. C. R. Symons, *J. Chem. Soc.* (*A*), 1968, 2696.
[217] J. E. Bennett, B. Mile, A. Thomas, and B. Ward, *Adv. Phys. Org. Chem.*, 1970, **8**, 1.
[218] D. E. Wood and R. V. Lloyd, *J. Chem. Phys.*, 1970, **52**, 3840.
[219] T. Richerzhagen and D. H. Volman, *J. Amer. Chem. Soc.*, 1971, **93**, 2062.

$CH_3CH{=}\dot{N}$ and $CH_3CH_2\dot{N}H$, were observed from ethylamine. It was concluded that $R^1R^2\dot{C}NH_2$ was the primary radical formed by photolysis, though in other studies it is found that $R^1R^2HC\dot{N}H$ is the primary radical.

Previous work [220] on the photolysis of tetramethyl-*p*-phenylenediamine (TMPD) in hydrocarbon glasses at 77 K has suggested that the radicals observed (*e.g.* $\dot{C}H_3$) originate from the hydrocarbon by energy transfer from TMPD. To check this hypothesis TMPD has been photolysed in perdeuterio-3-methylpentane.[221] The only radical apparently observed after photolysis was $\dot{C}H_3$, which must have originated from TMPD and not from 3-methylpentane. A broad background spectrum is attributed to the TMPD cation, but comparison of the spectra given in the paper suggests that a significant portion of this spectrum could arise from a deuteriated radical formed from the 3-methylpentane.

Alcohols. The photolysis [222] of 50% methanol–H_2O at 77 K yields $\dot{C}H_2OH$, $\dot{C}H_3$, $H\dot{C}O$, and $H^\bullet$ radicals. When the temperature is raised to 87 K the spectrum of the formyl radical increases whereas those of $\dot{C}H_3$ and $H^\bullet$ decrease. It is suggested that the growth of $H\dot{C}O$ is caused by the reactions:

$$H^\bullet + HCHO \longrightarrow H_2 + H\dot{C}O$$

or

$$H^\bullet + CO \longrightarrow H\dot{C}O$$

Unfortunately, for various reasons no attempt was made to follow the decays of $[\dot{C}H_3]$ or $[H^\bullet]$ for comparison with the growth of $[H\dot{C}O]$.

Electron-transfer Reactions Initiated by Photolysis.—*Organic Molecules.* The photo-oxidation of a range of organic molecules by ceric ions at 77 K has been studied extensively.[223–225] Primary alcohols gave $R\dot{C}HOH$ almost exclusively, whereas more C—C cleavage occurred for secondary alcohols, with propan-2-ol and butan-2-ol giving $\dot{C}H_3$ as well as $R^1R^2\dot{C}OH$. Tertiary alcohols gave mainly scission products, $\dot{C}H_3$ being formed from t-butyl alcohol and $\dot{C}_2H_5$ from 3-ethylpentan-3-ol. The mechanism for primary alcohols is expressed as:

$$\underset{\substack{\downarrow \\ Ce^{IV}}}{RCH_2OH} \xrightarrow{h\nu} R\dot{C}HOH + H^+_{solv} + Ce^{III}$$

whereas that for tertiary alcohols is considered to involve a concerted mechanism rather than the intermediate formation of an alkoxyl radical. However, in contrast, when ethanol is photolysed [226] in the presence of

[220] T. Tanei and H. Hatano, *Bull. Chem. Soc. Japan*, 1969, **42**, 3369.
[221] P. J. Bekowies and A. C. Albrecht, *J. Phys. Chem.*, 1971, **75**, 431.
[222] C. U. Morgan, *J. Phys. Chem.*, 1972, **76**, 494.
[223] D. Greatorex and T. J. Kemp, *Trans. Faraday Soc.*, 1971, **67**, 56.
[224] D. Greatorex and T. J. Kemp, *Trans. Faraday Soc.*, 1971, **67**, 1576.
[225] D. Greatorex and T. J. Kemp, *J. C. S. Faraday I*, 1972, **68**, 121.
[226] H. D. Burrows, D. Greatorex, and T. J. Kemp, *J. Amer. Chem. Soc.*, 1971, **93**, 2539.

Ce^{IV} in the liquid phase at 185 K, $\dot{C}H_3$ is formed exclusively. It thus appears that the environment has a marked influence on the course of the reaction.

In nearly all cases investigated,[224] photolysis of Ce^{IV} in carboxylic acids RCO_2H gives the corresponding alkyl radical ($R^{\bullet}$). This route seems to afford a useful method of preparing specific alkyl radicals. The authors prefer a concerted mechanism to one which involves the intermediacy of $RCO_2^{\bullet}$ because they do not observe this latter radical. However, it seems unlikely that $RCO_2^{\bullet}$ would be sufficiently stable, particularly in the presence of light, for a detectable concentration to be attained. Thus, the negative observation does not necessarily preclude the participation of $RCO_2^{\bullet}$ in the decomposition of RCO_2H.

The photo-oxidations of aldehydes, ketones, and esters by Ce^{IV} are not so clean, and the radicals formed depend on the type of solution used.[225] Aldehydes RCHO yielded mainly the alkyl radical ($R^{\bullet}$) or the acyl radical ($R\dot{C}O$). Ketones gave mainly radicals of the type $R^1\dot{C}HCOR^2$ by hydrogen abstraction, though in some cases, notably acetone, C—C cleavage occurred to yield an alkyl radical ($\dot{C}H_3$ from acetone). For esters both C—H and C—C bond cleavage occurred, the relative occurrence of the two modes depending markedly on the solvent.

In general the results suggest that in aqueous solutions C—C bond fission is preferred, whereas in non-aqueous solvents (*i.e.* the pure organic substrate + Ce^{IV}) C—H bond fission is dominant. However, this conclusion should be treated with caution, as in many cases the alkyl radical expected is $\dot{C}H_3$, and whereas $\dot{C}H_3$ is readily trapped at 77 K in aqueous systems, it is often not trapped in organic solvents.

Single crystals of glycine which have been doped with Cu^{II} exhibit a typical copper hyperfine spectrum of four lines. However, on u.v. irradiation [227] at 77 K a new spectrum appears, centred around $g = 2.0$, consisting of a major triplet with intensity ratio 1 : 2 : 1 and a hyperfine coupling of 25 G. From deuterium-substitution experiments it is concluded that the spectrum is due to the radical cation $(NH_2CH_2CO_2H)^+$.

Hydronium Radicals. The observation of the e.s.r. spectra of H_3O and D_3O, formed by the photolysis of a solution of ceric perchlorate–perchloric acid in H_2O or D_2O at 77 K, has been reported.[228] The hyperfine structure is virtually identical to that of $\dot{C}H_3$ or $\dot{C}D_3$, though the g-factors are slightly higher (2.0035) than those expected for methyl radicals (2.0026). However, two refutations of this assignment have been made recently. An experimental study [229] has failed to reproduce the results, and it is suggested that the observed spectra are due to $\dot{C}H_3$ and $\dot{C}D_3$ formed from trace organic impurities. A theoretical calculation [230] based on the INDO method, which

[227] Y. Tomkieicz and M. Fujimoto, *J. Chem. Phys.*, 1972, **56**, 3317.
[228] T. W. Martin and L. L. Swift, *J. Amer. Chem. Soc.*, 1971, **93**, 2788.
[229] J. A. Wargon and F. Williams, *Chem. Phys. Letters*, 1972, **13**, 579.
[230] N. D. Chuvylkin and G. M. Zhidomirov, *Chem. Phys. Letters*, 1972, **14**, 70.

gives reasonable agreement with experiment for $\dot{C}H_3$, $\dot{C}D_3$, and some other AB_3 radicals, shows that a hydrogen coupling constant of 130—150 G would be expected for H_3O depending on whether it has a pyramidal or planar configuration.

Trapped Electrons. A singlet (width 0.1—0.3 G; $g = 2.0001$) with characteristics expected for the spectrum of trapped electrons was observed[231] when Li, Na, K, Mg, or Cd in 3-methylpentane at 77 K was irradiated with light in the region of 254 nm. The metal 'solution' was formed by co-deposition of 3-methylpentane and the selected metal from the gas phase on to a cold finger maintained at 77 K. In the absence of light no e.s.r. spectrum was observed. The very narrow linewidth, the ease with which the singlet saturated, and the rapid decay (half-life 0.2 s) when the light was extinguished suggest that the trapping sites for the electrons are much weaker than those formed by cooling liquid 3-methylpentane to 77 K.

Charged Species. The co-deposition of sodium and HI in an excess of argon at 4 K yields a deposit which gives an e.s.r. spectrum arising solely from trapped Na atoms.[232] However, subsequent illumination of this deposit with light of wavelength longer than 500 nm causes the complete disappearance of the spectrum due to Na atoms and the appearance of that of H atoms. Light of wavelength longer than 600 nm has no effect and thus the sodium D-line transitions appear to be involved.

$$Na + HI \xrightarrow{h\nu} Na^+ + H^\bullet + I^-$$

When Cd, Mn, or Cr is used instead of Na, similar effects are obtained, but in addition the spectrum due to Cd^+, Mn^+, or Cr^+ is observed.

Electron-transfer reactions to organic molecules can also be induced by this method. For tetracyanoethylene (TCNE) the negative ion was formed, and the anisotropic coupling constants for ^{14}N were obtained ($A_\parallel = 5.7$ G; $A_\perp = 0.0$ G). For furan (21), the negative ion (22) is unstable and ring-opening occurs to give a β-substituted vinyl radical (23). The observed

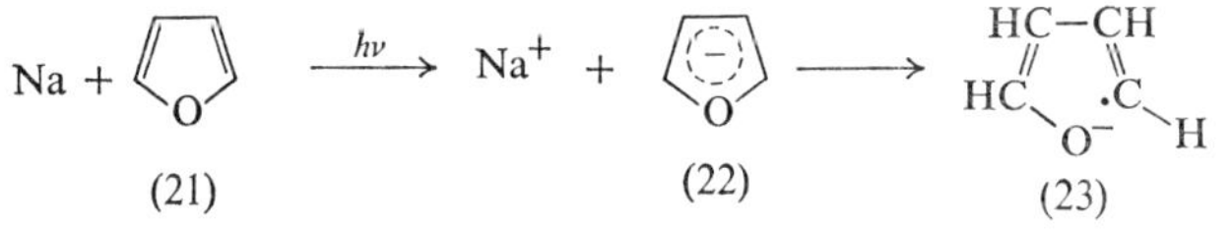

hyperfine-coupling constants are $A(H_\alpha) = 20$ G, $A(H_\beta) = 60$ G, which show that the vinyl radical remains exclusively in the configuration in which it was initially formed with the β-hydrogen atom *trans* to the unpaired electron.

The 2-, 3-, and 4-pyridyl radicals[233] have also been prepared by the corresponding reaction with pyridyl halides in an argon matrix. The

[231] F. W. Froben and J. E. Willard, *J. Phys. Chem.*, 1971, **75**, 35.
[232] P. H. Kasai, *Accounts Chem. Res.*, 1971, **4**, 329.
[233] P. H. Kasai and D. McLeod, *J. Amer. Chem. Soc.*, 1972, **94**, 720.

isotropic hyperfine-coupling constants have been obtained for all three pyridyl radicals. Hyperfine coupling to the nitrogen nucleus is only observed for 2-pyridyl (29 G), and the spectrum fully confirms the assignment of that observed in γ-irradiated pyridine to 2-pyridyl.[216] Subsequent photolysis with u.v. light causes ring-opening for both 2- and 3-pyridyl radicals, giving a β-substituted vinyl radical (24) and methyleneimino-radical (25), respectively. The coupling constants are comparable to those

2-pyridyl

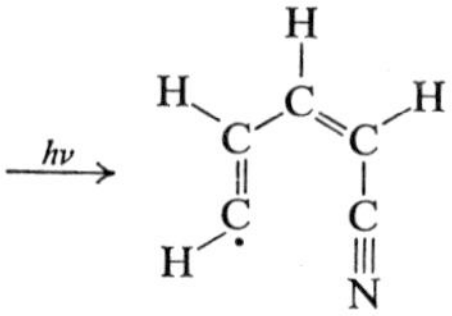

(24)

3-pyridyl

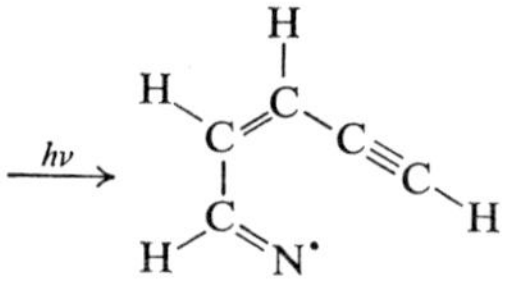

(25)

observed for the corresponding unsubstituted radicals, and again the vinyl radical retains the configuration with the β-hydrogen atom *trans* to the unpaired electron. The imino-radical can be photolysed further to give the butatrienyl radical $H_2C{=}C{=}C{=}\dot{C}H$. The 4-pyridyl radical is much more stable to photolysis, but after prolonged photolysis the spectrum of the butatrienyl radical is observed.

Direct Chemical Reaction.—The preparation of free radicals by direct chemical reaction at low temperature is a very limited field; however, it has permitted the observation of spectra of free radicals not observed by the other methods.

Acyl Radicals. The acetyl and benzoyl radicals have been prepared by the reaction of sodium atoms with the halides RCOCl at 77 K using a rotating cryostat.[234] Analysis of the e.s.r. spectra showed that both radicals are σ-type radicals, with the unpaired electron located primarily in a *sp*-hybrid orbital on the carbonyl-carbon atom. Comparison of the results with those for $H\dot{C}O$ and $HO_2CCH_2\dot{C}O$ shows that the free-radical centres for all four acyl radicals are very similar, with a spin distribution of $\rho_R = 0.25$, $\rho_C = 0.55$, and $\rho_O = 0.20$.

SCF MO calculations have been made for the acetyl radical,[235] and the calculated values of the isotropic coupling constants are in reasonable agreement with those found experimentally.

[234] J. E. Bennett and B. Mile, *Trans. Faraday Soc.*, 1971, **67**, 1587.
[235] H. Veillard and B. Rees, *Chem. Phys. Letters*, 1971, **8**, 267.

Aromatic Radicals. Further studies using the rotating cryostat[236] have been reported for the phenyl radical. In the light of results[237] for the phenyl radical trapped in xenon at 4 K, the earlier spectrum in benzene[238] at 77 K has been reinterpreted. Formerly it was postulated that two distinct trapping sites existed in benzene; however, subsequent comparison with the better-resolved spectrum in xenon shows that the observed effect is in fact due to the anisotropy of the *g*- and hyperfine-tensors. The two sets of values for the phenyl radical are in reasonable agreement.

The ^{13}C hyperfine couplings for the free valence carbon atom in the fully deuteriated phenyl radical have been observed.[236] The isotropic value (129 G) is smaller than that calculated by INDO theory[239] (151 G) with the assumption of sp^2 hybridization of the free valence orbital.

The 2-pyridyl and 2-pyrimidinyl radicals have also been prepared in the rotating cryostat.[236] The results for the 2-pyridyl radical trapped in deuteriated benzene at 77 K are in good agreement with those obtained in xenon[233] at 4 K. The ^{13}C hyperfine-coupling constants were observed for both radicals, the isotropic values being 171 and 241 G for 2-pyridyl and 2-pyrimidinyl, respectively. Thus, the spin density in the 2*s*-orbital of the valence carbon atom increases through the series phenyl (0.12), 2-pyridyl (0.15), and 2-pyrimidinyl (0.22).

The α-styryl radical $C_6H_5\dot{C}{=}CH_2$ has been prepared[97] by chemical reaction at 77 K. The e.s.r. spectrum displays a large triplet (1 : 2 : 1) with a coupling constant of 41.5 G, and a smaller quartet (1 : 3 : 3 : 1) of 6.1 G. It is concluded from the spectrum that the α-styryl radical has the structure of the α-methylenebenzyl radical, in which the unpaired electron is delocalized into the π-system of the aromatic ring and the two methylene protons lie in the antinodal plane of the π-orbital. Thus the α-styryl radical is a π-radical, in contrast to the benzoyl radical, which is a σ-radical. Confirmation of this conclusion is provided by the electronic spectrum of the α-styryl radical,[240] which is found to be very similar to those of the benzyl and α-methylbenzyl radicals, and also from INDO calculations.[67]

Hydrogen Atom Reactions. Free radicals have been formed[241] by the bombardment of solid uracil or dihydro-uracil with hydrogen atoms produced by flowing molecular hydrogen through an electric discharge. The dependence of the radical yield on the dose indicated that the free radicals were formed only in the surface layers of the powdered crystals.

Thermolysis.—Triphenylmethyl radicals are produced[206] when solid triphenylacetic acid is heated at 468 K or when sodium triphenylacetate is heated at 328 K.

236 J. E. Bennett and B. Mile, *J. Phys. Chem.*, 1971, **75**, 3432.
237 P. H. Kasai, E. Hedaya, and E. B. Whipple, *J. Amer. Chem. Soc.*, 1969, **91**, 4364.
238 J. E. Bennett, B. Mile, and A. Thomas, *Proc. Roy. Soc.*, 1966, **A293**, 246.
239 J. A. Pople, D. L. Beveridge, and P. A. Dobosh, *J. Amer. Chem. Soc.*, 1968, **90**, 4201.
240 B. Brocklehurst, J. S. Robinson, and D. N. Tawn, *Chem. Phys. Letters*, 1971, **12**, 610.
241 R. Bergene and T. Henricksen, *Acta Chem. Scand.*, 1972, **26**, 102.

Free radicals have also been detected[242] when malonic esters $R^1R^2C(CO_2Et)_2$ and cyanoacetic esters $R^1R^2C(CN)CO_2Et$ are heated at temperatures between 393 and 533 K.

Trapping from the Gas Phase.—A very elegant series of experiments has been reported on the trapping of small inorganic molecules in inert-gas matrices at 4 K. The e.s.r. spectra of the molecules MgF, CaF, SrF, and BaF, in their $^2\Sigma$ ground states in neon or argon matrices exhibited hyperfine coupling to the fluorine nucleus and, where applicable, to the metal nucleus as well.[243] The spectra from the neon matrices indicated that some preferential orientation of the molecules had occurred. In all cases less than 4% of the spin density is located on the F atom, showing that the molecules are highly ionic. The e.s.r. spectra of MgH, MgD, CaH, and SrH;[244] ZnH, CdH, HgH, and HgD;[245] BeH and BeD;[246] and PdH and PdD[247] have also been observed by this technique. The results for the alkaline-earth hydrides show that they are less ionic than the corresponding fluorides, but that the unpaired electron is still located mainly on the metal ion. The Group IIB hydrides (ZnH, CdH, and HgH) appear to be even less ionic, with about 35% of the spin density on the hydrogen atom in ZnH and CdH, and about 50% in HgH. Some extremely large metal hyperfine couplings are observed for ^{111}CdH, ^{113}CdH, ^{199}HgH, and ^{201}HgH; for the mercury compounds an effect of deuterium substitution on the metal-atom splitting is observed.

CdCN and HgCN have been prepared[248] and the results show that these cyanides are much more ionic than the corresponding hydrides, with a significant increase in the metal *s*-character relative to the hydrides.

The oxide molecules BO and AlO have been observed[249, 250] in their $^2\Sigma$ ground state in matrices of neon, argon, and krypton at 4 K. The BO molecule showed an extreme preferential orientation in neon. For BO the unpaired electron was located mainly in the 2*s*-orbital of boron ($\approx$ 51%) whereas for AlO the unpaired electron has 15—30% *s*-character and 40—70% *p*-character on Al. An additional species is observed with AlO which is tentatively attributed to a weakly bonded Al—X molecule (where X may be Al_2O), which may be considered as an isolated quenched aluminium atom.

The e.s.r. spectra of ScS and YS molecules in their $^2\Sigma^+$ ground states have also been observed,[251] and the earlier results on ScO and YO have been re-analysed. Again the unpaired electron is located mainly on

242 B. J. Kirkiacharian, *Bull. Soc. chim. France*, 1971, 1800.
243 L. B. Knight, W. C. Easley, W. Weltner, and M. Wilson, *J. Chem. Phys.*, 1971, **54**, 322.
244 L. B. Knight and W. Weltner, *J. Chem. Phys.*, 1971, **54**, 3875.
245 L. B. Knight and W. Weltner, *J. Chem. Phys.*, 1971, **55**, 2061.
246 L. B. Knight, J. M. Brom, and W. Weltner, *J. Chem. Phys.*, 1972, **56**, 1152.
247 L. B. Knight and W. Weltner, *J. Mol. Spectroscopy*, 1971, **40**, 317.
248 L. B. Knight and K. C. Lin, *J. Chem. Phys.*, 1972, **56**, 6044.
249 L. B. Knight, W. C. Easley, and W. Weltner, *J. Chem. Phys.*, 1971, **54**, 1610.
250 L. B. Knight and W. Weltner, *J. Chem. Phys.*, 1971, **55**, 5066.
251 N. S. McIntyre, K. C. Lin, and W. Weltner, *J. Chem. Phys.*, 1972, **56**, 5576.

the metal ion ($\approx$ 80%) and has about equal amounts of *s*- and p_σ-character (*i.e.* $\approx$ 40% each).

Copper, silver, and gold atoms [252] have been trapped in inert-gas matrices at 4 K. Matrix effects on the *g*-factors and hyperfine coupling constants of the trapped metal atoms are observed, and are similar to those previously reported for the alkali-metal atoms. However, in contrast to the multiple-trapping sites which occurred for the alkali-metal atoms,[253] no multiple-trapping sites were found for Cu, Ag, or Au in argon, krypton, or xenon, though some multiple-trapping sites exist in neon. In xenon a hyperfine interaction with the surrounding xenon atoms occurred for Cu, Ag, and Au atoms similar to that found for the alkali-metal atoms.

Over the past decade there have been a few reports of the trapping of polyatomic organic radicals from the gas phase, and the subsequent observation of their e.s.r. spectra at low temperature. However, the problems associated with the actual trapping process and the possibility of further reaction during condensation or after condensation are formidable, and no very extensive studies have been made. This technique has now been used to study the photochemical oxidation of hydrocarbons [254, 255] and the photolysis of alcohols [256] at low pressure in the gas phase. It is claimed [254] that $HO_2^{\bullet}$ can be distinguished from $RO_2^{\bullet}$ in the solid phase, but it seems unlikely that any definite assignment of the nature of the hydrocarbon moiety in $RO_2^{\bullet}$ could be made. Peroxyl radicals [255] attributed to $CH_3OO^{\bullet}$ and $CH_3CH_2OO^{\bullet}$ were trapped out from the photochemical oxidation of methane and ethane, respectively. For the photolysis [256] of methanol in the absence of oxygen, $\dot{C}H_2OH$ and $H\dot{C}O$ were trapped and identified; $CH_3\dot{C}HOH$ was trapped from ethanol, and it is claimed that $CH_3O^{\bullet}$ and $NO^{\bullet}$ were detected in the products from the photolysis of $MeNO_2$.

Neopentyl and t-butyl radicals have been trapped out at 77 K from neopentane which had been irradiated with low energy electrons (5—100 eV) in the gas phase.[257] Both neopentyl and t-butyl radicals are produced at electron energies in excess of the ionization potential of neopentane, whereas below this value only t-butyl radicals are observed. The dependence of the concentration of trapped radicals on electron energy and neopentane pressure was examined, and a reaction mechanism is given to explain the results.

[252] P. H. Kasai and D. McLeod, *J. Chem. Phys.*, 1971, **55**, 1566.
[253] C. K. Jen, V. A. Bowers, E. L. Cochran, and S. N. Foner, *Phys. Rev.*, 1962, **126**, 1749.
[254] T. A. Garibyan, A. A. Mantashyan, and A. B. Nalbandyan, *Armyan. khim. Zhur.*, 1971, **24**, 304.
[255] A. A. Mantashyan, M. A. Beibutyan, A. S. Saakyan, and A. B. Nalbandyan, *Doklady Akad. Nauk S.S.S.R.*, 1972, **202**, 120.
[256] T. A. Garibyan, A. A. Mantashyan, A. B. Nalbandyan, and A. S. Saakyan, *Armyan. khim. Zhur.*, 1971, **24**, 13.
[257] R. Marx, G. Mauclaire, and M. Wallart, *J. Chim. phys.*, 1971, **68**, 1680.

7
Transition-metal Ions

BY J. F. GIBSON

1 Introduction

Transition-metal e.s.r. is very much the realm of physics as well as that of chemistry, a fact which makes it a little difficult to prepare a report which is intended primarily for chemists. In order to keep the Report down to a reasonable size, the Reporter has omitted those papers, most of them concerned with impurity centres (many of these rare-earth ions) in host lattices, which seemed to be of interest primarily to physicists. Those papers which for full coverage were felt worthy of inclusion but did not justify a sentence in the text and those which are not readily available in a Western language are tabulated at the end.

The book by Wertz and Bolton [1] has appeared during the period of this review. It is comforting to find in one e.s.r. text a detailed account of transition metals as well as a good coverage of free radicals. Most authors shy away from the former. However, it is a pity that for simplicity the examples are confined mostly to ions in highly symmetric crystal fields and therefore are more pertinent to the physicist than to the chemist. Mention is made of the comprehensive review by Goodman and Raynor [2] since, although earlier than 1971, it was too late to be mentioned in the last Chemical Society Specialist Periodical Report on this topic.[3] The review on the stereochemistry of bis-chelate metal(II) complexes [4] includes much e.s.r. work from the past ten years or so, updating the 1968 review by Kuska and Rogers. It covers co-ordination number four only, excluding phthalocyanines, porphyrins, and the like. A short review of the past applications of e.s.r. and of n.m.r. to transition-metal complexes has appeared.[5] Another, of particular relevance to analytical chemists, covers the direct determination of transition-metal ions, alone or in mixtures, detection of titration end-points, equilibria, and cobalt–oxygen complexes.[6]

[1] J. E. Wertz and J. R. Bolton, 'Electron Spin Resonance. Elementary Theory and Practical Applications', McGraw-Hill, New York, 1972.
[2] B. A. Goodman and J. B. Raynor, *Adv. Inorg. Chem. Radiochem.*, 1970, **13**, 135.
[3] 'Spectroscopic Properties of Inorganic and Organometallic Compounds', ed. N. N. Greenwood, (Specialist Periodical Reports), The Chemical Society, London, 1970, vol. 3.
[4] R. H. Holm and M. J. O'Connor, *Progr. Inorg. Chem.*, 1971, **14**, 241.
[5] D. R. Eaton and K. Zaw, *Coordination Chem. Rev.*, 1971, **7**, 197.
[6] E. G. Janzen, *Analyt. Chem.*, 1972, **44**, 113R.

Some of the papers which were presented at the Fourth International Symposium on Magnetic Resonance, in Israel, August 1971, have been published together (see the introduction by Brey [7]).

In this Report, where e.s.r. parameters are given they refer to the spin hamiltonian

$$\mathscr{H} = \beta[g_z H_z S_z + g_x H_x S_x + g_y H_y S_y] + AS_z I_z + BS_x I_x + CS_y I_y + D[S_z^2 - \tfrac{1}{3}S(S+1)] + E(S_x^2 - S_y^2) + \tfrac{1}{6}a[S_x^4 + S_y^4 + S_z^4 - \tfrac{1}{5}S(S+1)(3S^2 + 3S - 1)] \quad (1)$$

The quartic term in F is not quoted, but where spin operator coefficients are given, the relationships are: $b_2^0 = D$; $b_2^2 = 3E$; $b_4^0 = a/2 + F/3$; $b_4^4 = 5/2a$. In axial symmetry: $g_z = g_\parallel$; $g_x = g_y = g_\perp$; $C = B$; $E = 0$. In cubic symmetry $D = 0$ and we use g and A.

2 General

Theory and Techniques.—More interest has been shown in the importance of core polarization to the contribution of hyperfine structure. The root of this phenomenon lies in the exchange interaction which, considering an unpaired α-spin in the valence shell, usually increases β-spin density at the nucleus by attracting α-spin away from that region thereby giving a negative contribution to the contact term. McMillan and Halpern [8] have chosen to estimate the experimental core polarization for some tetragonal (^2D) states, because these have no complicating zero-field splitting terms and because it is possible to estimate the extent of electron delocalization from the central metal atom. They used $K = \alpha^2 K_O + (1 - \alpha^2) K_L$, where K is the actual contact term derived from the experimental one with a knowledge of the g shift and $\langle r^{-3}\rangle$ from Hartree–Fock calculations, α is the coefficient of $|\psi_0\rangle$ the metal wavefunction, K_O is the expectation value of the core polarization operator over the metal orbitals (the experimental core polarization), and K_L is the corresponding value over the ligand orbitals. In this paper they have analysed numerous literature data to estimate K_O and K_L for series of $3d$ and $4d$ compounds, finding experimental values of core polarization within 20% of the theoretical estimates. Later papers from the same laboratory confirm these findings for $Cu^{2+}(3d^9)$ [9] and for $Ag^{2+}(4d^9)$ [10] which (in a series of frozen acid solutions) has a K_L almost as large as K_O, indicating that the electron density in the ligands can play an important role in polarizing the core.

A simple method for the analysis of large hyperfine splittings in $S = \frac{1}{2}$ systems has been given by Weil,[11] starting with Breit–Rabi formulation and

[7] W. S. Brey, *J. Magn. Resonance*, 1972, **6**, i.
[8] J. A. McMillan and T. Halpern, *J. Chem. Phys.*, 1971, **55**, 33.
[9] J. A. McMillan, N. S. Dixon, and T. Halpern, *J. Chem. Phys.*, 1971, **55**, 452; G. C. Schatz and J. A. McMillan, *ibid.*, p. 2342.
[10] J. A. McMillan and G. C. Munie, *J. Chem. Phys.*, 1972, **56**, 113.
[11] J. A. Weil, *J. Magn. Resonance*, 1971, **4**, 394.

bringing in anisotropies as perturbations. In this connection, a particularly useful feature of the paper by Booth *et al.*[12] on unstable intermediates containing mercury is the analysis of a dimer species in terms of the full Breit–Rabi equation.

Fackler *et al.*[13] have shown how anisotropic e.s.r. parameters of some copper(II) and oxovanadium(IV) complexes may easily be obtained from

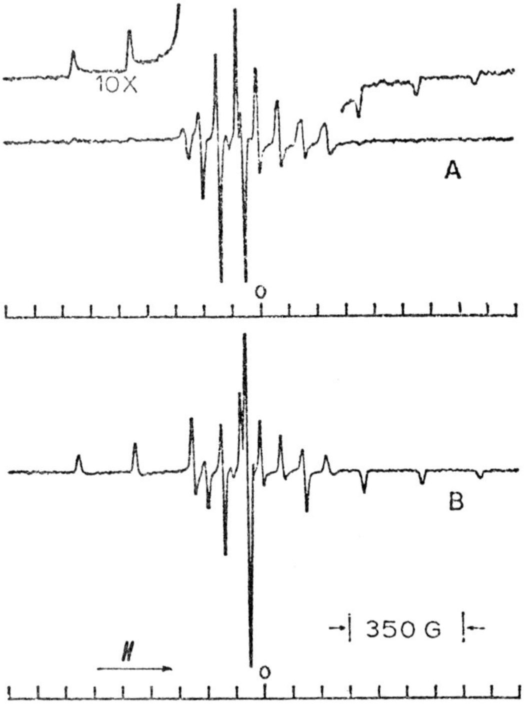

Figure 1 *E.s.r. spectra of* VO(sal)$_2$(pn) *oriented in a nematic glass at* − 150 °C: A, 0° *rotation;* B, 90° *rotation*
(Reproduced by permission from *J. Amer. Chem. Soc.*, 1972, **94**, 2436)

solutions in nematic glasses from liquid crystal solvents. The results compare favourably with diluted single-crystal measurements in the literature; they are better than results from frozen solution or powder spectra. However, one slight disadvantage is found in the weak interaction with the solvent. Such interaction is common for these complexes in basic solvents. The ease with which the parallel and perpendicular features for *NN'*-propylenebis(salicylaldiminato)oxovanadium(IV) may be seen is shown in Figure 1.

[12] R. J. Booth, H. C. Starkie, and M. C. R. Symons, *J. Chem. Soc.* (*A*), 1971, 3198.
[13] J. P. Fackler, jun., D. Levy, and J. A. Smith, *J. Amer. Chem. Soc.*, 1972, **94**, 2436.

E.s.r. spectroscopists should heed the warning [14] that the orientation of the **g** tensor is often not what might be expected from a knowledge of the first co-ordination sphere. Gerloch and Quested quote as an example *trans*-bis(mesityl)bis(triphenylphosphine)cobalt(II), which has a predominantly square-planar crystal field but with the unique axis of the **g** tensor

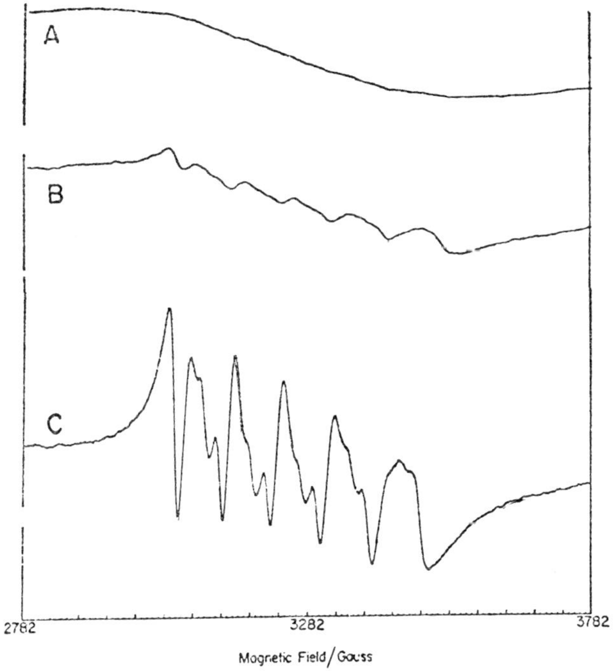

Figure 2 *E.s.r. spectra for* 5 × 10^{-4} mol l^{-1} *solution of* $MnCl_2$: A, *slowly frozen;* B, *rapidly frozen;* C, *frozen in Sephadex gel*
(Reproduced by permission from *J. Phys. Chem.*, 1971, **75**, 1203)

lying in the co-ordination plane rather than normal to it. The non-alignment of the **g** and fine-structure tensor axes of some copper compounds has also been reported.[15]

It is well known that solvent–solute segregation sometimes occurs in frozen solutions, resulting in high local concentrations of paramagnetic species which exhibit strong dipolar and exchange interactions. Cockle *et al.*[16] found that addition of a 1.0 mol l^{-1} of polar organic reagents containing such groups as OH, NH_2, or SH causes appreciable improvement in the spectra of cobalt(II) corrin adducts. The effect is thought to be

[14] M. Gerloch and P. N. Quested, *J. Chem. Soc.* (*A*), 1971, 2307.
[15] E. Buluggin, G. Dascola, D. C. Giori, and V. Varacca, *Phys. Status Solidi* (*B*), 1971, **45**, 217.
[16] S. Cockle, H. A. O. Hill, S. Ridsdale, and R. J. P. Williams, *J. C. S. Dalton*, 1972, 297.

due to structure formation of the solvent water. Leigh and Reed[17] point out that these effects are partially removed by rapid freezing but that embedding the aqueous solutions in a polydextran gel prior to freezing gives a spectacular improvement. Figure 2 shows such spectra for aqueous manganous chloride.

Equilibria and Reaction Rates.—Evidence for a short-lived 1 : 1 adduct between piperidine, n-hexylamine, or pyridine and the square-planar bis(di-n-butyldithiocarbamato)copper(II) has been produced.[18] Equilibrium constants for this formation reaction, and adduct lifetimes, were obtained from analysis of the e.s.r. line positions and linewidths. The e.s.r. spectrum of the vanadyl ion in the presence of lactic or glycollic acids indicates the existence of at least six different species, including one dimer, over the pH range 1—7. Complex formation equilibrium constants were estimated, and acid dissociation constants were measured for the hydroxyl protons in these species.[19]

Water-exchange rates with $[Mn(H_2O)_6]^{2+}$, $[Mn\,phen\,(H_2O)_4]^{2+}$, and $[Mn\,phen_2(H_2O)_2]^{2+}$ have been reported.[20] Substitution of phenanthroline on manganese(II) labilizes the co-ordinated water molecules, but the long relaxation times make precise kinetic studies rather difficult. The equilibria between $[Fe(NO)_2(EtOH)_2]^+$, $[Fe(NO)_2(EtOH)(EtO)]$, and $[Fe(NO)_2(EtO)_2]^-$ depend on pH and on bromide concentration.[21] The kinetics of the multistep aquation of $[Co(CN)_5NO]^{3-}$ have also been studied.[22]

A continuous flow method was used to monitor vanadyl concentration and that of an intermediate compound in its oxidation reaction with hydrogen peroxide.[23] Concerning such reactions as titanium(III) with peroxide, it has been shown incorrect to assume that the reaction responsible for radical formation in flow techniques is complete before the solution enters the cavity.[24]

Zeolites.—Copper(II) ions may be distributed over at least two different sites in X-type zeolites. When copper in one of these sites is reduced by but-1-ene, its e.s.r. signal disappears and does not reappear on reoxidation; instead, the appearance of a new e.s.r. signal shows that migration of copper ions has taken place during the but-1-ene treatment.[25] Bis(acetylacetonato)copper(II) forms adducts with aniline or ammonia when adsorbed on to X-type zeolite, but does not interact strongly with the zeolite surface.[26]

$[Cu(H_2O)_6]^{2+}$ becomes trapped in the centre of a supercage in Y-type zeolite. Dehydration causes migration to a sodalite cage, and subsequent

[17] J. S. Leigh, jun. and G. H. Reed, *J. Phys. Chem.*, 1971, **75**, 1202.
[18] B. J. Corden and P. H. Rieger, *Inorg. Chem.*, 1971, **10**, 263.
[19] R. R. Reeder and P. H. Rieger, *Inorg. Chem.*, 1971, **10**, 1259.
[20] M. Grant, H. W. Dodgen, and J. P. Hunt, *Inorg. Chem.*, 1971, **10**, 71.
[21] G. Martini and E. Tiezzi, *Trans. Faraday Soc.*, 1971, **67**, 2538.
[22] A. T. Nikitaev, A. Y. Sychev, and K. I. Zamarev, *Zhur. neorg. Khim.*, 1972, **17**, 985.
[23] H. B. Brooks and F. Sicilio, *Inorg. Chem.*, 1971, **10**, 2530.
[24] G. Czapski, *J. Phys. Chem.*, 1971, **75**, 2957.
[25] I. R. Leith, C. Kemball, and H. F. Leach, *Chem. Comm.*, 1971, 407.
[26] Y. Yamada, *Bull. Chem. Soc. Japan*, 1972, **45**, 60, 64.

reaction with ammonia, *etc.* causes co-ordination and migration back to the supercage.[27] Introduction of CO to Cu^{2+}-exchanged Y-zeolites causes reduction to Cu^{+},[28] and treatment of this with pyridine gives a centre containing four nitrogen atoms,[29] presumably $[Cu(py)_4]^{2+}$. Yamada has also reported formation of $[Cu(py)_4]^{2+}$ in X-zeolite by substitution.[26] Introduction of NO to Cu^{+} on Y-zeolite causes reappearance of e.s.r. signals for which the unpaired electron is essentially on NO.[30] Studies of chromium[31] and cobalt[32] ions on zeolites, and copper[33] ions on cation exchange resins, have also been reported.

Nitrosyl and Free-radical Complexes.—The saga of the irradiation product of sodium nitroprusside continues. Bloom *et al.*[34] found several species in the γ-irradiated hydrated salt: two of these are thought to be identical {essentially $[Fe^{2+}(CN)_5NO^{\bullet}]^{3-}$ with the unpaired electron mainly on the NO} but differing in the extent of libration; a third, originally thought to be $[Fe^{+}(CN)_5NO^{+}]^{3-}$, is now thought to be $[(CN{-})Fe^{+}({-}CN)_4NO^{+}]^{3-}$ or $[(HCN{-})Fe^{+}({-}CN)_4NO^{+}]^{3-}$ but still with the unpaired electron essentially in a d_{z^2} orbital. This last species was also thought to arise from γ-irradiation of $[Fe(CN)_5NO]^{2-}$ incorporated into an alkali-halide lattice.[35] The same authors found only two species in the γ-irradiated anhydrous salt;[36] the one of interest here is probably $[Fe^{3+}(CN)_5NO^{+}]^{-}$, a low-spin d^5 system with $g_{\parallel} = 2.0142$, $g_{\perp} = 2.0716$. In their study of compound (1),[37] the authors suggest $Cr^{2+}NO^{\bullet}$ might be a better formulation than $Cr^{+}NO^{+}$.

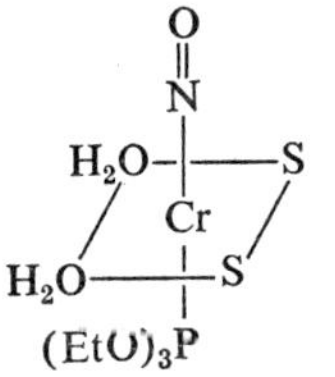

(1) S = S_2COEt or $S_2P(OEt)_2$

[27] C. M. Naccache and Y. Ben Taarit, *Chem. Phys. Letters*, 1971, **11**, 11.
[28] C. M. Naccache and Y. Ben Taarit, *J. Catalysis*, 1971, **22**, 171.
[29] J. Turkevich, Y. Ono, and J. Soria, *J. Catalysis*, 1972, **25**, 44.
[30] C. M. Naccache, M. Che, and Y. Ben Taarit, *Chem. Phys. Letters*, 1972, **13**, 109; C. C. Chao and J. H. Lunsford, *J. Phys. Chem.*, 1972, **76**, 1546.
[31] I. D. Mikheikin, O. I. Brotikovskii, and V. B. Kazanskii, *Kinetika i Kataliz*, 1972, **13**, 527.
[32] I. D. Mikheikin, O. I. Brotikovskii, G. M. Zhidomirov, and V. B. Kazanskii, *Kinetika i Kataliz*, 1971, **12**, 1442.
[33] K. Umezawa and T. Yamabe, *Bull. Chem. Soc. Japan*, 1972, **45**, 56.
[34] M. B. D. Bloom, J. B. Raynor, and M. C. R. Symons, *J. Chem. Soc.* (*A*), 1971, 3843.
[35] M. B. D. Bloom, J. B. Raynor, K. D. J. Root, and M. C. R. Symons, *J. Chem. Soc.* (*A*), 1971, 3212.
[36] M. B. D. Bloom, J. B. Raynor, and M. C. R. Symons, *J. Chem. Soc.* (*A*), 1971, 3209.
[37] N. S. Garif'yanov, A. D. Troitskaya, A. I. Razumov, Ts. A. Gurevich, and O. I. Kondrat'eva, *Russ. J. Inorg. Chem.*, 1971, **16**, 563.

Other compounds which have been studied include nitrosyls of iron,[38] chromium,[39] and manganese.[35]

Baird has reported a paramagnetic rhodium(II) nitrosyl compound $RhCl_3(PPh_2)_2NO$ which has $g_{\parallel} = 2.03$, $g_{\perp} = 2.01$; no hyperfine splitting was mentioned.[40] The photolysis of luminescent rhodium(III) 1,10-phenanthroline chelates gave metal-containing radicals with $g \approx 2.16$. No rhodium hyperfine splitting was reported, however, and the free electron is probably largely on the phenanthroline ligand.[41]

Radicals prepared by reduction of aromatic nitro- and nitroso-compounds with $[(CN)_5Co]^{3-}$ form species of the type $[^{\bullet}O(Ar)NCo(CN)_5]^{3-}$ in which the unpaired electron is largely in the aromatic π-antibonding orbital and d_{xz}. The d_{z^2} orbital is said to form a σ-bond to the nitrogen and a comparison with $[Fe(CN)_5NO]^{3-}$ is drawn.[42] Other systems which are more like free radicals chelated to metal ions have also been studied.[43]

Polynuclear Complexes.—Although not primarily intended for e.s.r., Griffith's theory of polynuclear compounds having two or three spins is of interest. Systems such as the nickel acetylacetonate trimer and the copper acetate dimer in which the constituent ions have spin degeneracy but not orbital degeneracy are covered. Weak interactions such as magnetic dipole–dipole are ignored.[44]

Dinuclear. First we consider two dinuclear systems which are not exchange coupled. The first is a dinuclear platinum species with one unpaired electron; its spectrum is shown in Figure 3. The spectrum arises from: (a) pairs with both nuclei having zero spin; (b) pairs with one nucleus only having spin ($S = \frac{1}{2}$); (c) pairs with both nuclei having spin ($S = \frac{1}{2}$). This is a good example of the need to consider second-order hyperfine interaction which causes the line with $|I, m_I\rangle = |1,0\rangle$ to be shifted from $|0,0\rangle$. The experimental relative intensities agree with the calculated ones within 5%. The species is believed to be $[(PtCl_4)_2]^{3-}$, formed by γ-irradiation of K_2PtCl_4.[45] The second is a diamagnetic sulphur-bridged cobalt(III) thioxanthate complex, which may be reduced by one electron. Although no e.s.r. is discussed, a rationale is suggested for the low and varying redox potentials in non-haem iron–sulphur proteins, many of which have been first characterized by their e.s.r. spectra.[46]

[38] J. P. Crow, W. R. Cullen, G. F. Herring, J. R. Sams, and R. C. Tapping, *Inorg. Chem.*, 1971, **10**, 1617; R. D. Feltham, W. Silverthorn, H. Wickman, and W. Wesolowski, *ibid.*, 1972, **11**, 676; W. V. Sweeney and R. E. Coffman, *J. Phys. Chem.*, 1972, **76**, 49.
[39] N. S. Garif'yanov and O. I. Kondrat'eva, *Teor. i eksp. Khim.*, 1971, **7**, 258.
[40] M. C. Baird, *Inorg. Chim. Acta*, 1971, **5**, 46.
[41] K. DeArmond and W. Halper, *J. Phys. Chem.*, 1971, **75**, 3230.
[42] J. Basters, *Rec. Trav. chim.*, 1972, **91**, 50.
[43] P. Clopath and Av. Zelewsky, *Chem. Comm.*, 1971, 47; D. G. Brown, T. Maier, and R. S. Drago, *Inorg. Chem.*, 1971, **10**, 2804; D. Jahr, K. E. Schwarzhaus, D. Nöthe, and P. K. Burkert, *Z. Naturforsch.*, 1971, **26b**, 1210; A. Tkáč, K. Veselý, and L. Omelka, *J. Phys. Chem.*, 1971, **75**, 2575.
[44] J. S. Griffith, *Structure and Bonding*, 1972, **10**, 87.
[45] T. Krigas and M. T. Rogers, *J. Chem. Phys.*, 1971, **55**, 3035.
[46] D. F. Lewis, S. J. Lippard, and J. A. Zubieta, *J. Amer. Chem. Soc.*, 1972, **94**, 1563.

Next we consider some dimers in which two $S = \frac{1}{2}$ centres (for example pairs of Cu^{2+}, VO^{2+}, Ti^{3+}, Mo^{5+}), although possibly weakly exchange-coupled, appear to interact only by a magnetic dipole–dipole mechanism. Dimer formation is usually confirmed by the observation of a $\Delta M = 2$ transition with appropriate hyperfine splitting at half-field. The Pilbrow

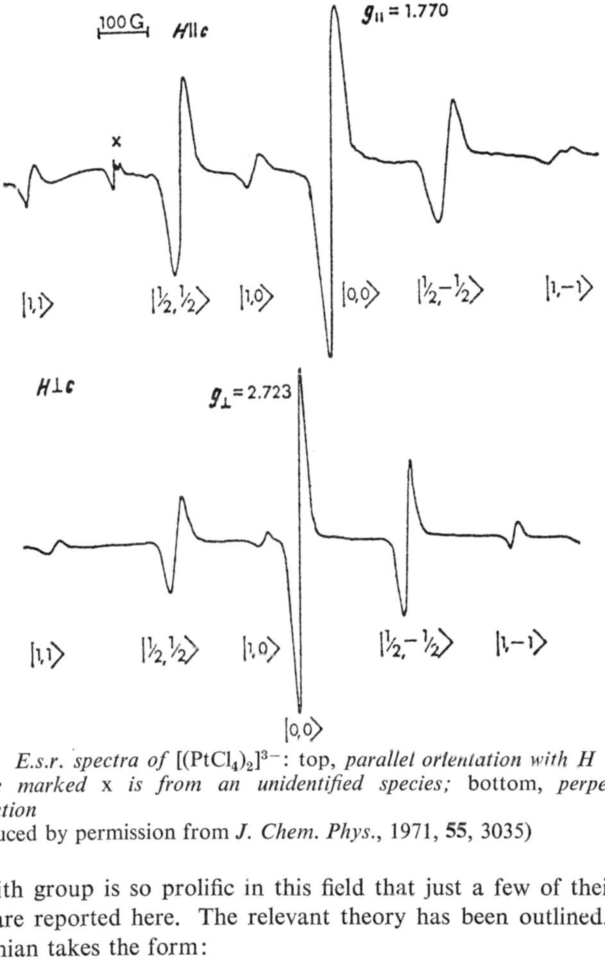

Figure 3 *E.s.r. spectra of* $[(PtCl_4)_2]^{3-}$: top, *parallel orientation with* $H \parallel c$; *the feature marked* x *is from an unidentified species;* bottom, *perpendicular orientation*
(Reproduced by permission from *J. Chem. Phys.*, 1971, **55**, 3035)

and Smith group is so prolific in this field that just a few of their latest papers are reported here. The relevant theory has been outlined.[47] The hamiltonian takes the form:

$$\mathscr{H} = \mathscr{H}_1 + \mathscr{H}_2 + \mathscr{H}_{dd} \quad (2)$$

in which $\mathscr{H}_1$ and $\mathscr{H}_2$, usually assumed axial with a common axis, refer to the isolated ions 1 and 2, and $\mathscr{H}_{dd}$ represents the magnetic dipole–dipole

[47] P. D. W. Boyd, T. D. Smith, J. H. Price, and J. R. Pilbrow, *J. Chem. Phys.*, 1972, **56**, 1253.

interaction (parameters D and E). Exchange interaction is assumed negligible and internuclear distances as low as 360 pm in dimeric vanadyl 1-hydroxycyclohexanecarboxylate,[48] or 400 pm in bis[Cu(monothio-β-diketone)$_2$],[49] are calculated on this basis. A [molybdenum(v)]$_2$ glutathione diamagnetic dimer has been reported to be in equilibrium with a paramagnetic dinuclear species at pH 8.[50] Enrichment with ^{95}Mo shows an eleven-line hyperfine splitting with $A = 0.0029\ cm^{-1}$. The Mo—Mo distance is calculated to be 600 pm. There is a puzzling reference to this system being a model system for xanthine oxidase in which, so far as the reporter is aware, e.s.r. evidence is for mononuclear molybdenum(v) only. Carr *et al.* have commented on this paper,[51] suggesting that a non-parallel axis alignment should be considered in the analysis. Hasegawa *et al.* have used the analysis to estimate the V—V distance, which increases from 420 to 650 pm as the pH is lowered, in dinuclear vanadyl tartrates.[52] This lengthening is interpreted as dissociation of a vanadium to (carboxyl) oxygen bond, caused by association (to the acidic form) of the carboxy-group. Two of the most recent reports from Smith and Pilbrow are concerned with some dimeric copper(II) and vanadyl chelates of 4,4′,4″,4‴-tetrasulphophthalocyanine[53] and polynuclear copper(II) chelates of tetrakis(aminomethyl)methane.[54]

Now we consider pairs of ions which are exchange-coupled. For short internuclear distances spin–spin exchange coupling becomes important and the term $JS_1 \cdot S_2$ is added to the hamiltonian of equation (2). Such systems are classified according to whether the exchange is ferromagnetic (J positive) or antiferromagnetic (J negative). Hatfield contends that for some oxygen-, sulphur-, and chlorine-bridged dinuclear copper complexes at least, the one is about as likely as the other. The Cu—Cu distance in all his examples is $<$ *ca.* 400 pm.[55] When exchange interaction is not negligible, it is no longer a simple matter to compute internuclear distances on the basis that the zero-field splitting D arises from $\mathscr{H}_{dd}$ alone. D may be the sum of two parts: D_{dd} arising from magnetic dipole–dipole interaction; and D_{pseudo} arising from exchange, through spin–orbit coupling. Chasteen, for the dinuclear copper n-butyrate monohydrate, finds $g_\parallel = 2.342$, $g_\perp = 2.063$, $|D| = 0.333\ cm^{-1}$, $|E| = 0.011\ cm^{-1}$, with D_{dd} and D_{pseudo} assuming roughly equal importance.[56]

One advantage of studying antiferromagnetically coupled pairs is that better resolution may be obtained at low temperatures, where the increased

[48] T. D. Smith, T. Lund, J. R. Pilbrow, and J. H. Price, *J. Chem. Soc.* (*A*), 1971, 2936.
[49] A. D. Toy, S. H. H. Chaston, J. H. Pilbrow, and T. D. Smith, *Inorg. Chem.*, 1971, **10**, 2219.
[50] T. J. Huang and G. P. Haight, jun., *J. Amer. Chem. Soc.*, 1971, **93**, 611.
[51] S. G. Carr, P. D. W. Boyd, and T. D. Smith, *J. C. S. Dalton*, 1972, 907.
[52] A. Hasegawa, Y. Yamada, and M. Miura, *Bull. Chem. Soc. Japan*, 1971, **44**, 3335.
[53] P. D. W. Boyd and T. D. Smith, *J. C. S. Dalton*, 1972, 839.
[54] T. D. Smith and A. E. Martell, *J. Amer. Chem. Soc.*, 1972, **94**, 3029.
[55] W. E. Hatfield, *Inorg. Chem.*, 1972, **11**, 216.
[56] N. D. Chasteen, *Inorg. Chem.*, 1971, **10**, 2339.

population of the singlet state has the effect of magnetic dilution of the remaining triplet species. Miners *et al.* made this point in reporting the e.s.r. in some nitrato-complexes of copper(II) with bidentate Schiff bases.[57] The latest news of the singlet–triplet–singlet model for dinuclear copper(II) carboxylates has been reported by Jotham *et al.*[58] who believe the spin coupling to be indirect, arising from mutual overlap of the two metal ions with suitable orbitals on the bridging ligands. Herring *et al.*, in a magnetic susceptibility and e.s.r. study of fifteen adducts of copper(II) arylcarboxylates with aniline and pyridine, found J in the range -290 to $-355\ cm^{-1}$. Their data are adequately accommodated by the Bleaney–Bowers singlet–triplet model but do not disprove the Jotham–Kettle model. The difficulty in finding a dinuclear species which is *completely* free of monomeric copper(II) was mentioned.[59] Belford *et al.* have reported $J = -18\ cm^{-1}$ in a stereospecific copper tartrate dimer which in solution is in simple equilibrium with a monomer.[60] Fournier *et al.* have reported Cr^{3+}–Cr^{3+} pairs in phosphate glass; J is about $-24\ cm^{-1}$ with zero-field splittings mostly about $0.03\ cm^{-1}$ but with some as high as $0.3\ cm^{-1}$ to account for absorption in the wings of the spectrum.[61]

We turn now to ferromagnetically coupled pairs. A series of hydroxo-bridged complexes of 2,2′-bipyridyl and copper(II) has been reported with $J > +20\ cm^{-1}$; for these a direct exchange mechanism is favoured because the internuclear distance is only 285 pm. Of several mechanisms discussed, the preferred one carries a β-spin from a filled orbital on atom A to a half-filled orbital on atom B, the initial and final half-filled orbitals having α-spin.[62] Three papers have been concerned with coupled pairs in bis(*NN*′-diethyldithiocarbamato)copper(II) diluted in the corresponding zinc compound. Cowsik *et al.* found $D = 0.0276\ cm^{-1}$, $E \approx 0\ cm^{-1}$ arising from dipole–dipole interaction, the principal axis of the **D** tensor being approximately along the Cu—Cu direction. The internuclear distance is 359 pm.[63] Villa and Hatfield[64a] also assumed dipole–dipole interaction only, but in their note[64b] reported $J = +12\ cm^{-1}$, with the two unpaired electrons in $d_{x^2-y^2}$ orbitals interacting *via* a σ-orbital mechanism. The same authors felt that the Cu—Cu distance of 380 pm in tetrakis(*NN*′-dimethylglyoximato)dicopper(II) is too large for either direct spin exchange or dipolar exchange; J in this case was larger at $+15\ cm^{-1}$ and a super-exchange mechanism using σ-orbitals was proposed.[64c] The first triplet

57 J. O. Miners, E. Sinn, R. B. Coles, and C. M. Harris, *J. C. S. Dalton*, 1972, 1149.
58 R. W. Jotham, S. F. A. Kettle, and J. A. Marks, *J. C. S. Dalton*, 1972, 428, 1133.
59 F. G. Herring, B. Landa, R. C. Thompson, and C. F. Schwerdtfeger, *J. Chem. Soc. (A)*, 1971, 528.
60 R. L. Belford, R. J. Missavage, I. C. Paul, N. D. Chasteen, W. E. Hatfield, and J. F. Villa, *Chem. Comm.*, 1971, 508.
61 J. T. Fournier, R. J. Landry, and R. H. Bartram, *J. Chem. Phys.*, 1971, **55**, 2522.
62 J. A. Barnes, D. J. Hodgson, and W. E. Hatfield, *Inorg. Chem.*, 1972, **11**, 144.
63 R. K. Cowsik, G. Rangarajan, and R. Srinivasan, *Chem. Phys. Letters*, 1971, **8**, 136.
64 (*a*) J. F. Villa and W. E. Hatfield, *Inorg. Chim. Acta*, 1971, **5**, 145; (*b*) *Chem. Comm.*, 1971, 101; (*c*) *J. Chem. Phys.*, 1971, **55**, 4758.

species containing silver has been reported; this is bis(*NN'*-di-isopropyldithiodicarbamato)silver(II) in the zinc salt. The **D** tensor components are much larger than those found in the corresponding Cu—Cu dimer in which the interaction is entirely dipolar. It follows that there are significant contributions from spin–orbit coupling, a proposal which is supported by the peculiar orientations of the principal axes of **D**, which is not axially symmetric around the Ag—Ag direction.[65] Ferromagnetic exchange has also been studied between Ni^{2+}–Ni^{2+} pairs[57] and between Co^{2+}–Co^{2+} pairs.[66]

A different kind of exchange interaction has been described in systems where S state ions have been doped into samarium chalcogenides.[67] The Sm^{2+} ion has a diamagnetic ($J = 0$) ground state with a paramagnetic excited state ($J = 1$) at about 300 cm^{-1}. The large g-shift for Mn^{2+} to 1.885 is interpreted as due to Mn^{2+}–Sm^{2+} antiferromagnetic exchange and a similar large shift for Eu^{2+} to 2.241 is due to Eu^{2+}–Sm^{2+} ferromagnetic exchange.

Trinuclear. Exchange-coupled triads of chromium ions in $[Cr_3O(carboxylate)_6(H_2O)_3]^+$ have been studied. In these compounds, each Cr^{3+} is in a tetragonally distorted octahedron with $S = \frac{3}{2}$. They have complex spectra above 77 K, but at 4.2 K a single, narrow, slightly asymmetric line is seen in the low region of $g = 1.90$. The preferred theory which explains these facts requires that the J_{ij} (exchange integral between a given pair in the triangular cluster) are not all the same. This could occur for very small differences in interatomic distances in the trimers, a postulate not yet confirmed by structural studies.[68] The paper by Cotton *et al.* on oxotriruthenium clusters might be of interest since, although it contains no e.s.r. data, a simple MO treatment is given.[69] (Vanadyl pyrophosphate)$_3{}^{6-}$ exists in solution as evidenced by its hyperfine splitting of 22 lines; it has $S = \frac{3}{2}$. The analysis was based upon that for a nitroxyl triradical and required that the resonance came mainly from the quartet state and partly from the two doublet states. The exchange term $|J|$ was much greater than the hyperfine term $|A|$.[70] Hudson has discussed exactly this case.[71] Tsukerblat *et al.* have also discussed the Heisenberg model of spin exchange in trimer clusters.[72]

[65] J. G. M. Van Reus, E. Van der Drift, and E. de Boer, *Chem. Phys. Letters*, 1972, **14**, 113.
[66] A. G. Anders, A. I. Zvyagin, and E. N. Khats'ko, *Soviet Phys. Solid State*, 1972, **13**, 2347.
[67] F. Mehran, K. W. H. Stevens, R. S. Title, and F. Holtzberg, *Phys. Rev. Letters*, 1971, **27**, 1368.
[68] Yu. V. Yablokov, V. A. Gaponenko, and V. V. Zelentsov, *Soviet Phys. Solid State*, 1971, **13**, 1081.
[69] F. A. Cotton, J. C. Norman, A. Spencer, and G. Wilkinson, *Chem. Comm.*, 1971, 967.
[70] A. Hasegawa, *J. Chem. Phys.*, 1971, **55**, 3101.
[71] A. Hudson, *Mol. Phys.*, 1971, **21**, 61.
[72] B. S. Tsukerblat, M. I. Belinskii, and A. V. Ablov, *Doklady Akad. Nauk S.S.S.R.*, 1971, **201**, 1410.

3 $S = \frac{1}{2}$

s^1 Configuration.—*Copper, Silver, and Gold Atoms.* Isotropic e.s.r. spectra have been observed from Cu, Ag, and Au atoms isolated in regular octahedral sites in Ne, Ar, Kr, and Xe matrices at 4.2 K.[73] The hyperfine splitting from electrons wholly in an *s*-orbital is naturally very large, and its analysis requires the general Breit–Rabi formula. Figure 4 shows how

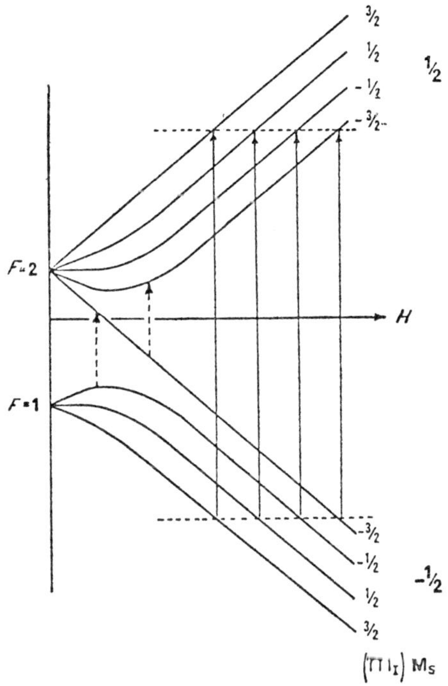

Figure 4 *Splitting of energy levels of a system with $S = \frac{1}{2}$ and $I = \frac{1}{2}$ in external magnetic field. Observable transitions are indicated for the two cases $h\nu > 2A$ and $h\nu < 2A$*
(Reproduced by permission from *J. Chem. Phys.*, 1971, **55**, 1566)

the energy levels corresponding to a copper atom with *A* isotropic would diverge. The dotted lines show the possibility of observing 'n.m.r.' transitions at X-band frequencies if *A* is large enough. Figure 5 shows the e.s.r. spectrum of Cu atoms in natural abundance isolated in a krypton matrix; it clearly does correspond to this situation. The signals below 2000 G are assigned to the 'n.m.r.' transition $M_s = -\frac{1}{2}$, $m_I = -\frac{3}{2} \leftrightarrow -\frac{1}{2}$ while the signals near 6000 G are assigned to the 'e.s.r.' transition $\Delta M_s = \pm 1$, $m_I = -\frac{3}{2}$; in each case the stronger line arises from ^{63}Cu, the weaker

[73] P. H. Kasai and D. McLeod, jun., *J. Chem. Phys.*, 1971, **55**, 1566.

from ^{65}Cu. This spectrum was run in dispersion mode and clearly shows up the opposite signs associated with the magnetic moments of the copper nucleus and its 4*s* electron. The A_{iso} values for ^{63}Cu, ^{107}Ag, and ^{197}Au are given as 6043, 1777, and 3096 MHz, respectively. The g_{iso} values are all close to 1.995. Gamma-irradiated $AgNO_3$ ices have been found to contain

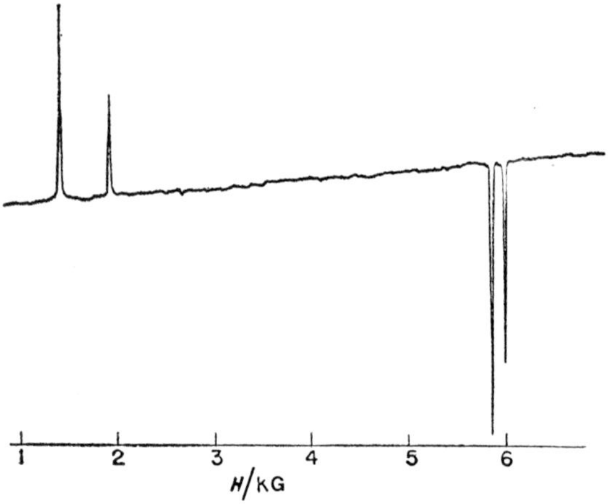

Figure 5 *E.s.r. spectrum of copper atoms in a krypton matrix. The spectrum was recorded in dispersion mode*
(Reproduced by permission from *J. Chem. Phys.*, 1971, **55**, 1566)

silver atoms in several interconvertible, magnetically distinct sites, some of which display anisotropy. One of these had ^{107}A = 1227 MHz, ^{107}B = 1211 MHz, $^{107}A_{iso}$ = 1216 MHz, $g_{\parallel}$ = 2.0032, and $g_{\perp}$ = 1.9930.[74] Starkie and Symons defend the suggestion that the unusual lineshape observed in silica gel treated with alcoholic $AgNO_3$ and then exposed to γ-rays is due to two isotropic species of Ag^0; that one of these species is highly anisotropic in **g** and ***A*** is defended by Gardner *et al.* in the note which follows.[75] Isotropic **g** and ***A*** tensors have been measured for the silver atoms which become trapped in cation sites after (Ag)KBr and (Ag)KI crystals have been *X*-irradiated. Super-hyperfine structure showed these sites to be regular octahedral.[76]

***d*¹ Configuration.**—*Tervalent Titanium.* The new compound $Ti[N(SiMe_3)_2]_3$ has been confirmed three-co-ordinate. At room temperature g = 1.911; at 135 K $g_{\parallel} > g_{\perp}$, confirming axial symmetry. This is compatible with a $^2A'$ ground state in D_{3h}.[77]

[74] B. L. Bales and L. Kevan, *J. Chem. Phys.*, 1971, **55**, 1327.
[75] H. C. Starkie and M. C. R. Symons, *J. Phys. Chem.*, 1971, **75**, 2705; C. L. Gardner, E. J. Casey, and C. W. M. Grant, *ibid.*, p. 2706.
[76] N. I. Mel'nikov, R. A. Zhitnikov, and P. G. Baranov, *Soviet Phys. Solid State*, 1971, **13**, 1117; *Izvest. Akad. Nauk S.S.S.R., Ser. fiz.*, 1971, **35**, 1384.
[77] D. C. Bradley and R. G. Copperthwaite, *Chem. Comm.*, 1971, 764.

Complexes of the general structure $[(\pi\text{-}C_5H_5)_2Ti(MR_n)_2]^-$, where M = N, P, As, Ge, or Sn, have been found by Kenworthy *et al.* to have a maximum *g* value of 1.988 ± 0.006. They postulated that the lower of these *g* values (assigned to g_y) is associated with the cyclopentadienyl groups, which must be similarly bonded in all these complexes, and the same as in $(\pi\text{-}C_5H_5)_2TiCl$. The third *g* value ($g_x$), which varies from 1.928 to 1.988 in

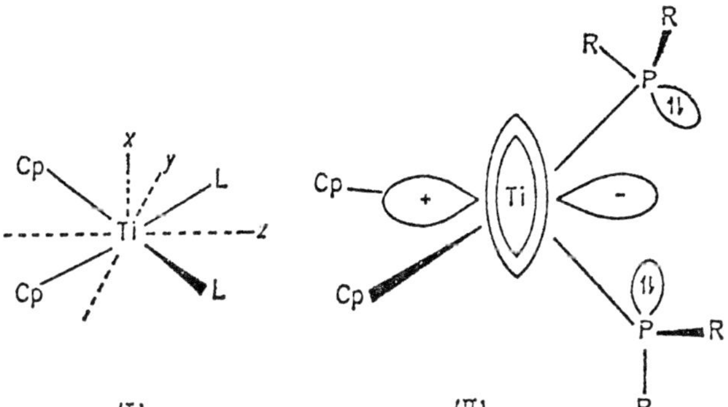

(I) (II)

Figure 6 *The orientation of:* (I), *the axes x, y, and z; and* (II), *the* d_{z^2} *orbital; with respect to the ligands in* $[(\pi\text{-}C_5H_5)_2TiL_2]^-$
[Reproduced from *J. Chem. Soc.* (*A*), 1971, 3428]

this series, must be connected with the other ligands. This is illustrated by Figure 6, which shows an approximately tetrahedral configuration for the titanium, and the unpaired electron in a d_{z^2} orbital. The momentum about the *y* axis results from coupling of d_{z^2} with d_{xz}, which is bonding to the cyclopentadienyl rings, whilst coupling of d_{z^2} with d_{yz}, which is bonding with respect to the other two ligands, generates momentum about the *x* axis. A series of hydrides $(\pi\text{-}C_5H_5)_2TiH_2M$ where M = Na, Li, MgBr, $AlCl_2$, or AlH_2 have been similarly studied by the same authors.[78] The solvent adducts $TiCl_3,L_3$ have been shown by their *g* anisotropy to have axial symmetry. Lower symmetry was expected and found for $TiCl_3$-(dioxan)$_2$L. In each case L = MeCN or THF.[79]

Quadrivalent Vanadium and Niobium. An important new series of alkyls has been reported;[80] a petroleum solution of $V(CH_2SiMe_3)_4$ has g_{iso} = 1.968, A_{iso} = 0.0050 cm^{-1}, $g_{\parallel}$ = 1.974, and $A_{\parallel}$ = 0.0111 cm^{-1}; these parameters are rather sensitive to concentration and temperature. It is noteworthy that monomeric, eight-co-ordinate compounds of general formula $V(RCS_2)_4$ have been prepared from aqueous solutions of vanadyl

78 J. G. Kenworthy, J. Myatt, and M. C. R. Symons, *J. Chem. Soc.* (*A*), 1971, 1020, 3428.
79 G. R. Hoff and C. H. Brubaker, jun., *Inorg. Chem.*, 1971, **10**, 2063.
80 W. Mowat, A. Shortland, G. Yagupsky, N. J. Hill, M. Yagupsky, and G. Wilkinson, *J. C. S. Dalton*, 1972, 533.

salts. Typical values of $g_{\parallel} = 1.969$, $g_{\perp} = 1.977$ if R = Me, for example, indicate that the unpaired electron is in a $d_{x^2-y^2}$ orbital and that the arrangement of sulphur atoms is dodecahedral; such an arrangement for R = $C_6H_4CH_2$ was later confirmed by *X*-ray diffraction methods.[81]

An analysis of 27 vanadyl compounds revealed the core polarization parameters $K_O = 160$ kG, $K_L = 0$. A few ionic V^{4+} compounds showed a lower value of K_O which was possibly caused by a (positive) 4*s* contribution resulting from partial covalence.[8] Some 4*s* character in the σ-bonding was thought also to be the cause of a reduced A_{iso} in certain vanadyl salicylaldimines.[13] Single crystals of diamagnetic $(NH_4)_3[VO_2(C_2O_4)_2]2H_2O$ when irradiated with *X*-rays or u.v. light gave stable centres, presumed vanadyl, for which the unpaired electron is in d_{xy}, there being a covalent in-plane σ-bond and an ionic out-of-plane π-bond.[82]

A solution spectrum of $VOCl_2,2PEt_3$ gives a 1 : 2 : 1 triplet for each vanadium line. These are not ^1H splittings because for the corresponding complex with arsine the triplets are replaced by septets. By studying the linewidth variation, the authors favoured a trigonal-bipyramidal structure rather than a square-pyramidal structure for this type of complex.[83]

Vanadyl ions have been shown to substitute for Al^{3+} in tanzanite, a newly found, gem-quality zoisite from Tanzania. The e.s.r. parameters recorded were $g_z = 1.948$, $g_x = 1.942$, $g_y = 1.968$, $A = 0.0163$, $B = 0.0049$, and $C = 0.0053$ cm^{-1}.[84]

Vanadyl ions have been studied in less well-defined environments including: alkali borate glasses [85] which contain VO^{2+} in a C_{4v} site and not V^{3+} or V^{5+} as previously reported; petroleum residuals which contain square-pyramidal complexes;[86] and zirconium dioxide surfaces [87] upon which adsorption of oxygen leads to the superoxide ion O_2^-. The vanadium in this last system is clearly catalytic since, although O_2^- is formed when there is no vanadium present, its e.s.r. signal is two or three orders of magnitude less intense.

The tetrakis(dialkylamido)niobium(IV) compounds (dialkyl = Et_2, $MeBu^n$, $Pr^n{}_2$, or C_5H_{10}) are monomeric, air-sensitive, purple-brown liquids.[88] Toluene solutions gave well-resolved e.s.r. spectra with $A > B$ and $g_{\parallel} < g_{\perp}$. These parameters are typical of a D_{2d} distortion from tetrahedral symmetry, a configuration partly due to nitrogen–metal π-bonding. The hydrochloric acid solution spectra of $[NbOCl_4(H_2O)]^{2-}$ and $[NbOCl_5]^{3-}$

[81] O. Piovesana and C. Furlani, *Chem. Comm.*, 1971, 256; M. Bonamico, G. Dessy, V. Fares, P. Porta, and L. Scaramuzza, *ibid.*, p. 365.
[82] R. Srinivasan and C. K. Subramanian, *Indian J. Pure Appl. Phys.*, 1971, **9**, 24.
[83] G. Henrici-Olivé and S. Olivé, *J. Amer. Chem. Soc.*, 1971, **93**, 4154.
[84] D. R. Hutton, *J. Phys. (C)*, 1971, **4**, 1251.
[85] H. Toyuki and S. Akagi, *Phys. and Chem. Glasses*, 1972, **13**, 15.
[86] F. E. Dickson, C. J. Kunesh, E. L. McGinnis, and L. Petrakis, *Analyt. Chem.*, 1972, **44**, 978.
[87] V. M. Fenin, V. A. Shvets, and V. B. Kazanskii, *Kinetics and Catalysis*, 1971, **12**, 1113.
[88] D. C. Bradley and M. H. Chisholm, *J. Chem. Soc. (A)*, 1971, 1511.

were also well resolved. Second-order corrections were employed to estimate g and hyperfine parameters, which were then used to calculate coefficients in a simple MO scheme.[89]

Quinquevalent Chromium, Molybdenum, and Tungsten; Sexivalent Rhenium. When strontium titanate is doped with chromium, the impurity ion is usually Cr^{3+}. However, oxidation above 1000 K by oxygen or chlorine gives Cr^{5+} in a Ti^{4+} site. E.s.r. at room temperature gave $g_{\parallel} = 1.961$, $g_{\perp} = 1.920$, $A = 0.0035$, and $B = 0.0010\,cm^{-1}$.[90] The radius of this impurity ion is smaller than that of its host, and tetragonal distortion results from the displacement of Cr^{5+} along the cubic axis. This distortion destroys the orbital degeneracy giving d_{xy} lowest. Likewise, the low-intensity e.s.r. spectrum from chromium ions in rutile is believed to be that of Cr^{5+} in a substitutional site. The e.s.r. parameters reported are $g_z = 2.105$, $g_x = 1.928$, $g_y = 1.976$, $A = 0.0038$, $B = 0.0027$, and $C = 0.0024\,cm^{-1}$. These hyperfine values are much bigger than $0.0015\,cm^{-1}$ found for Cr^{3+} in the same crystal.[91]

When added to dioxan in contact with solid Na_3CrO_4, various acids such as malic, malonic, salicylic, *o*-hydroxytoluic, maleic, and citric, and also various other hydroxylic substances and amines, gave paramagnetic solutions which, from their e.s.r. spectra, were assumed to contain Cr^V complexes; no e.s.r. data were given.[92] Complexes believed to be $[(dt)_2MO]^-$ (M = Cr or Mo; dtH_4 = diphenylthiocarbazone) were produced by heating DMF solutions of $K_2Cr_2O_7$ or $(NH_4)_2MoO_4$ with dtH_4. The hyperfine structure shows two equivalent ^{14}N with $A \approx 2$ G. The g values for both metals were about 1.985.[93]

Mo^{5+} ions occupy germanium sites in tetragonal GeO_2 and are charge-compensated by oxygen.[94]

Removal of oxygen atoms from the surface of a Bi–Mo mixed oxide, which catalyses the oxidation of simple olefins, led to Mo^V in a tetragonal-pyramidal structure, with approximately C_{4v} symmetry. The e.s.r. parameters ($g_{\parallel} = 1.935$, $g_{\perp} = 1.872$) were compared with those of $[MoOX_5]^{2-}$, X = F, Cl, or Br.[95] The same structure was proposed for Mo^V on the surface of a molybdenum silicate catalyst. However, in this latter case, cooling to 77 K yielded another e.s.r. signal assumed to arise from Mo^V in a tetrahedral site. This site could be modified by adsorption of water or ammonia.[96]

[89] D. P. Johnson and R. D. Bereman, *J. Inorg. Nuclear Chem.*, 1972, **34**, 679.
[90] A. Lagendijk, R. J. Morel, M. Glasbeek, and J. D. W. Van Voorst, *Chem. Phys. Letters*, 1972, **12**, 518.
[91] L. S. Sochava and Yu. N. Tolparov, *Soviet Phys. Solid State*, 1972, **13**, 1802.
[92] M. Mitewa, P. R. Boutchev, and V. Bojinov, *Inorg. Nuclear Chem. Letters*, 1972, **8**, 51.
[93] N. S. Garif'yanov, *J. Struct. Chem.*, 1971, **12**, 148.
[94] T. Purcell and R. A. Weeks, *Phys. Letters (A)*, 1972, **38**, 473.
[95] L. Burlamacchi, G. Martini, and E. Ferroni, *Chem. Phys. Letters*, 1971, **9**, 420.
[96] I. N. Marov, V. K. Belyaeva, Yu. N. Dubrov, and A. N. Ermakov, *Russ. J. Inorg. Chem.*, 1971, **16**, 81; V. M. Vorotyutsev, V. A. Shevts, and V. B. Kazanskii, *Kinetics and Catalysis*, 1971, **12**, 1108.

The best-fit values of the core polarization parameters for a series of $[MoOX_5]^{2-}$, X = halogen or SCN, were found to be K_O = 385 kG, K_L = 12.5 kG.[8]

Paramagnetic centres believed to be WO_4^{3-} were found in early, defective scheelite crystals but not in later ones, temperatures of 77 K or higher being used for the e.s.r. experiments.[97] Hyperfine structure from ^{183}W nuclei in the next neighbouring position was observed and the **g** tensor axes were not coincident with the crystal axes. E.s.r. parameters were given as g_z = 1.610, g_x = 1.816, g_y = 1.807, A = 0.0140, B = 0.0074, and C = 0.0076 cm^{-1}. However, a temperature of 20 K or lower was required to observe the e.s.r. spectra of W^{5+} in single crystals of Cs_2ZrCl_6, in which the point-group of Zr^{4+} is cubic with a perfect octahedron of Cl^-. The e.s.r. showed tetragonal symmetry with $g_\parallel$ = 1.8113, $g_\perp$ = 1.7674, A = 0.0125, and B = 0.0063 cm^{-1}. The core polarization K_O was estimated to be about 1200 kG, which is higher than that estimated by Low.[98]

The core polarization K_O for another $5d^1$ ion, Re^{6+} in $ReOCl_4$, was. estimated to be 822 kG.[8]

***d^5* Configuration.**—*Tervalent Iron, Ruthenium, and Osmium.* Griffith has pointed out that, in some instances, difficulty is found in fitting experimental g values to his low-spin d^5 equations [99] which imply that $g_x^2 + g_y^2 + g_z^2 = S$ must not exceed 16, and he has quoted experimental data which violate this condition. Since S_{max} = 16 when k = 1 and always $S_{max} < 16$ when $k < 1$, we need an explanation of why k must exceed unity for the low-spin d^5 equations to hold in these instances. As a personal communication from DeSimone, he quoted k = 1.06, 1.20, or 2.45 for ferrihaemoglobin azide, adding that the last of these values is obviously unreasonably high. The present Reporter agrees, but further finds that this solution does not even correspond to a resonance in the ground state! However, the first two solutions do. Thornley had already shown that configuration interaction can lead to a substantial increase in the effective orbital angular momentum of the ground $^2T_2$ term by three mechanisms: (i) states of t_2^4e mixed in by electrostatic interaction; (ii) states of t_2^4e mixed in by spin–orbit interaction; and (iii) charge-transfer states mixed in by spin–orbit interaction. He found that for $[IrX_6]^{2-}$ only the first of these was important.[100] Griffith extended this calculation, which still applies only to octahedral symmetry, and found [101]

$$\begin{aligned} k &= k_1 + 6Bk_2(a_1 E_1^{-1} + a_2 E_2^{-1}) \\ E_1 &= \Delta + (7 - \sqrt{34})\, B - C \\ E_2 &= \Delta + (7 + \sqrt{34})\, B - C \end{aligned} \qquad (3)$$

[97] V. P. Solntsev and M. Ya. Shcherbakova, *J. Struct. Chem.*, 1971, **12**, 369; M. D. Kapitonov, M. Ya. Shcherbakova, and V. P. Solntsev, *Geokhimiya*, 1972, 205.
[98] S. Maniv, W. Low, and A. Gabay, *J. Phys. and Chem. Solids*, 1971, **32**, 815.
[99] J. S. Griffith, 'The Theory of Transition-Metal Ions', Cambridge University Press, 1964.
[100] J. H. M. Thornley, *J. Phys.* (*C*), 1968, **1**, 1024.
[101] J. S. Griffith, *Mol. Phys.*, 1971, **21**, 135.

E_1 and E_2 may be estimated from the crystal field and Racah parameters, a_1 and a_2 from the secular equations. The k symbols are as follows: k is the quantity calculable from the low-spin d^5 equations in Griffith's book;[99] k_1 is the quantity pertinent to t_2 orbitals only; and k_2 is the quantity pertinent to t_2 e mixing. Cotton has used equation (3) to correct some earlier data. For example, for $K_3Fe(CN)_6$, he corrected Bleaney and O'Brien's higher, reasonable value of $k = 0.87$ to $k_1 = 0.72$.[102] (But see also Merrithew and Modestino who, in accounting for the temperature-dependence of the Mössbauer quadrupole splitting, rejected $k = 0.87$ in favour of Bleaney and O'Brien's *lower* value of $k = 0.56$.[103])

Clearly, to analyse low-spin d^5 e.s.r. data properly so as to reject all unwanted solutions, electronic spectral assignments and a greater degree of refining are required. A good example of this is the analysis of the **g** tensor measured for single crystals of osmium-doped trichlorotris(diethylphenylphosphine)rhodium(III). In his treatment, Hill[104] first reformulated the theory in terms of five-electron states rather than Griffith's one-hole states. He then considered two contributions to configuration interaction, (i) and (ii) above, and gave the new wavefunctions of the ground sextet, the new matrix of low-symmetry field and spin–orbit coupling within this sextet, and the new equations for g_i which are:

$$\begin{aligned} g_x &= 2[-2ac - b^2 - \sqrt{(2)}\, kb(a + c)] \\ &\qquad + (2\lambda k/E)\,[4ac - a^2 - 3c^2 + \sqrt{(2)}\, b(a - c)] \\ g_y &= 2[-2ac + b^2 + \sqrt{(2)}\, kb(a - c)] \\ &\qquad + (2\lambda k/E)\,[4ac + a^2 + 3c^2 - \sqrt{(2)}\, b(a + c)] \\ g_z &= 2[a^2 - b^2 + c^2 + k(a^2 - b^2)] + (2\lambda k/E)\,[4b^2 - 2\sqrt{(2)}\, ab] \end{aligned} \qquad (4)$$

E is the single excitation energy assumed for all the $t_2^4 e$ states and was estimated from the optical spectrum. Starting with rough values of a, b, c, and k (from Griffith's equations), an iterative method, using the corrected g_i from equation (4), was used to find the single set of g_i which gave consistent satisfactory results. The parameters are $g_x = -1.44$, $g_y = 0.32$, $g_z = 3.32$, $\lambda = 2650\ cm^{-1}$, and $k_1 - 0.80$. The relationship between the principal axes of the **g**-tensor and the molecular octahedron was also found experimentally and is shown in Figure 7. It is seen that d_{xy} will π-bond with three phosphines and one chloride, d_{yz} with two phosphines and two chlorides, and finally d_{xz} with one phosphine and three chlorides. The e.s.r. calculation puts these orbitals in the order $d_{xy} < d_{yz} < d_{xz}$, which is in agreement with the conventional thinking of t_{2g} orbitals being depressed by π-bonding with high-energy phosphine acceptors.

There is disagreement between the coefficients a, b, and c for the low-spin complex $(Ph_4P)_3[Fe\{S_2C_2(CN)_2\}_3]$ published by Cotton and Gibson[105]

[102] S. A. Cotton, *Inorg. Nuclear Chem. Letters*, 1972, **8**, 371.
[103] P. B. Merrithew and A. J. Modestino, *J. Amer. Chem. Soc.*, 1972, **94**, 3361.
[104] N. J. Hill, *J. C. S. Faraday, II*, 1972, **68**. 427.
[105] S. A. Cotton and J. F. Gibson, *J. Chem. Soc. (A)*, 1971, 803.

and those obtained by Rickards *et al.*[106] The answer to this anomaly lies in the fact that of their eight possible physical solutions, Cotton and Gibson reported as the most reasonable, only the one with the k value closest to 1 (k = 1.08). This corresponds to a resonance from the $|-1\rangle$ state, there being only a small, low-symmetry distortion. The Reporter, using Griffith's equations, finds, however, that one of the seven other solutions, the one

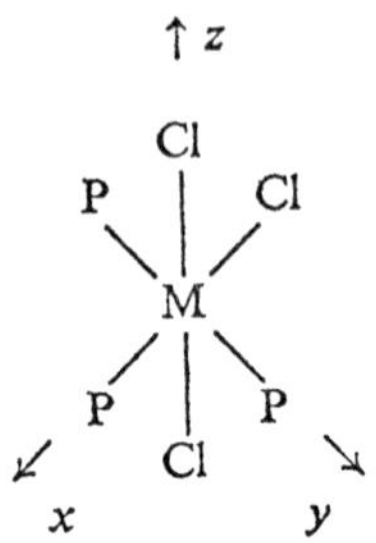

Figure 7 *Relationship of* **g**-*tensor eigenvectors to the molecular structure of* $OsCl_3(PEt_2Ph)_3$
(Reproduced from *J. C. S. Faraday, II*, 1972, **68**, 427)

which has k = 1.29 and which corresponds to the resonance in a d_{xy} state, well separated (11.57λ) from the other two states, has coefficients very close ($a = -0.048$, $b = -0.999$, $c = 0.015$) to those of Rickards *et al.* The coefficients are not identical because the latter authors used a simpler analysis with $k = 1$; they found only the one real solution in their analysis.

Other reports on low-spin d^5 systems have included: complexes of iron with persulphide ligands;[107] Ru^{3+} in MgO, for which g is almost isotropic at 1.8294, ^{101}A is almost isotropic at 0.00498 cm^{-1}, k = 0.87, and core polarization = − 658 kG;[108] a study of some Ru^{III} and Os^{III} compounds;[109] and studies of equilibria between low-spin d^5 and high-spin d^5.[110, 105]

***d*^{7} Configuration.**—*Bivalent Cobalt.* Engelhardt and Green have performed some ligand-field calculations for tetragonal low-spin cobalt(II) complexes to try to explain the very wide range of g for cobalt phthalocyanine. The reported g values are $g_\parallel \approx 1.90$, $g_\perp \approx 2.90$, whereas most complexes, for example corrins and vitamin B_{12} model compounds, have 'normal' g values of $g_\parallel \approx 2.01$, $g_\perp \approx 2.2$. They concluded that all of these compounds have a Γ_6 or $^2A_1$ ground state with the unpaired electron essentially in a d_{z^2}

106 R. Rickards, C. E. Johnson, and H. A. O. Hill, *J. Chem. Soc. (A)*, 1971, 797.
107 R. Rickards, C. E. Johnson, and H. A. O. Hill, *J. Chem. Soc. (A)*, 1971, 1755.
108 A. Reizman, J. T. Suss, and S. Szapiro, *Solid State Comm.*, 1971, **9**, 1799.
109 R. S. Abdrakhmanov, N. S. Garif'yanov, and E. I. Semenova, *Zhur. neorg. Khim.*, 1972, **17**, 1177.
110 V. V. Zelentsov, G. M. Larin, E. V. Ivanov, and N. V. Gerbelen, *Teor. i eksp. Khim.*, 1971, **7**, 798; P. C. Healy and A. H. White, *Chem. Comm.*, 1971, 1446.

orbital and that the phthalocyanine is different from the rest because of the absence of axial co-ordination.[111] A marked change in the e.s.r. parameters was observed when cobalt complexes of tetrasulphonated phthalocyanine were allowed to react with DMSO, pyridine, or imidazole; the reported g values were in the 'normal' range. Efforts to make simple 1 : 1 oxygen adducts of these compounds were unsuccessful.[112] Abel *et al.*, however, recorded similar 'normal' g values for cobalt(II) tetrasulphophthalocyanine in the presence of nitrogenous base (ammonia) and were able to observe reversible oxygenation even in the absence of such base. The g values of the oxygen adduct are $g_{\parallel} = 2.075$, $g_{\perp} = 2.004$ and the complex was described as Co^{III} with superoxide anion O_2^-.[113]

There is continued interest in the oxygenation of complexes with Schiff bases such as *NN'*-ethylenebis(salicylideneiminato)cobalt(II).[114] Yablokov *et al.* have oxygenated both finely powdered crystalline material and solutions. They concluded that the Co^{II} remains tetrahedral in the crystal, but that in solution the equilibrium between tetrahedral and planar is shifted towards planar as the basicity of the co-ordinated solvent is increased; oxygen reacts with the planar configuration to give an adduct which is written as $Co^{III}–O_2^-$. Busetto *et al.* have proposed two forms of a 1 : 1, O_2 : Co adduct. On the basis of e.s.r. intensity measurements, they have proposed a diamagnetic, oxygen-bridged, dinuclear intermediate in the evolution of one form from the other. Ochiai has shown that oxygenation readily takes place in pyridine, DMSO, or DMF but not in toluene, chloroform, or acetone. He argued that a $(d_{xy})^1$ configuration does not permit oxygenation, but co-ordination of one strong axial ligand results in a $(d_{z^2})^1$ configuration which is required for oxygenation.

Thioethers, thiolate anions, thiols, and nitrogenous bases bind to cobalt(II) corrins axially as 1 : 1 complexes.[16] Likewise, pyridine and piperidine bind to bis(dithioacetylacetonato)cobalt(II).[115] In the latter case, a square-pyramidal structure with a $(d_{z^2})^1(d_{xy})^0$ configuration was assumed and it was reasoned that, on co-ordination of the nitrogenous ligand, d_{z^2} is raised in energy while d_{xy} is lowered by the cobalt atom moving out-of-plane.

E.s.r. studies of $[Co(CN)_5]^{3-}$ have been popular.[116] Tsay *et al.* studied ethyleneglycol–water solutions and electron-irradiated powder samples of

[111] L. M. Engelhardt and M. Green, *J. C. S. Dalton*, 1972, 724.
[112] L. D. Rollmann and S. J. Chan, *Inorg. Chem.*, 1971, **10**, 1978.
[113] E. W. Abel, J. M. Pratt, and R. Whelan, *Chem. Comm.*, 1971, 449.
[114] Yu. V. Yablokov, V. K. Voronkova, V. F. Shishkov, and O. A. Osipov, *Russ. J. Inorg. Chem.*, 1971, **17**, 214; C. Busetto, C. Neri, N. Palladino, and E. Perrotti, *Inorg. Chim. Acta*, 1971, **5**, 129; E. Ochiai, *J. C. S. Chem. Comm.*, 1972, 489.
[115] K. M. Erck and B. B. Wayland, *Inorg. Chem.*, 1972, **11**, 1141.
[116] F. D. Tsay, H. B. Gray, and J. Danon, *J. Chem. Phys.*, 1971, **54**, 3760; W. C. Lin, C. A. McDowell, and D. J. Ward, *ibid.*, 1972, **56**, 1018; A. O. Caride, H. Panepucci, and S. I. Zanette, *ibid.*, 1971, **55**, 3651; D. Guenzburger, A. O. Caride, and E. Zuleta, *Chem. Phys. Letters*, 1972, **14**, 239; M. C. R. Symons and J. G. Wilkinson, *J. Chem. Soc. (A)*, 1971, 2069.

$K_3Co(CN)_6$ and reported the structure of the $[Co(CN)_5]^{3-}$ ion to be a slightly distorted square-pyramid with no solvent bound in the sixth co-ordination site. A computer simulation showed the two strong, well-spaced hyperfine components at the low-field end of the *X*-band spectrum to be so-called angular anomalies. The e.s.r. data are inconsistent with a previous assignment of $^2A_1 \rightarrow {}^2B_1$ for the band at 10 350 cm^{-1}; it was re-assigned $^2A_1 \rightarrow {}^2E$. Lin *et al.* have corrected their earlier assumption that $A_\parallel$ and $A_\perp$ are positive; they now agree with Tsay *et al.* that these parameters are probably of opposite sign. It is amusing that there is a formula error in their correction. Caride *et al.* have assigned optical bands and calculated the correct g values assuming $k = 0.6$. Their assignment places the orbitals in the order $b_2 < e < a_1 \ll b_1$, with e higher than b_2, contrary to expectations from a crystal-field treatment in the absence of a sixth ligand. A possible explanation is that the cobalt atom is expected to lie above the plane so that the ligand-function orbitals overlap with d_{xz} and d_{yz}. Symons and Wilkinson have shown that the five-co-ordination of this ion is preserved when $[Co(CN)_5]^{3-}$ is incorporated into alkali-metal halides and γ-irradiated. They deduced $A_\parallel$ to be positive but $A_\perp$ and A_{iso} to be negative. They also interpreted the low-field, strong, well-spaced hyperfine components as non-principal turning points. However, their assumed electron configuration while placing the unpaired electron essentially in d_{z^2} required an order $d_{xz}, d_{yz} < d_{xy} < d_{z^2}$. The same authors have incorporated $K_3Co(CN)_6$ into alkali-metal halides and although $[Co(CN)_5]^{3-}$ is the main paramagnetic species after γ-irradiation, $[Co(CN)_4(NC)_2]^{4-}$ was reported to be formed but lost on annealing.[117]

Univalent Ruthenium, Bivalent Rhodium, and Tervalent Palladium and Platinum. $[Ru(CN)_4(NC)_2]^{5-}$ is produced by γ-irradiation of $K_4Ru(CN)_6$,-$3H_2O$ at 77 K. A small ($\approx$ 3 G) slightly anisotropic hyperfine splitting from two equivalent nitrogen atoms is seen. On warming to room temperature, $[Ru(CN)_4(NC)]^{4-}$ or $[Ru(CN)_5(NC)]^{5-}$ is formed; no preference could be made for just one of these species.[118] Labilization of axial cyanides was also apparent from the formation of $[Rh(CN)_4Cl_2]^{4-}$ when $K_3Rh(CN)_6$ in a KCl lattice was electron irradiated.[119]

Molecular oxygen oxidation of *NN*-dimethylacetamide, cumene, or cyclo-octene (= cot), using the dinuclear Rh^{I} catalyst $[RhCl(cot)_2]_2$, gives paramagnetic monomeric species. Their e.s.r. spectra have been analysed as superoxide ion and a Rh^{II} species which are weakly interacting.[120] Addition of excess triphenylphosphine to an ethanolic solution of $RhCl_3,3H_2O$ and NO gives an orange precipitate containing, apparently, the bivalent species $RhNOCl_3(PPh_3)_2$.[40]

[117] M. C. R. Symons and J. G. Wilkinson, *J. C. S. Dalton*, 1972, 1086.
[118] R. S. Eachus and F. G. Herring, *Canad. J. Chem.*, 1972, **50**, 162.
[119] R. P. A. Muniz, N. V. Vugman, and J. Danon, *J. Chem. Phys.*, 1971, **54**, 1284.
[120] B. R. James, F. T. T. Ng, and E. Ochiai, *Canad. J. Chem.*, 1972, **50**, 590.

The $4d^7$ ions Rh^{2+} and Pd^{3+} have been doped into MgO. For each ion two spectra are seen at helium temperatures, one of these being isotropic and the other anisotropic; the lineshape of this latter spectrum is consistent with a strain-split ground state. On warming to 77 K only one isotropic spectrum is seen, its g value being the average of those of the lower temperature anisotropic spectrum. For example, for Pd^{3+}, $g_{\parallel} = 2.00$, $g_{\perp} = 2.17$ at 4.2 K becomes $g_{iso} = 2.1136$ at 77 K.[121]

In their γ-irradiation studies of K_2MCl_4, M = Pt or Pd, Krigas and Rogers have identified various species, including $[MCl_5]^{2-}$, which have low-spin d^7 configurations.[122] A four-line pattern from one chlorine nucleus, with $g_{\parallel} = 2.012$, $g_{\perp} = 2.149$, is found for $[PdCl_5]^{2-}$; the species $[PtCl_5]^{2-}$ has virtually the same hyperfine splitting tensor but additionally shows hyperfine splitting from a platinum nucleus. Both supposedly form by a chlorine atom attaching itself to $[MCl_4]^{2-}$ in a fifth co-ordination position along the fourfold axis. The ground state in each case is $(d_{x^2-y^2})^2(d_{z^2})^1$. Mixing in with some σ-bonded, almost pure, chlorine p_z accounts for the axially symmetric chlorine hyperfine splitting, whereas the hyperfine field at the platinum nucleus results mainly from polarization induced by the $5d$ electron, with little or no $6s$ involvement. A large hyperfine field K_O = 800 kG has also been quoted for Pt^{3+} in yttrium aluminium garnet.[8]

Forbes *et al.* have reported an interesting series of electron-transfer compounds $[Pt\text{–}N_2S_2]^n$, where N_2S_2 are chelates containing unsaturated five-membered rings and $n = -2, -1, 0, 1,$ or 2. The two species with $n = \pm 1$ are paramagnetic and are formed respectively by bromine oxidation or sodium amalgam reduction of the $n = 0$ species. E.s.r. data are scanty, the technique being used simply to confirm the existence of the complete series.[123]

d^9 Configuration.—Those ions with one positive hole in a d-shell usually give rise to easily observable e.s.r. spectra. The cupric ion is the most studied member of this class, and several references have been relegated to the Table at the end for no better reason than to economize on space in the text. Hathaway has recently explained the principles used in many of his papers.[124] The most important single piece of information comes from the lowest g value: if approximately 2.0 then the configuration is probably $(d_{z^2})^1$ and the structure trigonal-bipyramidal; if greater than 2.0, the configuration is $(d_{x^2-y^2})^1$ or, less commonly, $(d_{xy})^1$, indicating that the structure is possibly square-pyramidal.

The first observation of ^{17}O hyperfine splitting has been claimed for 0.5% ^{63}Cu doped into $KZnSO_4,6H_2O$. The hyperfine splitting tensor is approximately axial with 22 G along the Cu—O bond and with 10 and 12 G normal to this. The unpaired electron is therefore in a σ-orbital to oxygen.[125]

121 J. T. Suss, A. Raizman, and S. Szapiro, *J. Magn. Resonance*, 1972, **6**, 438.
122 T. Krigas and M. T. Rogers, *J. Chem. Phys.*, 1971, **54**, 4769; **55**, 3035.
123 C. E. Forbes, A. Gold, and R. H. Holm, *Inorg. Chem.*, 1971, **10**, 2479.
124 B. J. Hathaway, *J. C. S. Dalton*, 1972, 1196.
125 B. L. Silver, D. Getz, and Z. Luz, *Chem. Phys. Letters*, 1972, **13**, 113.

Bivalent Copper: Structural Interest. The two species earlier identified as *cis*- and *trans*-bis-(2,2′-bipyridyl)copper(II), have been shown to be disproportionation products, $[Cu(bipy)(OH_2)_4]^{2+}$ and $[Cu(bipy)_3]^{2+}$.[126]

Nitrogen hyperfine splitting has been observed in 1 : 1 adducts between copper(II) β-diketonates and pyridine or quinuclidine. The e.s.r. data are not compatible with the expected axial ligation. The authors suggest a square-pyramidal structure with the single nitrogen donor in the basal plane.[127]

Of the several papers [128] from the Hathaway group, which are concerned with the electronic properties and stereochemistry of the cupric ion, four will be picked out. A $d_{x^2-y^2}$ ground state was suggested for the approximately square-pyramidal CuN_3O_2 chromophore of μ-formatomono-(diethylenetriamine)copper(II) formate because $g_x = 2.034$ and $g_y = 2.094$ are along the $d_{x^2-y^2}$ lobe directions, and $g_z = 2.244$. On the other hand, a d_{xy} ground state was suggested for the approximately square-pyramidal CuN_4S chromophore in thiocyanatotriethylenetetraminecopper(II) thiocyanate. This ground state is unusual for a non-π-bonding chelate ligand and in this case is thought to derive from the accumulative distortion of the quadridentate ligand. A d_{z^2} ground state was proposed for the tetragonally compressed CuX_4L_2 chromophore of $Cu(NH_3)_2Cl_2$ diluted in NH_4Cl. In this case the Cu^{2+} ion was in a square of Cl^- ions, with NH_4^+ and a vacancy along the tetragonal axis. Notice $g_\parallel < g_\perp$ with $g_\parallel = 2.000$, $g_\perp = 2.219$. Finally, for the compressed tetrahedral CuN_4 chromophore of bis-(2,2′-bipyridylamine)copper(II) diperchlorate, $g_z = 2.244$, $g_x = 2.059$, and $g_y = 2.069$. The lower two of these g values correlate well with the copper–nitrogen directions, suggesting a $d_{x^2-y^2}$ ground state rather than the more usual d_{xy} which has been assumed previously for this environment.

Several near-tetrahedral copper(II) dipyrromethene complexes were studied to investigate the effect of distortion along a dyad axis.[129] Both $g_\parallel$ and $g_\perp$ were found to decrease over the range $g_\parallel = 2.279$ to 2.224, $g_\perp = 2.070$ to 2.039, as the tetrahedron was distorted towards square-planar; the anisotropic ^{63}Cu hyperfine parameters increased in the same direction. A most interesting, but unexplained, reversible disappearance of the ^{63}Cu hyperfine splitting was observed with most of these complexes as the temperature was varied over the range + 122 to − 50 °C.

Two other tetrahedral cupric ions have been studied.[130] Cu^{2+} in ammonium fluoride has a very high g_z of 2.4702, while $g_x = 2.0771$, and

[126] F. A. Walker and H. Sigel, *Inorg. Chem.*, 1972, **11**, 1162.

[127] B. B. Wayland and M. D. Wisniewski, *Chem. Comm.*, 1971, 1025.

[128] M. J. Bew, B. J. Hathaway, and (in part) R. J. Fereday, *J. C. S. Dalton*, 1972, 1229; R. J. Dudley and B. J. Hathaway, *ibid.*, p. 882; D. E. Billing, R. J. Dudley, B. J. Hathaway, and A. A. G. Tomlinson, *J. Chem. Soc.* (*A*), 1971, 691; G. Davey, R. J. Dudley, and B. J. Hathaway, *ibid.*, p, 1446; M. J. Bew, R. J. Dudley, R. J. Fereday, B. J. Hathaway, and R. C. Slade, *ibid.*, p. 1437; R. J. Dudley and B. J. Hathaway, *ibid.*, p. 1442; R. J. Fereday and B. J. Hathaway, *ibid.*, p. 2758; D. E. Billing, B. J. Hathaway, and A. A. G. Tomlinson, *ibid.*, p. 2839.

[129] Y. Murakami, Y. Matsuda, and K. Sakata, *Inorg. Chem.*, 1971, **10**, 1734.

[130] R. A. Vaughan, *Phys. Status Solidi* (*B*), 1972, **49**, 247; I. H. Parker, *J. Phys.* (*C*), 1971, **4**, 2967.

$g_y = 2.1040$. The flattened $[CuBr_4]^{2-}$ has $g_z = 2.220$, $g_x = 2.053$, and $g_y = 2.068$. In each case, absence of copper hyperfine splitting is probably due to admixture of the $3d^84p$ configuration into the ground state.

Bivalent Copper: Theoretical Interest. MO coefficients have been calculated in several cases. Well-resolved hyperfine splitting from the nitrogen ligands is observed when $^{63}Cu^{II}$ protoporphyrin IX dimethyl ester is diluted in the metal-free base, and it is concluded that delocalization of the unpaired electron is within 5% of that of the corresponding ferric compound.[131] Calculations have been based on axially symmetric data [13] and on rhombic data.[132] In the latter case, both the anisotropic presence of unpaired spin at the ligands in $CuX_2,2H_2O$ (X = Cl or F), and the mixing of d_{z^2} into the $d_{x^2-y^2}$ ground state, are evaluated. On the other hand, metal $4p_z$ was thought to mix into the ground state of copper(II) doped into tetracetatocadmium(II) hexahydrate. Also the large $4p_x$, $4p_y$ admixture into the lowest excited state $|E\rangle$ is the reason for the large, anisotropic orbital reduction; the parameters are $k_\parallel = 0.879$, $k_\perp = 0.546$.[133]

The $g_\parallel$ value of 2.52 is the highest yet reported for a copper compound. This was for $Cu(OMPA)_3(ClO_4)_2$ where OMPA is (2). Low temperatures are

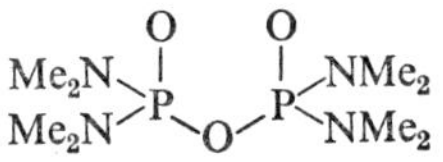

(2)

required to observe the anisotropy in this complex because, at room temperature, the dynamic Jahn–Teller effect results in the observation of a single isotropic line.[134] The relationship between correlation time and linewidth for aqueous solutions of $[Cu(dipyridyl)_3]^{2+}$ and $[Cu(H_2O)_6]^{2+}$ is invalid. This is because fast reorientation of the axis of the Jahn–Teller distortion occurs in such complexes where all orientations are equivalent. The Jahn–Teller effect is not dynamic in $[Cu(py)(H_2O)_4]^{2+}$ or when ethanol is added to $[Cu(H_2O)_6]^{2+}$.[135]

A least-squares analysis of cupric compounds with $A_\parallel > A_\perp$ gave best-fit parameters for core polarization of $K_O = 204$ kG, $K_L = 63$ kG.[8] A lower value of $K_O = 150$ kG was recorded for a $d_{x^2-y^2}$ ground state with both $A_\parallel$ and $A_\perp$ negative, and a still lower value, $K_O = -107$ kG, for a d_{z^2} ground state with $A_\parallel$ positive and $A_\perp$ negative.[9]

Bivalent Copper Chelates. When cupric ions were doped into L-alanine, hyperfine splitting was observed from only one amino-group ^{14}N nucleus,

[131] Y. Hsu, *Mol. Phys.*, 1971, **21**, 1087.
[132] E. Buluggiu, G. Dascola, D. C. Giori, and A. Vera, *J. Chem. Phys.*, 1971, **54**, 2191.
[133] R. S. Eachus, F. G. Herring, and Bo-Long Poh, *J. Chem. Soc.* (*A*), 1971, 614.
[134] M. D. Joeston, R. C. Koch, T. W. Martin, and J. H. Venable, jun., *J. Amer. Chem. Soc.*, 1971, **93**, 1138.
[135] M. Noack, G. F. Kokoszka, and G. Gordon, *J. Chem. Phys.*, 1971, **54**, 1342.

the copper being essentially three-co-ordinate. U.v. irradiation at 77 K formed two new centres which, on warming, underwent an irreversible change to another centre, with quintet ligand hyperfine splitting.[136] A compressed environment with d_{z^2} ground state was shown by its g values of $g_z = 2.013$, $g_x = 2.262$, and $g_y = 2.153$ to be appropriate for copper(II)-doped L-histidine hydrochloride monohydrate.[137]

Two important differences were observed between dithio- and diselenocarbamates, diethyldiselenocarbamatocopper(II) having the very low value of $g_{iso} = 2.022$, and no linewidth variation due to slow tumbling. No frozen solution spectra could be obtained.[138] Diethyl diselenophosphatocopper(II) has also been studied.[139] The unusual values of $g_{\parallel} = 2.036$ and $g_{\perp} = 1.981$ have been reported for an NN'-dibenzylethylenediamine complex,[140] and the bonding parameters in some amines, β-ketoenolates, dioximates, and Schiff-base complexes have been discussed.[141]

Adducts with Bivalent Copper Chelates. The $A_{\perp}$ value of bisacetylacetonatocopper(II) has been shown to exhibit a marked dependence on solvent. Results correlated well with Gutmann's donor number and with Selbin's $D_{II,I}$ parameter.[142] However, earlier workers were mistaken in supposing that this solute forms two distinct toluene and chloroform adducts when dissolved in a 2 : 3 toluene–chloroform mixture; Wisniewski and Wayland have shown that one of these is an ethanol adduct arising because ethanol is used as a stabilizer for chloroform.[143] Zamaraev *et al.* have also studied the adducts of various electron donor molecules, including methanol, with β-diketone complexes of copper(II).[144]

A large (132 G) isotropic hyperfine splitting from a single ^{31}P nucleus was observed in chloroform solutions of $Cu(hfac)_2PPh_3$, where hfac = hexafluoroacetylacetone. Other parameters were $g_{iso} = 2.13$, $A_{Cu} = 59$ G. Similar adducts were formed with PEt_3, PBu_3, PMe_2Ph, $PMePh_2$, and $AsPh_3$.[127, 145]

Some amine adducts, for example bis(p-toluidine)diacetatocopper(II) trihydrate, have planar CuN_2O_2 chromophores which are oxygen-bridged

[136] M. Fujimoto and Y. Tomkiewicz, *J. Chem. Phys.*, 1972, **56**, 749; M. Fujimoto and J. Janecka, *ibid.*, 1971, **55**, 1152; R. A. Wilkinson, jun. and I. Miyagawa, *ibid.*, 1971, **55**, 2177; H. C. Allen, jun., M. I. Mandrioli, and J. W. Becker, *ibid.*, 1972, **56**, 997.

[137] R. Hirasawa and H. Kon, *J. Chem. Phys.*, 1972, **56**, 4467.

[138] G. M. Larin, P. M. Solozhenkin, M. E. Dyatkina, and N. I. Kopitsya, *J. Struct. Chem.*, 1971, **12**, 715.

[139] T. M. Larin, P. M. Solozhenkin, M. E. Dyatkina, E. V. Semenov, and N. I. Zemlyanskii, *Izvest. Akad. Nauk S.S.S.R., Ser. khim.*, 1972, 680.

[140] K. C. Patel and D. E. Goldberg, *Inorg. Chem.*, 1972, **11**, 759.

[141] I. Adato, A. H. I. Ben-Bassat, and S. Sarel, *J. Phys. Chem.*, 1971, **75**, 3828; M. D. Alexander, P. C. Harrington, and A. Van Heuvelen, *ibid.*, p. 3355.

[142] I. Adato and I. Eliezer, *J. Chem. Phys.*, 1971, **54**, 1472.

[143] M. D. Wisniewski and B. B. Wayland, *J. Chem. Soc.* (*A*), 1971, 2727.

[144] K. I. Zamaraev, U. N. Nizel'skii, and T. E. Lipatova, *Teor. i eksp. Khim.*, 1971, **7**, 789; K. I. Zamaraev, U. N. Nizel'skii, T. E. Lipatova, V. A. Vonsyatskii, N. Y. Shteinshneider, and G. M. Zhidomirov, *ibid.*, p. 776.

[145] R. A. Zelonka and M. C. Baird, *Chem. Comm.*, 1971, 780.

thus: —Cu—O—Cu—O—Cu—O—, to give magnetically concentrated octahedral configurations. E.s.r. at *Q*-band frequencies gave a single line which was analysed to yield $g_z = 2.252$, $g_x = 2.103$, and $g_y = 2.060$, where *x* and *y* bisect the copper ligand directions.[146]

Univalent Nickel, Palladium, and Platinum, and Bivalent Silver and Gold. Ni^{+} doped into alkali-metal halides showed a compressed structure with $g_{\parallel} = 2.048$ and $g_{\perp} = 2.270$ and no dynamic Jahn–Teller effect. Cu^{2+} in the same crystals had an elongated environment and did show a dynamic Jahn–Teller effect above 130 K.[147]

Gamma irradiation of K_2MCl_4, M = Pt or Pd, yields d^9 centres which have been identified by their e.s.r. spectra. The palladium centre is $[PdCl_4]^{3-}$, as indicated by the 13-line hyperfine splitting from four equivalent chlorine nuclei.[122, 148] The anisotropic hyperfine splitting from ^{105}Pd is believed to have both $A_{\parallel}$ and $A_{\perp}$ positive on the basis that a positive A_{iso} arises from core polarization of the negative moment of the ^{105}Pd nucleus. However, its 10-line hyperfine splitting, characteristic of three equivalent chlorine nuclei, shows that the platinum centre is $[PtCl_3]^{2-}$.[122] In each case the unpaired electron is in a $d_{x^2-y^2}$ orbital, but in the latter case the D_{2h} symmetry may permit some of the metal hyperfine splitting to arise from admixture of the 6*s* orbital; this is not permitted for the former case, for which the symmetry is D_{4h}.

Crystals of KI supposedly containing Pd^{I} showed an e.s.r. spectrum which looks suspiciously like that of Mn^{II}.[149] The Pt^{I} species of the PtN_2S_2 type has already been mentioned.[123]

Surprisingly, K_O for Ag^{2+} in five frozen acidic solutions is less (421 kG) than it is for Ag^{2+} co-ordinated to some nitrogen or sulphur donors (450 kG); it is likewise surprising that K_L is more for the former (274 kG) than it is for the latter (155 kG).[8] Expanding the population of frozen acidic solution examples to forty-nine only deepened the mystery, when it was found that K_O = 390 kG and K_L = 307 kG.[10]

Tetragonal elongation of an otherwise octahedral site was found when Ag^{2+} was doped into $CdCl_2$; however, $g_{\parallel}$ was at least 5° off the Ag—Cl direction.[150] A similar elongation was reported for $Ag^{2+}(D_{4h})$ in $AgMF_6$, where M = Hf, Zr, Sn, or Pb.[151] In each, $g_{\parallel} > g_{\perp}$ but the *g* values were much closer to 2.0 for the former.

146 R. C. Komson, A. T. McPhail, F. E. Mabbs, and J. K. Porter, *J. Chem. Soc. (A)*, 1971, 3447.
147 G. Kuwabara, *J. Phys. Soc. Japan*, 1971, **31**, 1074.
148 S. Fujiwara and M. Nakamura, *J. Chem. Phys.*, 1971, **54**, 3378; M. Nakamura and S. Fujiwara, *J. Coordination Chem.*, 1971, **1**, 221.
149 G. D. Sootha, T. C. Tripathi, and D. C. Parashar, *Phys. Status Solidi (B)*, 1971, **48**, 45K.
150 T. Miyanaga, K. Kan'no, S. Naoé, and H. Matsumoto, *J. Phys. Soc. Japan*, 1971, **30**, 1669.
151 G. C. Allen, R. F. McMeeking, R. Hoppe, and B. Müller, *J. C. S. Chem. Comm.*, 1972, 291.

Both **g** and ***A*** are rhombic for Ag^{2+} in the bis(dialkyldiselenocarbamate) complexes. Low *g* values (2.081, 1.928, and 1.997) result from the important contribution of the spin–orbit interaction in the selenium atoms and from the strong π covalence of the Ag—Se bond. In the series Cu, Ag, Au, the contribution of A_{iso} increases strongly, in agreement with an increased admixture of *s*-orbitals into the ground state.[152]

The first example of the reaction gold(I) + gold(III) → 2 gold(II) has been detected using the *cis*-1,2-dicyanoethylenedithiolate (= mnt) ligand. The anion $[Au(mnt)_2]^{2-}$ had the same e.s.r. parameters as the previously reported $(Bu^n{}_4N)_2Au(mnt)_2$.[153]

4 $S = 1$

Tervalent Vanadium.—$[V(H_2O)_6]^{3+}$ in $CsAl(SO_4)_2,12H_2O$ is trigonally distorted. *X*-Band measurements in the {100} and {111} planes show that $g_\parallel$ (1.975) and $A_\parallel$ (0.0102 cm^{-1}) are along the ⟨111⟩ direction.[154] No measurement of *D* was reported but it was assumed to be about 10 cm^{-1}. The *D* value for $[V(H_2O)_6]^{3+}$ in guanidinium vanadium(III) sulphate hexahydrate was estimated to be 3.74 cm^{-1} using single crystal magnetic susceptibility methods, and the *D* value for V^{III} in hexaureavanadium(III) bromide trihydrate was found to be 5.82 cm^{-1} by low-temperature specific-heat measurements.[155]

Quadrivalent Chromium.—An important class of elimination-stabilized alkyls has been announced.[80] One of these compounds, $Cr(CH_2SiMe_3)_4$, gave a spectrum, typical of a triplet ($S = 1$) state with axial symmetry, which passed through various stages on cooling, as the zero-field splitting parameter, *D*, gradually increased. The resolution was best at about 150 K (see Figure 8). The spectrum is compatible with the proposed tetrahedral chromium(IV) state.

Bivalent Nickel.—An entirely theoretical treatment concerned with metal ions in fluoride lattices defines covalency factors in terms of metal- and ligand-orbital admixture coefficients and metal-to-ligand overlap integrals. These are related to *g* values, metal hyperfine splitting, and the ^{19}F hyperfine splitting tensor ***T***. Since in general the absolute signs of T_x, T_y, T_z, and T_{iso} are unknown and because there are insufficient data about 10 *Dq*, the characterization of the bonding in these complexes is limited. At present the **g** tensor and metal hyperfine splitting tensor are more reliable criteria than ^{19}F superhyperfine structure. Examples are given for Ni^{2+} ($t_{2g}^6\, e_g^2$).[156]

152 R. Kirmse, B. Lorenz, W. Windsch, and E. Hoyer, *Z. anorg. Chem.*, 1971, **384**, 160; *Z. Chem.*, 1971, **11**, 27.
153 J. H. Waters, T. J. Bergendahl, and S. R. Lewis, *Chem. Comm.*, 1971, 834.
154 J. G. Clarke and J. A. MacKinnon, *Canad. J. Phys.*, 1971, **49**, 1539.
155 J. N. McElearney , R. W. Schwartz, S. Merchant, and R. L. Carlin, *J. Chem. Phys.*, 1971, **55**, 466; J. N. McElearney, R. W. Schwartz, A. E. Siegel, and R. L. Carlin, *J. Amer. Chem. Soc.*, 1971, **93**, 4337.
156 L. Shields, *J. Chem. Soc.* (*A*), 1971, 1048.

Methods of evaluating g, D, and E from powder spectra have been discussed, with particular reference to N-donor ligands such as pyrazole, MeCN, and imidazole. D *versus* H graphs for $\lambda(= E/D) = 0.0$, 0.03, 0.10, 0.17, 0.25, and 0.333 are given. The full range of D reported is represented by the two compounds Ni(pyrazole)$_6$(ClO$_4$)$_2$ and Ni(N-methylimidazole)$_6$(ClO$_4$)$_2$ for which, respectively, $D = 0.07$ cm^{-1}, $\lambda = 0$, $g = 2.19$, and $D = 0.90$ cm^{-1}, $\lambda = 0$, $g = 2.185$.[157]

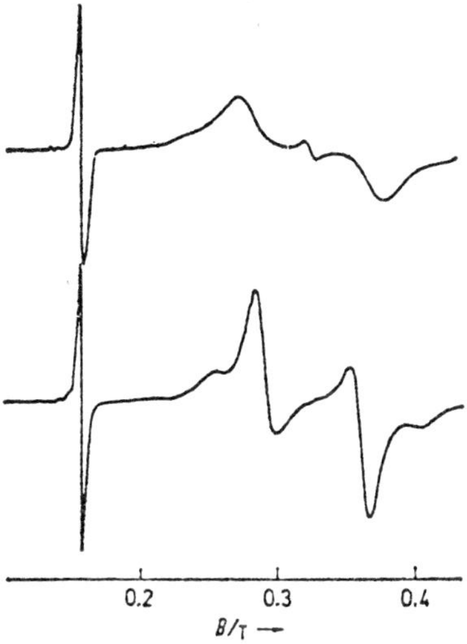

Figure 8 *E.s.r. spectrum of* $Cr(CH_2SiMe_3)_4$ *in toluene:* top, 100 K; bottom, 165 K. *E.s.r. parameters at* 150 K *were* $g = 1.989$, $D = 0.053$ cm^{-1} (Reproduced from *J. C. S. Dalton*, 1972, 533)

Above a certain temperature, T_c, usually between 4 and 77 K, the ligand field around Ni^{II} in $Zn(NH_3)_6X_2$ and $Cd(NH_3)_6X_2$ (where X = Cl, Br, or I) was found to be perfectly octahedral; only one e.s.r. line was seen. Below T_c, rotation of the NH_3 molecules was quenched and a three-line, anisotropic spectrum was seen. Parameters reported were $D = 0.265$—0.3 cm^{-1} and $g \approx 2.18$.[158]

A very strong zero-field splitting was found for $[Ni(H_2O)_6]^{2+}$ in $ZnSeO_4$,-$6H_2O$; its parameters were $g_{\parallel} = 2.219$, $g_{\perp} = 2.238$, and $D = +4.20$ cm^{-1}. The tetragonal crystal field stemming from decreased negative charge along the C_4 direction was thought to arise because the two axial water molecules have three hydrogen bonds, whereas the four equatorial water molecules have only two hydrogen bonds.[159]

[157] J. Reedijk and B. Nieuwenhuijse, *Rec. Trav. chim.*, 1972, **91**, 533.
[158] C. Trapp and C.-I. Shyr, *J. Chem. Phys.*, 1971, **54**, 196.
[159] A. Jiudo and R. J. Myers, *J. Magn. Resonance*, 1972, **6**, 633.

5 $S = \frac{3}{2}$

Calculations for various metal ions in fluoride lattices showed the (t_{2g}^3) configuration, exemplified by V^{2+} and Cr^{3+}, to have the smallest covalency factors.[156]

Bivalent Vanadium.—V^{2+} occupies orthorhombic sites in MgO as shown by $g_{iso} = 1.991$, $|D| = 0.1819\,cm^{-1}$, and $|E| = 0.0419\,cm^{-1}$. The local distortion from O_h is due to cation-vacancy nearest neighbours along [110]. 'Forbidden' transitions were also seen.[160] V^{2+} in octahedral sites in MgF_2 or $KMgF_3$ also showed non-cubic splittings.[161]

The e.s.r. spectra of V^{2+} in two sites, one of these a Ca^{2+} site, has been claimed[162] for a gem-quality crystal of blue zoisite, $Ca_2Al_3Si_3O_{12}(OH)$, but Hutton has declared that these spectra arise from the vanadyl ion VO^{2+}.[84]

Tervalent Chromium.—The metal oxidation states in the three-co-ordinate $Cr(NPr^i_2)_3$ which is trigonal planar,[163] and the four-co-ordinate $[Cr(CH_2SiMe_3)_4]^-$ which is tetrahedral,[80] were confirmed as Cr^{3+} by their e.s.r. spectra, which were typical of $S = \frac{3}{2}$ with a large zero-field splitting. In each case the frozen solution e.s.r. spectra gave $g_\parallel \approx 2.0$ and $g_\perp \approx 4.0$. These spectra were difficult to obtain because, after exposure to the merest trace of oxygen the solutions gave, instead, intense e.s.r. spectra in the region of $g = 1.98$.

Lineshape reconstructions for multi-electron centres having anisotropy in $\mathbf{g}$ and $\mathbf{D}$ can become rather involved. Figure 9 shows a very good computer simulation of an axially symmetric Cr^{3+} ion in $[Co(Cr)(NH_3)_6][CdCl_5]$. Note that one of the most prominent features in the experimental spectrum comes from transitions for which the field is not along one of the principal axes. In this example, the magnetic susceptibility was calculated from a knowledge of the zero-field splitting and it was deduced that there must be an antiferromagnetic interaction between the hexamminechromium(III) centres.[164]

Triglycine fluoroberyllate, when doped with chromic ions, contains the octahedral complex $[(glycine)_2Cr(BeF_4)]^{3+}$. There is a deformation along [111] which lies in the *yz* plane, the only plane in which fluorines bind. The e.s.r. parameters are $g_\parallel = 1.9805$, $g_\perp = 1.9815$, $D = 0.0620\,cm^{-1}$, and $E = 0.0117\,cm^{-1}$, which should be compared with $D = 0.430\,cm^{-1}$, $E = 0.070\,cm^{-1}$, and $g = 1.987$ for the corresponding complex in which SO_4^{2-} replaces BeF_4^{2-}. The low value of D is said to arise because fluorine, being more electronegative than oxygen, is more strongly attracted to the Cr^{3+} ion.[165]

160 A. J. B. Codling and B. Henderson, *J. Phys. (C)*, 1971, **4**, 1242.
161 M. D. Sturge, F. R. Merritt, L. F. Johnson, H. J. Guggenheim, and J. P. Van der Ziel, *J. Chem. Phys.*, 1971, **54**, 405.
162 T. Tsang and S. Ghose, *J. Chem. Phys.*, 1971, **54**, 856.
163 D. C. Bradley, M. B. Hurtshouse, and C. W. Newing, *Chem. Comm.*, 1971, 411.
164 W. E. Estes, D. Y. Jeter, J. C. Hempel, and W. E. Hatfield, *Inorg. Chem.*, 1971, **10**, 2074.
165 J. Stankowski and S. Waplak, *Bull. Acad. polon. Sci., Sér. Sci. chim.*, 1971, **19**, 243.

Wolberg has given a résumé of MO theory applied to octahedral complexes of $3d^3$ configuration, following this with an attack on Symons' contention that the large isotropic ligand hyperfine splitting of $[Cr(CN)_6]^{3-}$ and $[Cr(CN)_5NO]^{3-}$ is primarily due to core polarization. He feels that dipolar overlap is more important.[166]

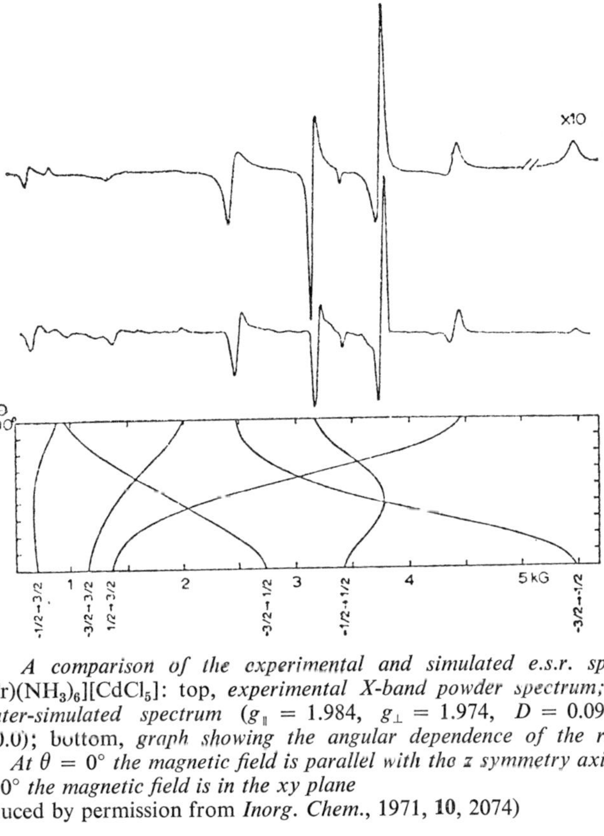

Figure 9 *A comparison of the experimental and simulated e.s.r. spectra of* $[Co(Cr)(NH_3)_6][CdCl_5]$: top, *experimental X-band powder spectrum;* middle, *computer-simulated spectrum* ($g_{\parallel} = 1.984$, $g_{\perp} = 1.974$, $D = 0.0949\ cm^{-1}$, $E = 0.0$); bottom, *graph showing the angular dependence of the resonance fields. At* $\theta = 0°$ *the magnetic field is parallel with the z symmetry axis, and at* $\theta = 90°$ *the magnetic field is in the xy plane*
(Reproduced by permission from *Inorg. Chem.*, 1971, **10**, 2074)

Bivalent Manganese and Tervalent Iron.—Bis(diethyldithiocarbamato)-manganese(II) is believed to be the first Mn^{II} complex of intermediate symmetry and predominantly axial field in which the $^4T_1$ state has crossed beneath the $^6A_1$ and $^2T_2$ states to become the ground state.[167] Acetone or dichloromethane solutions of $[Ph_4P][Fe(py)\{S_2C_2(CN)_2\}_2]$ are further examples of the spin quartet ground state for ferric ions.[105] In all cases $g_{\parallel} \approx 2.0$ and $g_{\perp} \approx 4.0$, typical of a ground state with $S = \frac{3}{2}$ and a large

[166] A. Wolberg, *J. Chem. Phys.*, 1971, **54**, 1428.
[167] S. Lahiry and V. K. Anaud, *Chem. Comm.*, 1971, 1111.

zero-field splitting, the resonance being observed in the $M_s = \pm \frac{1}{2}$ Kramers doublet.

Bivalent Cobalt.—Effective g values spread from below 2.0 to above 6.0 are not uncommon for cobalt(II) compounds; they arise from an effective spin $S' = \frac{1}{2}$. Results have been reported for several tetrahedral environments,[168] some with $g_\parallel > g_\perp$, others with $g_\parallel < g_\perp$, and from the tetragonally distorted hexa-aquo-ion with $g_\parallel = 5.975$, $g_\perp = 3.45$.

6 $S > \frac{3}{2}$

$S = 2$.—*Bivalent Iron.* As shown by Griffith,[99] the $^5T_2$ state may be assumed isomorphous with ^5P, with the result that there are three spin–orbit states of importance; these are a singlet, ψ_0, and a doublet, $\psi_{\pm1}$. When the separation between them is larger than about 10 cm^{-1}, and if the doublet is lowest, the magnetic properties may be described by an effective spin $S' = \frac{1}{2}$. However, for Fe^{2+} doped into a single crystal of $CdCl_2$, Trapp *et al.* found an extremely broad, strain-sensitive, allowed transition. Its behaviour indicated that the crystal field in $CdCl_2$ is very nearly octahedral, that ψ_0 is quite close to $\psi_{\pm1}$ (about 3 cm^{-1}) and that the magnetic properties are best represented by an effective spin $S' = 1$.[169] Similar conclusions were reached for Fe^{2+} in $KMgF_3$ where, in addition to the broad $\Delta M = 1$ transition, a $\Delta M = 2$ transition, and a double quantum transition were also seen.[170]

$S = \frac{5}{2}$.—Calculations concerned with covalence between metal ions and fluoride lattices have appeared. Mention has been made of the $(t_{2g})^3\,(e_g)^2$ configuration exemplified by Mn^{2+} and Fe^{3+}.[156]

Bivalent Manganese. Sharma has shown that for Mn^{2+} in the trigonal site in $CdCl_2$, the spin–spin contribution to D (+ 0.001 535 cm^{-1}) largely cancels the spin–orbit contribution (– 0.001 043 cm^{-1}), to give a value which agrees well with experiment (+ 0.0005 cm^{-1}).[171] The intensities, and the doubling, of the 'forbidden' hyperfine transitions of Mn^{2+} in calcite have been used to show that D, A, and B all have the same sign; the transitions arise through admixture of the quadratic term in D.[172] For Mn^{2+} in MgO it is the admixture of the quartic terms in a which permits the 'forbidden' transitions.[173]

On exposure of CaS to explosive shock the two natural Mn^{2+} impurities diminished at the expense of a new manganous centre, which was curious

[168] Yu. V. Yablokov, V. K. Voronkova, V. F. Shishkov, A. V. Ablov, and Zh. Yu. Vaisbein, *Soviet Phys. Solid State*, 1972, **13**, 831; G. E. Shankle, J. N. McElearney, R. W. Schwartz, A. R. Kampf, and R. L. Carlin, *J. Chem. Phys.*, 1972, **56**, 3750.
[169] C. Trapp, D. Smith, and R. Johnson, *J. Chem. Phys.*, 1971, **55**, 5195.
[170] J. T. Vallin and W. W. Piper, *Solid State Comm.*, 1971, **9**, 823.
[171] R. R. Sharma, *Phys. Rev.* (*B*), 1971, **3**, 76.
[172] P. Mialke, A. Briguet, and B. Tribollet, *J. Phys. Chem. Solids*, 1971, **32**, 2639; P. A. Narayana, *J. Chem. Phys.*, 1971, **55**, 4283.
[173] P. A. Narayana and K. V. L. N. Sastry, *J. Chem. Phys.*, 1971, **54**, 2281.

in that the cubic symmetry of its site was maintained, there being no evidence for zero-field splitting. At the same time A_{iso} increased, suggesting a fall in the delocalization of the electrons on to the sulphide ligands. In a single-crystal study of the same material, Wagner *et al.* observed very narrow lines (ΔH = 0.05—0.3 G) which permitted the measurement of what they claim to be the smallest *a* value reported for Mn^{2+} in any host. The parameters were $g = 2.001\ 75$, $A = -0.0077\ cm^{-1}$, and $a = 5.2 \pm 0.3 \times 10^{-6}\ cm^{-1}$.[174]

Attempts to prepare Co(Mn)(β-picoline)$_4$I$_2$ from acetone solution were unsuccessful. Analysis of e.s.r. spectra of the reaction product gave $D = 0.057\ cm^{-1}$, $E \approx 0.0$, consistent with the *trans*-octahedral structure [Co(Mn)(β-picoline)$_4$(H$_2$O)$_2$]I$_2$. If the iodine were co-ordinated to the metal, a D value of about 0.87 cm^{-1} would be expected.[175] Complex formation of Cl^- and of SO_4^{2-} in water–methanol mixtures has been reported,[176] as has also the substitution of Mn^{2+} for Ni^{2+} in [Ni(MeCO$_2$)$_2$,-4H$_2$O] and [NiSO$_4$,7H$_2$O].[177]

Naturally occurring blodite, [Na$_2$Mg(SO$_4$)$_2$,4H$_2$O], contains two differently oriented but otherwise equivalent Mn^{2+} centres, with g = 1.9993, $b_2^0 = 0.0334\ cm^{-1}$, $b_2^2 = 0.0260\ cm^{-1}$, and $b_4^0 = 0.0001\ cm^{-1}$.[178] Blue zoisite contains Mn^{2+} substituted for Ca^{2+}, the b_2^0 value being 0.01 cm^{-1}.[162]

Tervalent Iron. Theoretical studies of Fe^{3+} in 'II–VI tetrahedral' crystals such as ZnO, ZnS, ZnSe, and ZnTe have shown that electron transfer within a chromophore, *e.g.* FeS$_4$, is adequate to explain the positive g shift and the magnitude of D.[179] However, Das *et al.*, who have used self-consistent, charge extended Hückel wavefunctions to calculate the spin–spin and spin–orbit contributions to zero-field splitting, find that the spin–orbit mechanism is by far the more important of the two, for most haemin type molecules, by virtue of a low-lying $^4A_2$ state.[180]

In a series of papers, Cotton and Gibson have used a band assignment method to estimate D and E values, and have correlated them with the structures of several classes of ferric complex. For example: distortions in some formally octahedral complexes with oxygen donors have been shown to be present, but not attributable to crystal packing;[181a] the relatively high D of 0.16 cm^{-1} for [Al(Fe)(H$_2$O)$_6$Cl$_3$] is probably due to the iron taking up a site [Fe(H$_2$O)$_4$Cl$_2$]$^+$;[181b] *trans*-octahedral cations, [Fe(H$_2$O)$_4$X$_2$]$^+$, and tetrahedral anions, [FeX$_4$]$^-$, were proposed for the

174 M. C. R. Symons, *J. Chem. Soc.* (*A*), 1971, 1648; G. R. Wagner, J. Murphy, and M. Rubinstein, *J. Phys. Chem. Solids*, 1972, **33**, 273.
175 M. Goodgame and P. J. Hayward, *J. Chem. Soc.* (*A*), 1971, 3406.
176 J. R. Bard and J. O. Wear, *Z. Naturforsch.*, 1971, **26b**, 1091.
177 R. Janakiraman and G. C. Upreti, *J. Chem. Phys.*, 1971, **54**, 2336.
178 B. Kiggins and A. Manoogian, *Canad. J. Phys.*, 1971, **49**, 3174.
179 H. Watanabe and H. Kishiahita, *Progr. Theor. Phys. Suppl.*, 1970, **46**, 1.
180 P. S. Han, T. P. Das, and M. F. Rettig, *J. Chem. Phys.*, 1972, **56**, 3861.
181 (*a*) S. A. Cotton and J. F. Gibson, *J. Chem. Soc.* (*A*), 1971, 1690; (*b*) *ibid.*, p. 1693; (*c*) *ibid.*, p. 859; (*d*) *ibid.*, p. 1696.

compounds FeX_2L_3 where L = Ph_3PO or Ph_3AsO, and X = Cl or Br;[181c] and finally, the meridional (C_{2v}) configuration was preferred for [In(Fe)-(methylpyrazole)$_3$Cl$_3$], which had a near rhombic e.s.r. signal, whereas the facial (C_{3v}) configuration was suggested for [In(Fe)(pyrazole)$_3$Cl$_3$], whose e.s.r. parameters were between rhombic and axial.[181d]

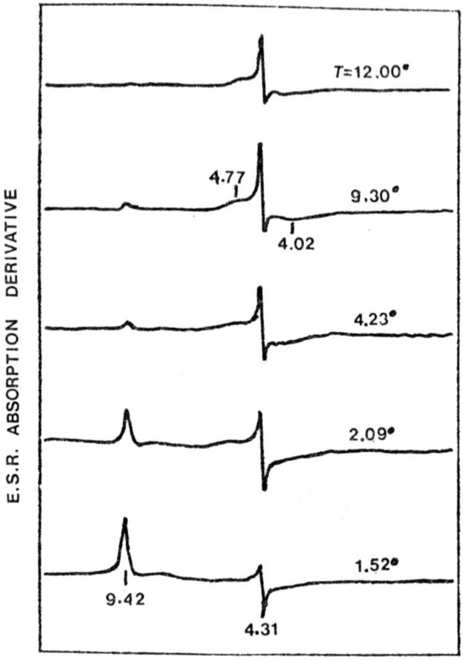

Figure 10 *E.s.r. spectra of* Pseudomonas oleovorans *rubredoxin containing one iron atom per molecule of protein. The spectrometer gains for the different traces have been adjusted arbitrarily*
(Reproduced by permission from *J. Biol. Chem.*, 1971, **246**, 5877)

Measurements of D have also been made by depopulation experiments. If the D value is large enough, lowering the temperature will depopulate the middle Kramers doublet, which in rhombic symmetry gives a signal at $g_{eff} \approx 4.3$, to the benefit of the lowest Kramers doublet, which gives a signal near $g_{eff} \approx 9.4$. By intensity measurements over the temperature range 1.4—40 K, Peisach *et al.* have thus shown ferric pyrrolidone dithiocarbamate to have $D = 1.68\ cm^{-1}$ and $E = 0.55\ cm^{-1}$. They therefore regarded this sulphur-ligated complex as a model compound for the rubredoxin isolated from *Pseudomonas oleovorans*, whose e.s.r. spectra as a function of temperature are shown in Figure 10.[182] Other sulphur-ligated

[182] J. Peisach, W. E. Blumberg, E. T. Lode, and M. J. Coon, *J. Biol. Chem.*, 1971, **246**, 5877.

complexes have much lower D values; for example, the values of $D = 0.16\ cm^{-1}$ and $E = 0.016\ cm^{-1}$ have been assigned to In(Fe){$S_2P(OEt)_2$}$_3$.[105]

Lang *et al.* found a good correlation between Mössbauer and e.s.r. measurements for polycrystalline, iron-doped [Ga(edtaH)H_2O], but some incompatibilities between the g_{eff} values near 4.3 and those near 9.7, when Fe(edtaH) was dissolved in 83% glycerol.[183] Cox *et al.*, however, reported curious behaviour for some compounds of the type FeCl($R^1COCHCOR^2$)$_2$. For example, if $R^1 = R^2$ = Me, Mössbauer measurements indicated a diamagnetic ground state with pairwise antiferromagnetic coupling, but the e.s.r. signal at $g_{eff} = 4.22$ clearly indicated paramagnetism with rhombic symmetry! Again, if $R^1 = R^2$ = Ph, a simple paramagnetic ground state was indicated by the Mössbauer measurement, yet no e.s.r. signal could be seen![184]

Hutton[84] has opposed the suggestion that ferric ions (with $D = 0.14\ cm^{-1}$) substitute for Al^{3+} in blue zoisite,[162] and instead has assigned the spectrum to Cr^{3+}.

***S* = 3.**—*Univalent Manganese.* When manganese is doped into NaCl, a centre with $g = 1.9992$ and $A = 0.0231\ cm^{-1}$ may be observed. This very large hyperfine splitting for Mn^+, $3d^54s$ (^7S), has been ascribed to a $4s$ contribution which is much larger than core polarization.[185]

Lanthanides and Actinides.—Ce^{3+} ($4f^1$) has been assigned to a C_{3i} site in $La_2(SO_4)_3,9H_2O$. This twelve-co-ordinate site is unusual because La^{3+} was previously thought to be the only lanthanide large enough for it.[186]

It is rare to find ligand hyperfine splitting among the lanthanide ion centres because the f electrons are deeply buried. However, fluorine hyperfine splitting was observed when Yb^{3+} ($4f^{13}$) was doped into CdF_2; the ion was at the body centre of a cube, with eight fluorine neighbours.[187] It was also observed when Eu^{2+} ($4f^7$) was doped into the same site of the same host,[188] but was said to arise more from the fluorine nuclear Zeeman term, than from overlap between the europium $4f^7$ wavefunction and the ligand fluorine nuclei.

Gd^{3+} ($4f^7$) has been doped into many host lattices: a generalized hamiltonian has been derived using tensor decomposition algebra appropriate to the C_{3h} site symmetry in lanthanum ethyl sulphate;[189] similar overall features are exhibited by the room-temperature spectra of some

183 G. Lang, R. Aasa, K. Garbett, and R. J. P. Williams, *J. Chem. Phys.*, 1971, **55**, 4539.
184 M. Cox, B. W. Fitzsimmons, A. W. Smith, L. F. Larkworthy, and K. A. Rogers, *J. Chem. Soc.* (*A*), 1971, 2158.
185 M. Ikeya and N. Itoh, *J. Phys. and Chem. Solids*, 1971, **32**, 2569.
186 R. N. Rogers, P. I. Biagi, and A. T. Long, *J. Chem. Phys.*, 1971, **55**, 5417.
187 T.-T. Chang, M. I. Cohen, and W. R. Hosler, *J. Chem. Phys.*, 1971, **54**, 4278.
188 T. C. Ensign, *J. Chem. Phys.*, 1971, **54**, 5188; A. Tzalmona and P. S. Pershan, *ibid.*, 1971, **55**, 4804.
189 H. A. Buckmaster, R. Chatterjee, and Y. H. Shing, *J. Magn. Resonance*, 1971, **4**, 85.

isostructural fluorides, which may be diamagnetic as in LaF_3, or paramagnetic as in CeF_3, PrF_3, or NdF_3;[190a] only five of the seven $\Delta M_s = 1$ transitions were within the range of the spectrometer for the C_3 Gd^{3+} ion in a $LiNbO_3$ host;[190b] all seven lines were seen in $Y(OH)_3$ and $La(OH)_3$ for which there was a slight variation in parameters with temperature;[190c] and in $LaCl_3$, previously unresolved structure has been observed.[190d]

Nd^{3+} ($4f^3$) has been doped into LaH_x ($2 < x < 3$). The e.s.r. results were interpreted in terms of a $\Gamma_8^{(2)}$ quartet lowest, favouring hydride anions, rather than the protonic model in which hydrogen atoms lose their electrons to the conduction band.[191]

Two distinct types of defect site, of C_{2v} or lower symmetry rather than the normal C_{3h} symmetry, were found for the $4f^{12}$ ion in thulium ethyl sulphate.[192]

Es^{2+} ($5f^{11}$) has been stabilized in CaF_2. A cubic field splits the ^4I ground term leaving a Γ_6 doublet lowest, in which resonance is observed, with an effective spin $S' = \frac{1}{2}$; $|g|$ was found to be 5.809; the theoretical value is $-$ 6.0. This reduction in g was ascribed to covalent bonding, which is more important here than in the lanthanides, because the $5f$ orbitals have better overlap with the ligands than do the $4f$.[193]

$^{237}Np^{4+}$ in strained crystals of ThO_2 was characterized by a large zero-field splitting and a $S' = \frac{1}{2}$ ground state. On the other hand, in unstrained crystals the impurity site is cubic and has a quartet ($S' = \frac{3}{2}$) ground state. Much hyperfine splitting from the neptunium nucleus ($I = \frac{5}{2}$) was seen.[194]

7 Table of Compounds not Included in the Text

The classification used here follows fairly closely that in the text.

Paramagnetic molecule, centre, or ion	*Reference*
Ti^{3+} in hexagonal CdS single crystals	195
Interstitial Ti^{3+} defects in TiO_2	196
Several Ti^{3+} centres in X-irradiated α-quartz	197
V^{4+} in cristobalite and quartz glass	198
Thio-oxinates of Mo^{V}, W^{V}, and Cu^{II}	199
Ni^{3+} in $MgO–NiO–Li_2O$ solid solutions	200

[190] (*a*) V. K. Sharma, *J. Chem. Phys.*, 1971, **54**, 496; (*b*) P. F. McDonald, C. P. Tam, and Y. W. Mok, *ibid.*, 1972, **56**, 1007; (*c*) P. D. Scott, *ibid.*, 1971, **54**, 5385; (*d*) R. W. Reynolds and L. A. Boatner, *ibid.*, 1971, **54**, 1827.

[191] B. Stalinsky and H. Drulís, *Bull. Acad. polon. Sci., Sér. Sci. chim.*, 1971, **19**, 739.

[192] D. C. Baker and H. H. Dearman, *J. Chem. Phys.*, 1972, **56**, 3664.

[193] N. Edelstein, *J. Chem. Phys.*, 1971, **54**, 2488.

[194] R. P. Richardson and J. B. Gruber, *J. Chem. Phys.*, 1972, **56**, 256.

[195] R. Böttcher and J. Dziesiaty, *Phys. Status Solidi* (*B*), 1972, **50**, 413.

[196] P. Meriaudeau, M. Che, P. C. Gravelle, and S. J. Teichner, *Bull. Soc. chim. France*, 1971, 13; T. Purcell and R. A. Weeks, *J. Chem. Phys.*, 1971, **54**, 2800.

[197] H. Rinneberg and J. A. Weil, *J. Chem. Phys.*, 1972, **56**, 2019; M. Okada, H. Rinneberg, J. A. Weil, and P. M. Wright, *Chem. Phys. Letters*, 1971, **11**, 275.

[198] V. S. Grubin, *Izvest. Akad. Nauk S.S.S.R., Neorg. Materialy*, 1971, **7**, 2113.

[199] I. N. Marov, V. K. Belyaeva, U. N. Dubrov, and A. N. Ermakov, *Zhur. neorg. Khim.*, 1972, **17**, 991.

[200] A. Cimino, D. Cordischi, G. Guarino, and A. Micheli, *Trans. Faraday Soc.*, 1971, **67**, 1776.

Paramagnetic molecule, centre, or ion	*Reference*
MO calculations on some copper complexes	201
Cu^{2+} exchange cations in kaolinite	202
Cu^{2+} in adamite	203
A new Cu^{2+} centre in NaCl	204
Copper acetate in acetic acid solutions	205
X-Irradiated crystals of copper acetate monohydrate	206
Unimolecular layers of copper stearate	207
Eight-co-ordination in tetra-(6-aminohexanoic acid)copper(II) perchlorate	208
Tetrahedral Zn(Cu){di-(2-pyridyl)amine}$_2$X$_2$, X = Cl, ClO_4, or PF_6	209
Bis-[3-phenyl-5-pyridyl(2)-pyrazolato]copper(II)	210
Influence of axial ligands on copper(II) dithiophosphates and dithiocarbamates	211
Substitution in copper(II)-β-thioxoketones	212
Copper(II) dialkenyl dithiophosphate	213
Anodic formation and cathodic reduction of copper(II) pyrophosphates	214
Ni^{II}, V^{II}, and Mn^{II} in single crystals of $CsMgCl_3$	215
Angular variation of linewidth of Ni^{2+} in CaO	216
Ni^{2+} doped into ZnS and CdS crystals	217
Cr^{3+} in magnesium chromite spinel	218
Cr^{3+} in a rhombic site in lithium gallium spinel	219
Cr^{3+} in vanadium pentoxide single crystals	220
Cr^{3+} in yttrium and europium orthoaluminate	221
Cr^{3+} and Fe^{3+} in sillimanite	222
Photochromic Fe^{5+} ($3d^3$) in $SrTiO_3$	223

[201] N. J. Trappeniers, G. de Brouckère, and C. A. Ten. Seldam, *Chem. Phys. Letters*, 1971, **8**, 327.
[202] V. A. Sil'chenko, N. N. Pafomov, U. I. Tarasevich, I. V. Matyash, and Z. E. Suyunova, *Ukrain. khim. Zhur.*, 1971, **37**, 1238.
[203] T. J. Seed, J. Le-Marshall, D. R. Hutton, and G. J. Troup, *Phys. Status Solidi* (*A*), 1972, **10**, 255.
[204] S. I. Farcas, A. Darabant, and A. Nicula, *Phys. Status Solidi* (*B*), 1972, **50**, 755.
[205] G. Nyberg, *J. Phys. Chem.*, 1971, **75**, 2228.
[206] A. Kawamori and I. Miyagawa, *J. Chem. Phys.*, 1971, **55**, 1336.
[207] J. Messier and G. Marc, *J. Phys. Paris*, 1971, **32**, 799.
[208] C. D. Garner, P. Lambert, F. E. Mabbs, and J. K. Porter, *J. C. S. Dalton*, 1972, 320.
[209] J. C. Lancaster, W. R. McWhinnie, and P. L. Welham, *J. Chem. Soc.* (*A*), 1971, 1742.
[210] H. Hennig and R. Kirmse, *Z. anorg. Chem.*, 1972, **388**, 238.
[211] N. D. Yordanov and D. Shopov, *Inorg. Chim. Acta*, 1971, **5**, 679.
[212] D. Rehorek, R. Kirmse, and Ph. Thomas, *Z. anorg. Chem.*, 1972, **389**, 199.
[213] E. V. Semenov, P. M. Solozhenkin, N. I. Zemlyanskii, and Y. I. Mel'nik, *Doklady Akad. Nauk Tadzh. S.S.R.*, 1971, **14**, 28.
[214] M. S. Shapnik, K. A. Zinkicheva, A. V. Il'yasov, and G. S. Vozdvizherskii, *Elektrokhimiya*, 1972, **8**, 295.
[215] G. L. McPherson, T. J. Kistenmacher, and G. D. Stucky, *J. Chem. Phys.*, 1971, **54**, 1432.
[216] R. Calvo, M. C. G. Passeggi, and C. Fainstein, *Phys. Letters* (*A*), 1971, **37**, 201.
[217] M. Schulz and G. G. Wepfer, *Solid State Comm.*, 1972, **10**, 405.
[218] E. I. Abdrashitova and R. Y. Khodakovskaya, *Fiz. Tverd. Tela*, 1972, **14**, 1234.
[219] T. J. Glynn, L. Kelleher, G. F. Imbusch, D. M. Larkin, F. R. Merritt, and M. J. Berggren, *J. Chem. Phys.*, 1971, **55**, 2925.
[220] G. Sperlich and K. Jansen, *Phys. Status Solidi* (*B*), 1971, **48**, 339.
[221] A. Pinto, N. Z. Sherman, and M. J. Weber, *J. Magn. Resonance*, 1972, **6**, 422.
[222] J. Le-Marshall, D. R. Hutton, G. J. Troup, and J. R. W. Thyer, *Phys. Status Solidi* (*A*), 1971, **5**, 769.
[223] K. A. Müller, Th. von Waldkirch, W. Berlinger, and B. W. Faughnan, *Solid State Comm.*, 1971, **9**, 1097.

Paramagnetic molecule, centre, or ion	*Reference*
Co^{2+} in triclinic $ZnMoO_4$ crystals	224
Co^{2+} in potassium tantalate	225
Co^{2+} combined with oxygen vacancies in TiO_2	226
Mn^{2+} in a trigonal site in single crystals of BaO	227
Orthorhombic Mn^{2+} with superhyperfine splitting in LiF	228
Very detailed analysis of Mn^{2+} in CaF_2	229
Mn^{2+} in lithium metavanadate	230
Mn^{2+} in natural crystals of newberyite $MgHPO_4,3H_2O$	231
Simulation of spectrum from Mn^{2+} in strontium borate glasses	232
Fe^{3+} in an axial Nb^{5+} site in $LiNbO_3$	233
Fe^{3+}in natural single crystals of anatase (TiO_2)	234
Surprisingly large value of D (3.8 cm^{-1}) for Fe^{3+} in SnO_2	235
Improved e.s.r. spectra of Fe^{3+} in alum	236
Fe^{3+} doped into potassium 12-tungstoborate(III), $K_5BW_{12}O_{40}$	237
Fe^{3+} ions in cubic sites in CaF_2	238
Fe^{3+}–Ag^+ centres in CdS	239
Dynamic Jahn–Teller effect of La^{2+} ($5d^1$) in $SrCl_2$	240
Trigonal centres in Ce^{3+} ($4f^1$) activated CaF_2	241
Nd^{3+} (f^3), Dy^{3+} (f^9), Er^{3+} (f^{11}), and Yb^{3+} (f^{13}) centres in LYF_4, a scheelite	242
Bivalent Eu ($4f^7$) in cadmium chloride and lead bromide	243
Gd^{3+} in bis-(1,10-phenanthroline)lanthanum trinitrate single crystals	244
Hyperfine splitting of $^{155}Gd^{3+}$ in cubic sites in ThO_2	245
Gd^{3+} in a tetrahedral site in $SrMoO_4$	246
Gd^{3+} in highly luminescent La_2O_2S single crystals	247

[224] A. G. Anders, A. I. Zvyagin, N. M. Nesterenko, and L. N. Pelikh, *Soviet Phys. Solid State*, 1972, **13**, 1624.

[225] D. M. Hannon, *Phys. Status Solidi* (*B*), 1971, **43**, 21K.

[226] Y. Miyako and Y. Kazumata, *J. Phys. Soc. Japan*, 1971, **31**, 1727.

[227] P. Weightman, D. E. Dugdale, and L. V. Holroyd, *J. Phys.* (*C*), 1971, **4**, 3291.

[228] C. G. Bousquet, A. S. Valls, and M. G. Sucre, *J. Magn. Resonance*, 1972, **6**, 117.

[229] R. J. Richardson, S. Lee, and T. J. Menne, *Phys. Rev.* (*B*), 1971, **4**, 3837.

[230] G. L. Bir, L. V. Dmitrieva, and S. A. Kizhaev, *Fiz. Tverd. Tela*, 1972, **14**, 24.

[231] A. Manoogian and B. Kiggins, *Amer. Mineral.*, 1972, **57**, 52.

[232] P. C. Taylor and P. J. Bray, *J. and Phys. Chem. Solids*, 1972, **33**, 43.

[233] H. H. Towner, Y. M. Kim, and H. S. Story, *J. Chem. Phys.*, 1972, **56**, 3676.

[234] M. Horn and C. F. Schwerdtfeger, *J. Phys. and Chem. Solids*, 1971, **32**, 2529.

[235] V. F. Annfrienko, L. G. Yandralova, and D. V. Tarasova, *Soviet Phys. Solid State*, 1972, **13**, 1970.

[236] M. A. Collins, J. F. Duncan, and K. Trotter, *J. Inorg. Nuclear Chem.*, 1971, **33**, 3943.

[237] H. So and M. T. Pope, *J. Chem. Phys.*, 1971, **55**, 2786.

[238] S. Zenbutu, R. Nakata, M. Sumita, and E. Higuchi, *Jap. J. Appl. Phys.*, 1971, **10**, 1497.

[239] R. C. Kemp, *J. Phys.* (*C*), 1972, **5**, 716.

[240] J. R. Herrington, T. L. Estle, and L. A. Boatner, *Phys. Rev.* (*B*), 1971, **3**, 2933.

[241] D. N. Chambers, *Phys. Letters* (*A*), 1971, **36**, 77.

[242] J. P. Sattler and J. Nemarich, *Phys. Rev.* (*B*), 1971, **4**, 1.

[243] D. Geist, E. Gerstenhauer, and G. Rackebrandt, *Z. Naturforsch.*, 1971, **26a**, 1589; B. Willemsen and W. C. Hommels, *Phys. Status Solidi* (*A*), 1972, **10**, 183.

[244] I. A. Gavrilov and V. B. Kravchenko, *Zhur. strukt. Khim.*, 1972, **13**, 334.

[245] G. Bacquet, J. Dugas, C. Escribe, L. Vassilieff, and B. M. Wanklyn, *Phys. Rev.* (*B*), 1972, **5**, 2419.

[246] M. L. Meil'man, *Soviet Phys. Solid State*, 1972, **13**, 2663.

[247] K. Aiki, K. Narita, M. Tanabe, and S. Takaoka, *J. Phys. Soc. Japan*, 1971, **31**, 1595.

Paramagnetic molecule, centre, or ion	*Reference*
Gd^{3+} (f^7), Dy^{3+} (f^9), Er^{3+} (f^{11}), Tm^{2+} (f^{13}), and Yb^{3+} (f^{13}) centres in the cubic perovskite $KMgF_3$	248
Tetragonal Ho^{3+} (f^{10}) centres in CaF_2 crystals	249
Powder e.s.r. spectra for some rare-earth trifluorides including HoF_3 and YbF_3	250
Dynamic effects of ion–lattice coupling for Er^{3+} and Yb^{3+} in ZnSe	251
Paramagnetic and paraelectric resonance of the $\mu = \pm 1$ state of Tm^{3+} ($4f^{12}$) in thulium ethyl sulphate	252
The $4f^{13}$ ions Tm^{2+} and Yb^{3+}, in cubic sites	253
The $5f^5$ ions Am^{4+} and Pu^{3+} in cubic single crystals of ThO_2	254

[248] M. M. Abraham, C. B. Finch, J. L. Kolopus, and J. T. Lewis, *Phys. Rev. (B)*, 1971, **3**, 2855.
[249] L. S. Kornienko and A. O. Rybaltovskii, *Soviet Phys. Solid State*, 1972, **13**, 1785.
[250] S. L. Carr and W. G. Moulton, *J. Magn. Resonance*, 1971, **4**, 400.
[251] W. E. Hagston, *Mol. Phys.*, 1971, **21**, 373.
[252] D. C. Baker and H. H. Dearman, *J. Chem. Phys.*, 1971, **54**, 5427.
[253] J. E. Lowther, *Mol. Phys.*, 1971, **22**, 899.
[254] M. M. Abraham, L. A. Boatner, C. B. Finch, and R. W. Reynolds, *Phys. Rev. (B)*, 1971, **3**, 2864.

8

Inorganic Radicals

BY T. A. CLAXTON

1 Introduction

This chapter selectively reports on those small radicals which are directly pertinent to the understanding of structure and reactivity in inorganic chemistry. It is not felt necessary for all radicals discussed here to be themselves inorganic. There are many radicals, which could be labelled inorganic, which are primarily of interest in the processes accompanying radiation; these have not been covered here, particularly if the radicals are not well characterized. One of the difficulties in this subject is in the identification of the species since, for example, oxygen (^{16}O) is non-magnetic and its presence cannot always be unequivocally proved. This means that an important aspect of current research is revision of previous assignments. Here it has been generally accepted that the current assignments are correct for the purpose of discussion. Some attempt has been made to divide this chapter into sections to improve clarity but the divisions are naturally, in some instances, arbitrary.

The isotropic hyperfine coupling constant will be denoted by a(X) or a_X for the nucleus X and the anisotropic coupling constant by B_{ii}(X) where $i = x, y, z$. For molecules of axial symmetry, real or apparent, $B_{\parallel}$ and $B_{\perp}$ may be used for the cases B_{zz} and B_{xx},B_{yy}, respectively. The combined hyperfine coupling constant $A_{ii}(X) = a(X) + B_{ii}(X)$ will also be used.

Gauss will be the unit used for all coupling constants (10 G $\equiv$ 1 mT).

2 Gas-phase Radicals

The number of radicals observed by e.s.r. in the gas phase continues, albeit slowly, to increase. The reaction of hydrogen atoms and NF_2 in the gas phase has given rise to transitions of NF in the $^3\Sigma^-$ ground state and also for the $J = 2$ level in its metastable $^1\Delta$ excited state.[1]

Of particular interest is the e.s.r. study of the Renner effect.[2] Gaseous NCO exists in a two-fold degenerate electronic state in the equilibrium linear nuclear configuration. The degeneracy is lifted during the doubly

[1] A. H. Curran, R. G. Macdonald, A. J. Stone, and B. A. Thrush, *Chem. Phys. Letters*, 1971, **8**, 451.

[2] A. Carrington, A. R. Fabris, B. J. Howard, and N. J. D. Lucas, *Mol. Phys.*, 1971, **20**, 961.

degenerate bending vibration and the resulting interaction is known as the Renner effect. The electron resonance spectrum is described for the lowest rotational levels of the $^2\Pi_{\frac{3}{2}}$, $n = 1$ ($J = \frac{3}{2}$), $^2\Delta_{\frac{5}{2}}$, $n = 2$ ($J = \frac{5}{2}$), $^2\Phi_{\frac{7}{2}}$, $n = 3$ ($J = \frac{7}{2}$) vibronic states.

The ^{17}O hyperfine coupling constant has been measured for $O_2(^1\Delta_g)$ in its $J = 2$ state.[3] The metastable $^1\Delta_g$ has considerable importance in oxidation reactions.

3 The Methylene Radical

Methylene, CH_2, is discussed separately because it is very important from the theoretical viewpoint at the present time, as it is the smallest known polyatomic radical existing in the triplet state, although this has only been very recently confirmed.[4] All isotopic possibilities have now been studied, the most significant being CD_2 in a xenon matrix [$a(^{13}C)$ = 89 G].[5] The angle of bending was estimated to be 137.7° from this value, in good accord with theoretical expectation. Although the geometry of CH_2 is similar in every environment in which it has been prepared, there is evidence of significant environmental interaction in xenon,[6] in spite of the fact that in this medium the radical seems to be freely rotating about a fixed long axis. Photolysis of diazomethane condensed with xenon at 4 K produced two radicals, both thought to be the methylene radical, with identical geometries but located at different crystalline sites. It is estimated that the zero-field parameters are increased by *ca.* 50% owing to intermolecular interactions with the xenon atoms. The contributions from zero-point vibrations can be compensated by comparing isotopic derivatives. The parameters for CH_2 in SF_6 and krypton are almost identical but differ significantly from the xenon results. Since there is no obvious reason to suppose that a substantial spin–orbit interaction with the krypton matrix does not occur the results are not necessarily those for 'free' CH_2.

4 Diatomic Hydrides

Alkali-metal Hydrides.—One of the interesting features in barium sulphate is the existence of two types of 'lattice' hydrogen centres.[7] One of these shows a splitting only slightly lower than that for a hydrogen atom in the gas phase, *i.e.* 503.9 G compared with 507.3 G. The other centre is associated with a sodium ion and, although some spin density (estimated to be about 5%) is on the sodium cation, the hyperfine coupling is very large

[3] C. A. Arrington, A. M. Falick, and R. J. Myers, *J. Chem. Phys.*, 1971, **55**, 909.
[4] R. A. Bernheim, H. W. Bernard, P. S. Wang, L. S. Wood, and P. S. Skell, *J. Chem. Phys.*, 1970, **53**, 1230.
[5] R. A. Bernheim, H. W. Bernard, P. S. Wang, L. S. Wood, and P. S. Skell, *J. Chem. Phys.*, 1971, **54**, 3223.
[6] E. Wasserman, V. J. Kuck, R. S. Hutton, E. D. Anderson, and W. A. Yager, *J. Chem. Phys.*, 1971, **54**, 4120.
[7] M. B. D. Bloom, R. S. Eachus, and M. C. R. Symons, *J. Chem. Soc.* (*A*), 1971, 833.

[$a(^1H)$ = 511.5 G]. *Ab initio* unrestricted Hartree–Fock (UHF) calculations [8] have already predicted the large values for NaH^+ but the situation is not as simple as this. The anisotropic coupling constants [$B_{\parallel}(^1H)$ = 1.7 G], although small, are in disagreement with UHF calculations using only *s*-type valence orbitals [$B_{\parallel}(^1H)$ = − 0.42 G].[9] It was concluded that there must be some polarization of the 1*s* hydrogen atomic orbital, achieved by mixing in the 2*p* orbital [which on its own would give $B_{\parallel}(^1H)$ = 3.17 G]. The extraordinarily large values for $a(^1H)$ arise from coefficients larger than unity for the hydrogen atomic orbital in the molecular orbital for the unpaired electron, which is only possible if orbital overlap is not ignored. In these circumstances it is inappropriate to discuss the amount of electron on the atoms. The smaller value for KH^+, $a(^1H)$ = 500 G, compared to NaH^+, $a(^1H)$ = 511.5 G, can be partly associated with a smaller orbital overlap.

Group IIa Hydrides.—The radicals MgH, CaH, SrH, and BaH in their ground electronic $^2\Sigma^+$ state have been studied in solid argon matrices at 4 K.[10] Hyperfine coupling to a metal was observed only in the case of ^{25}MgH. Since these are neutral radicals the unpaired electron can be imagined to be in a weakly antibonding orbital. According to expectations the greater the electronegativity difference of the two atoms, Δ, the smaller is the proton hyperfine coupling constant (Table 1). It is interesting to

Table 1 *Comparison of properties for the Group II metal hydrides*

	$a(^1H)$	Δ	IP/eV[a]	D_0^0/eV[b]
MgH	78.5	0.87	7.64	< 2.49
CaH	48.2	1.06	6.11	< 1.70
SrH	43.5	1.11	5.69	< 1.68
BaH	16.8	1.13	5.21	< 1.82
ZnH	171	0.44	9.39	0.850
CdH	184	0.64	8.99	0.678
HgH	254	0.66	10.43	0.376

[a] Ionization potential of the metal.
[b] Molecular dissociation energy.

note that the observed g_H shift is associated with the environment rather than with spin–orbit coupling effects.

The hydrogen atoms trapped in CaF_2 [$a(^1H)$ = 521 G][11] are to be regarded as interstitial hydrogen atoms. The temperature dependence of the e.s.r. spectrum is explained in terms of a three-dimensional harmonic oscillator.

Group IIb Hydrides.—The HgH^{2+} radical ion has been assigned to the species prepared by annealing frozen beads of mercury(II) salts in

8 T. A. Claxton and D. McWilliams, *Trans. Faraday Soc.*, 1970, **66**, 513.
9 T. A. Claxton and N. A. Smith, *Trans. Faraday Soc.*, 1971, **67**, 1859.
10 L. B. Knight and W. Weltner, *J. Chem. Phys.*, 1971, **54**, 3875.
11 S. G. Sligar and H. Blum, *Phys. Rev. B*, 1971, **3**, 3587.

7M-$HClO_4$ after γ-irradiation at 77 K.[12] In addition to evidence from deuteriation the analogy to AgH^+ [13] is recalled. Qualitatively the bonding is similar to the alkali-metal hydrides. The almost identical proton coupling constants [$a(^1H)$ = 300 and 302 G for HgH^{2+} and AgH^+, respectively] can be correlated with the almost identical electronegativities of Hg and Ag.

Two other hydrides have been observed which are best described as ion pairs; $Gd^{3+} - H^-$ [14] and $Ce^{3+} - H^-$.[15]

The neutral radicals ZnH, CdH, and HgH, analogous to the Group IIa hydrides, have also been studied trapped in solid argon matrices at 4 K (Table 1).[16] The correlation with electronegativity of the metal atom is not reproduced in the same way as for the Group IIa hydrides. Better correlations are found by using the first ionization potential of the metal atom or the molecular dissociation energy.

Group VI Hydrides.—The hydroxyl radical, claimed to be the first example in hydrated organic crystals, has been observed in two types of environment after γ-irradiation of single crystals of sodium hydrogen maleate trihydrate.[17] From the anisotropies of the g-values it was concluded that the radical with the larger average g-value interacts more weakly with the environment (which is consistent with the environmental interaction observed for the methyl radical as shown in Table 3). The species OH(I) (see Table 2) resembles the OH radical observed many times in irradiated

Table 2 *g and A values (in G) for the two types of OH radicals observed in sodium hydrogen maleate trihydrate*

	g_1	g_2	g_3	A_1	A_2	A_3
OH(I)	2.0034	2.0081	2.0618	− 27	− 44.7	3.5
OH(II)	1.9947	2.0065	2.1198	− 29.4	− 44.8	4.8

Table 3 *Isotropic proton hyperfine coupling constants and g-values for species having similarities with the methyl radical*

Assumed radical	$a(^{13}C)$	$a(^1H)$	g_{av}	Ref.
CH_3		23.04	2.00255	16
H_3O		22.84	2.0035	26
CH_3/Na	39.3	22.10	2.0023	23
CH_3/Vycor		19.3	2.002	24

ice, and OH(II) is more like that found in hydrated sulphates. (In the abstract of the paper [17] it is stated that the relative signs of the hyperfine coupling tensors are determined by forbidden transitions but this is not elaborated upon in the text.)

[12] R. J. Booth, H. C. Starkie, and M. C. R. Symons, *J. Chem. Soc. (A)*, 1971, 3198.
[13] R. S. Eachus and M. C. R. Symons, *J. Chem. Soc. (A)*, 1970, 1333.
[14] K. Zdansky and A. Edgar, *Phys. Rev. (B)*, 1971, **3**, 2133.
[15] I. T. Jacobs, G. D. Jones, K. Zdansky, and R. A. Satten, *Phys. Rev. (B)*, 1971, **3**, 2888.
[16] L. B. Knight and W. Weltner, *J. Chem. Phys.*, 1971, **55**, 2061.
[17] K. Toriyama and M. Iwasaki, *J. Chem. Phys.*, 1971, **55**, 1890.

The other Group VI hydrides, SH, SeH, and TeH are known and have been extensively discussed again.[18] In particular, the complex e.s.r. spectrum obtained by Gunning *et al.*[19] assigned to SH is disproved. Results for SD are given.

5 AB_3 Radicals

Although by itself the methyl radical cannot be considered as an inorganic radical it forms part of the important isoelectronic series, BH_3^-, CH_3, NH_3^+ and H_3O^{2+}. Most observations of the methyl radical are well understood in a variety of environments but a new explanation has been suggested[20] for the observation of Garbutt *et al.*[21] in 1968 that below 150 K on Vycor glass the ^{13}C hyperfine coupling constant decreased below the generally accepted value, $a(^{13}C) = 38$ G, such that the extrapolation to 0 K predicted $a(^{13}C) = 10.6$ G. Symons and Rao[20] have suggested that under these circumstances the radical is not tumbling freely, which leads to a misinterpretation of the e.s.r. spectrum by assigning the measured values to $a(^{13}C)$. At 0 K the extrapolated value would be $A_\perp$. It is not easy to test this explanation, owing to the well-known influence of the out-of-plane vibration. It has recently been estimated[22] that $B_\perp(^{13}C) = -30$ G after taking the zero-point vibration energy into account. Therefore $A_\perp(^{13}C) \sim 8$ G predicted in this way is in good agreement with the extrapolated value.[21]

Two other examples of the methyl radical interacting with its immediate environment are of more significance. The quartet of quartets observed after γ-irradiation of sodium acetate trihydrate has been assigned to a methyl radical interacting with a neighbouring sodium (^{23}Na) nucleus, which itself determines the orientation of the radical.[23] It is suggested that the small differences (Table 3) between the isotropic hyperfine coupling constants compared with a 'free' methyl radical can be understood in terms of a slight bending. If so this would have to be very small to account for a change in $a(^{13}C)$ of only 1 G. A modification of the out-of-plane zero-point vibration due to the interaction with the sodium ion may be a sufficient explanation for the small discrepancy. If this is the case it is inappropriate to estimate spin density losses to the sodium ion. A less well-defined but equally significant interaction has been observed using porous Vycor glass as the supporting medium for methyl radicals.[24] Two different species are assigned to methyl radicals in different environments; one of these species is quite normal with no obvious interaction. The abnormal species

[18] Y. Kotake, M. Ono, and K. Kuwata, *Bull. Chem. Soc. Japan*, 1971, **44**, 2056.
[19] D. A. Stiles, W. J. R. Tyerman, O. P. Strauss, and I. E. Gunning, *Canad. J. Chem.*, 1966, **44**, 2149.
[20] M. C. R. Symons and K. V. S. Rao, *J. Chem. Soc.* (*A*), 1971, 263.
[21] G. B. Garbutt, H. D. Gesser, and M. Fujimoto, *J. Chem. Phys.*, 1963, **48**, 4605.
[22] T. A. Claxton, M. J. Godfrey, and N. A. Smith, *J. C. S., Faraday II*, 1972, **68**, 181.
[23] J. Janecka, H. M. Vyas, and M. Fujimoto, *J. Chem. Phys.*, 1971, **54**, 3229.
[24] T. Katsu, M. Yanagita, and Y. Fujita, *J. Phys. Chem.*, 1971, **75**, 4064.

has been found to be preferentially formed from photolysis of methane at 77 K in the presence of certain metal oxides, such as V_2O_5 and MoO_3, and the method has been used to observe CD_3 and C_2H_5 in both forms. There is a *g*-value shift (Table 3) between the two forms and a reduction in a_H for the abnormal form.

In 1963 [25] the possible existence of H_3O and NH_4 was predicted although a number of searches since then have either been negative or inconclusive. 0.01M Aqueous ceric perchlorate–perchloric acid solutions at − 180 °C were irradiated with u.v. light and then annealed at − 120 °C. The resulting e.s.r. spectrum was assigned to the radical H_3O.[26] The distribution of the unpaired electron in CH_3 and H_3O is expected to be so vastly different that it is hard to conceive that the spectrum of H_3O could be so similar to CH_3 (see Table 3). Apart from simple qualitative considerations *ab initio* UHF calculations suggest that $a(^1H) \sim 200$ G for H_3O.[27] The comparison in Table 3 of these results and others assigned to methyl radicals in differing environments immediately suggests that the species is, in all probability, the methyl radical also. Martin and Swift [26] have realized this similarity and have attempted to show experimentally that CH_3 radicals are not responsible for the observed spectrum. Clearly experiments with ^{17}O are essential. The high *g*-value for H_3O is also unexpected but may indicate an association with ceric ions. This would also apply equally well to a methyl radical as indicated above.[23, 24] $(H_3O)^{2+}$, which was rejected because the observed species can be produced in basic media,[26] could conceivably have a very similar spectrum to CH_3, from qualitative reasoning and *ab initio* UHF calculations.[27]

Whereas the temperature dependence of $a(^{13}C)$ is well understood in terms of the Boltzmann population of the out-of-plane vibrational energy levels [28] no change in $a(^{14}N)$ or $a(^1H)$ has been observed over the range 77—300 K for the isoelectronic radical ion NH_3^+.[20] This is expected since the force constant for the out-of-plane vibration in NH_3^+ is calculated to be at least twice that for CH_3.[22] The well-resolved spectrum assigned to the BH_3^- radical ion obtained using tetramethylammonium borohydride is also independent of temperature between 84 and 300 K.[29] This is surprising since for an isoelectronic series one might have expected a smooth trend in the force constants for the out-of-plane vibrations, leading to a more marked temperature dependence of $a(^{11}B)$ compared to $a(^{13}C)$. Apart from this difficulty the assignment is qualitatively satisfactory but theoretically other difficulties exist. Again one looks for an explanation involving an interaction with the environment.[22] It is not unexpected that, if BH_3^- and CH_3 interact with the environment, similar evidence has

[25] H. J. Bernstein, *J. Amer. Chem. Soc.*, 1963, **85**, 484.
[26] T. W. Martin and L. L. Swift, *J. Amer. Chem. Soc.*, 1971, **93**, 2788.
[27] T. A. Claxton, M. J. Godfrey, and M. C. R. Symons, unpublished work.
[28] D. M. Schrader and K. Morokuma, *Mol. Phys.*, 1971, **21**, 1033.
[29] E. D. Sprague and F. Williams, *Mol. Phys.*, 1971, **20**, 375.

become available for NH_3^+. Janecka and Fujimoto[30] have studied NH_3^+ in NH_4ClO_4 and concluded, in agreement with Symons and Rao,[20] that the motion of the radicals is not random. In addition it is postulated that the random rotational diffusion occurs about two distinct directions in the unit cell jumping from one orientation to another. Analogous conclusions were obtained for NH_3^+ prepared by γ-irradiating adenosine diphosphate.[31]

A more particular type of environmental interaction involves hydrogen bonding.[32] The increase in $a(^{14}N)$ for NH_2 and H_2CN in environments where hydrogen bonding can be expected is explained by assuming that NH_2 becomes more like NH_3^+ and H_2CN more like H_2CNH^+, which is isoelectronic with the vinyl radical and probably has a similar structure. The NH_2 radial has also been observed in a medium which does not contain ammonia or an amino-group, that is, in a γ-irradiated frozen (77 K) aqueous azide matrix.[33]

Trends in the data for NH_3^+, CH_3, and BH_3^- suggest that all these radicals are planar.[34] For the second-row elements the results are consistent with the expected bent configuration. The absence of resolved proton splitting for AlH_3^- is theoretically reasonable.[35] SiH_3 is well established with resolved proton splittings, and an e.s.r. spectrum attributed to PH_3^+ has been claimed,[36] again without resolved proton hyperfine coupling. Theoretical[37] and experimental estimates of the degree of bending from planarity indicate that PH_3^+ is the most planar of the series. Two mechanisms must be at play in determining the spin density at the protons because only resolvable splitting is observed for the middle member of the series. If it is assumed that the proton coupling constant increases positively the larger the bending, then the charge effect must produce an increase in the positive proton coupling constant the more positive the radical charge, in order to rationalize the results. The theoretical results for AlH_3^- and SiH_3 agree quite well with experiment, but $a(^{31}P) = 195.5$ G from theory, compared with $a(^{31}P) = 517$ G from experiment for PH_3^+. Significant environmental interactions cannot be ruled out, particularly since such interactions are expected to be larger for charged radicals than for neutral radicals.

This work has been supplemented by studying the alkyl derivatives of these radicals.[36] γ-Irradiation of frozen sulphuric acid solutions of the alkyl derivatives of phosphorus produced radical species described as PEt_3^+ and $PBu^n{}_3^+$. It was estimated that these alkyl derivatives were more planar than PH_3^+. On the other hand the corresponding alkyl derivatives

[30] J. Janecka and M. Fujimoto, *J. Magn. Resonance*, 1971, **4**, 47.
[31] I. Suzuki and R. Abe, *J. Phys. Soc. Japan*, 1971, **31**, 951.
[32] K. V. S. Rao and M. C. R. Symons, *J. Chem. Soc.* (*A*), 1971, 2163.
[33] P. N. Moorthy, K. N. Rao, and J. Shankar, *Nature*, 1971, **234**, 17.
[34] R. C. Catton, M. C. R. Symons, and H. W. Wardale, *J. Chem. Soc.* (*A*), 1969, 2622.
[35] T. A. Claxton and N. A. Smith, *Theor. Chim. Acta*, 1971, **22**, 378.
[36] A. Begum, A. R. Lyons, and M. C. R. Symons, *J. Chem. Soc.* (*A*), 1971, 2290.
[37] N. A. Smith, Ph.D. Thesis, University of Leicester, 1971.

of aluminium including $AlMe_3^-$, $Al(n\text{-}octyl)_3^-$, $Al(n\text{-}decyl)_3^-$, and $AlEt_2Cl^-$, obtained by γ-irradiation at 77 K, were estimated to be more bent than AlH_3^-. Electronegativity difference arguments would require an increase in bending for the alkyl derivatives. Although steric effects favour planar structures, hyperconjugation is suggested to be dominant in the explanation of the tendency of phosphorus compounds towards planarity. Significant hyperconjugation is invoked only for PR_3^+, owing to the effective positive charge on the central atom.

Associated with this work has been the observation, simultaneously with the above alkyl derivatives, of radicals of the general type $R_2C{-}X$ (R is H or alkyl, X is an alkyl-substituted P or Al).[38] This is possibly a result of the non-selective nature of damage by γ-irradiation or a subsequent complex reaction sequence. It is concluded [38] that, in contrast to the results of recent *ab initio* SCF-MO calculations,[39] d_π–p_π bonding between phosphorous and oxygen is not of chemical significance. Also contrary to other work on phosphobetaine radicals,[40] the sign of $a(^{31}P)$ is concluded to be negative.

6 Other Group V Radicals

For the first time, evidence for formaldiminoxyl radicals (CH_2NO) has been obtained by photolysing diazomethane in an argon matrix at 4.2 K in the presence of NO.[41] The existence of CH_2NO had been invoked as an intermediate in the gas-phase reaction of CH_2 and NO. The hyperfine coupling constants: $a(^{14}N) = 30$ G, *cis* $a(^1H) = 2.8$ G, *trans* $a(^1H) = 26.2$ G, agree very well with calculated values:[42] $a(^{14}N) = 33$ G, *cis* $a(^1H) = 6$ G, *trans* $a(^1H) = 27$ G. Under similar conditions, photolysing diazomethane in the presence of D_2O produced deuteriated methyl radicals, thereby providing further evidence for the ability of methylene radicals to abstract hydrogen atoms.

γ-Irradiation at 77 K of phosphoryl chloride, phosphorus trichloride, and phosphorus pentachloride produced e.s.r. spectra subsequently assigned to the radicals $POCl_3^-$, $POCl_2$, PO_2Cl_2, PCl_2, and PCl_4.[43] The first three are respectively isostructural with PO_4^{4-}, PO_3^{2-}, and PO_4^{2-} and use was made of the characteristic ^{31}P hyperfine tensors in the assignments. One related aspect concerns the radical PF_2 for which varying results have been obtained. The results of Gordy *et al.*[44] are supported [43] after comparison with PCl_2.

38 A. Begum, A. R. Lyons, and M. C. R. Symons, *J. Chem. Soc. (A)*, 1971, 2288.
39 I. H. Hillier and V. R. Saunders, *Chem. Comm.*, 1970, 316; *J. Chem. Soc. (A)*, 1970, 2475.
40 E. A. C. Lucken and C. Mazeline, *J. Chem. Soc. (A)*, 1966, 1074; 1967, 439.
41 J. B. Farmer, C. L. Gardner, M. C. L. Gerry, C. A. McDowell, and P. Raghunathan, *J. Phys. Chem.*, 1971, **75**, 2448.
42 G. Berthier, H. Lemaire, A. Rassat, and A. Veillard, *Theor. Chim. Acta*, 1965, **3**, 213.
43 A. Begum and M. C. R. Symons, *J. Chem. Soc. (A)*, 1971, 2065.
44 W. Gordy, G. Jackel, and W. Nelson, *J. Chem. Phys.*, 1970, **52**, 4572.

Almost identical results have been obtained by Kerr and Williams [45] who concluded that the radicals $POCl_3^-$ and PCl_4 must be tumbling almost freely in their respective matrices: phosphoryl chloride and phosphorus trichloride. The structure of $POCl_3^-$ seems best described by a distorted trigonal bipyramid with the two identical chlorines placed on the trigonal axis.

γ-Irradiation of potassium difluorophosphate gave a species which was identified as PO_2F^-;[46] one of the few isoelectronic radicals with one or more of the oxygen ligands replaced by fluorine. This spectrum has similarities with that assigned to the $PF^{\pm}$ radical by Morton.[47]

γ-Irradiation at 77 K of potassium antimonate gave a spectrum assigned to SbO_4^{2-}; sodium hexafluoroantimonate gave a similar spectrum, suggesting a species of the type $\dot{O}$—SbX where X is an undetermined collection of ligands.[48] Similar species were observed using analogous hexafluorarsenate crystals. The detailed natures of these radicals are discussed and a number of possibilities are proposed.

The e.s.r. technique is well known to provide results which are readily explained directly in terms of hyperconjugation. Evidence for greatly enhanced conjugation has been observed in alkyl radicals containing phosphorus, arsenic, or tin [49] of the general type $H_2C^{\bullet}—CH_2X$ (where X is a group containing P, As, or Sn). The delocalization of the unpaired electron is unexpectedly large which results in an estimated orbital population of *ca.* 6% for *s* (compared with 1% commonly observed in similar situations) and 15% for *p*. Certain mechanistic repercussions are indicated but p_π–d_π interactions are ruled out as being responsible.

The NO_2 radical has been studied once again,[50] this time using ^{17}O over a range of temperature in single crystals of $NaNO_3$. At room temperature the NO_2 radicals are thought to be rotating about the C_3 rotation axis of the parent NO_3^- ion. At 77 K the radical is trapped at three distinct sites in the crystal.

7 Other Group VI Radicals

The paramagnetic species obtained by dissolving sulphur in oleum has been studied again using sulphur samples enriched with 10 and 25% ^{33}S.[51] The spectra are not resolved and a completely unambiguous assignment of the spectra to the species S_8^+ is not possible. Although indications were that S_8^+ fitted the observed spectra better than S_4^+ a more recent communication from Beaudet and Stephens [52] prefers S_4^+ using sulphur

[45] C. M. L. Kerr and F. Williams, *J. Phys. Chem.*, 1971, **75**, 3023.
[46] A. Begum, S. Subramanian, and M. C. R. Symons, *J. Chem. Soc.* (*A*), 1971, 700.
[47] J. R. Morton, *Canad. J. Phys.*, 1963, **41**, 706.
[48] M. C. R. Symons, *J. Chem. Soc.* (*A*), 1971, 2393.
[49] A. R. Lyons and M. C. R. Symons, *Chem. Comm.*, 1971, 1068.
[50] B. L. Silver, J. Koresh, and S. Schlick, *Mol. Phys.*, 1971, **22**, 225.
[51] M. Stillings, M. C. R. Symons, and J. G. Wilkinson, *J. Chem. Soc.* (*A*), 1971, 3201.
[52] R. A. Beaudet and P. J. Stephens, *Chem. Comm.*, 1971, 1083.

enriched to 91.8% ^{33}S. There are some discrepancies in the *g*-values (2.0134 [51] and 2.0163 [52]) and hyperfine coupling constants (7.7 G [51] and 8.87 G [52]) which require explanation before any firm conclusions can be arrived at. The current situation is that the species observed is neither S_4^+ or S_8^+. Evidence for S_5^+ is now under consideration.[53]

8 Group VII Radicals

Monatomic Ions.—A brief report has appeared describing the possible observation of the Br^{2-} and I^{2-} ions in halide-doped barium sulphate solutions which have been γ-irradiated at 77 K.[54]

Diatomic Ions.—The existence of Cl_2^- is well established. It has recently proved useful as a probe with regard to the nature of some zeolites.[55] Two distinct adsorption sites are observed in the case of faujasite. One of these sites is analogous to a 'free' Cl_2^- (as observed in KCl) but the other site has been interpreted as being under the influence of a strong electric field, the radical being so oriented that the field is perpendicular to the inter-nuclear axis. This field will mix the low-lying $^2\Pi_g$ state with the $^2\Sigma_u$ ground state. In mordenite the interpretation of the results requires a large component of the electric field to be oriented parallel to the molecular axis so that the possibility of significant mixing with the $^2\Sigma_g$ excited state is considered.

The V_k centres in alkali-halide crystals correspond to the same species which are found by γ-irradiation at 77 K of small beads of aqueous solutions of alkali-metal halides. I_2^- occurs in saturated aqueous solutions with slightly different e.s.r. parameters from those found in dilute solutions.[56] An environmental interaction is indicated. In addition the presence of strong electric fields is possible, so that mixing in of excited electronic states cannot be excluded. A clear example of such mixing may occur with nitrogen atoms trapped in sodium azide at 77 K where the ^4P state could mix in with the ^4S state to explain the observed spectra.[57]

Hydroxyhalides.—Evidence for $BrOH^-$ and IOH^- has now been provided [56] so that in the complete series only FOH^- has not been observed. The estimated spin density on the halogen in this series varies linearly with electronegativity of the halogen, predicting a very low spin density on fluorine in FOH^-. FO_2^- is known and fits on the plot. There seems to be no need to suggest that FO_2^- correlates well since a factor of at least two in the spin density on fluorine between FO_2^- and FOH^- would not affect the linearity. If FOH^- has a small spin density on fluorine then it would be correspondingly smaller for FO_2^-.

53 M. C. R. Symons, personal communication.
54 K. V. S. Rao and M. C. R. Symons, *Chem. Comm.*, 1971, 268.
55 J. A. R. Coope, C. L. Gardner, C. A. McDowell, and A. I. Pelman, *Mol. Phys.*, 1971, **21**, 1043.
56 I. Marov and M. C. R. Symons, *J. Chem. Soc. (A)*, 1971, 201.
57 A. Begum and M. C. R. Symons, *J. Chem. Soc. (A)*, 1971, 2062.

Oxides.—The long list of oxides of the non-metals is now extended by the assignment of e.s.r. spectra for γ-irradiated $KBrO_4$ crystals at 10 or 26 K to the species BrO_4^-, BrO_4, and $(BrO_4)_2^-$.[58] These observations, including the formation of BrO_2 above 20 K, are quite analogous to the results previously obtained using $KClO_4$, and the conclusions regarding the types of primary centres and the observed thermal processes are the same. The radicals observed by γ-irradiation of sodium chlorate at room temperature have been restudied at temperatures as low as 19 K. The species, which clearly contained a single strongly interacting nucleus, was assigned by Gamble to be ClO_3.[59] This was not in accord with other assignments for ClO_3 and Symons[60] has produced evidence suggesting that Gamble's radical is ClO_2 undergoing restricted rotational or librational motion. ClO_2 has also been studied while adsorbed on zeolites. As with Cl_2^- two distinct adsorption sites are apparent on faujasite with very different hyperfine splitting constants.[55] One interesting feature is that one of the adsorbed species can be pumped off preferentially.

IO_2F^- has been identified as a product in γ-irradiated single crystals of potassium oxyfluoroiodate.[61] CNDO calculations of the unpaired spin density distribution led to the conclusion that the most probable structure is similar to a slightly non-planar T-shaped radical. The large quadrupole coupling constant [$e^2qQ(^{127}I)$ = 1096 MHz] and the asymmetry parameter (η = 0.11) are consistent with this geometry. Iodine orbitals of d symmetry are not necessary in the qualitative description of this radical.

XeF.—This radical is interesting since it is isoelectronic with IF^-. It is prepared by γ-irradiating XeF_2 at 77 K.[62] Since both XeF and IF^- fit reasonably well, on an electronegativity basis, with other V_k centres which could not conceivably involve significant d-orbital participation in the bonding, it was concluded that in XeF only the xenon $5s$ and $5p_z$ orbitals were used in the bonding. A recent calculation[63] on XeF_2 made a major use of the d_{z^2} orbital but a similar calculation on XeF agreed with the conclusion outlined above from an interpretation of the e.s.r. results.

[58] J. R. Byberg, *J. Chem. Phys.*, 1971, **55**, 4867.
[59] F. T. Gamble, *J. Chem. Phys.*, 1965, **42**, 3542.
[60] M. C. R. Symons, *J. Chem. Soc.* (*A*), 1971, 698.
[61] S. Subramanian and M. T. Rogers, *J. Phys. Chem.*, 1971, **75**, 3479.
[62] R. S. Eachus and M. C. R. Symons, *J. Chem. Soc.* (*A*), 1971, 304.
[63] R. C. Catton and K. A. R. Mitchell, *Chem. Comm.*, 1970, 457.

9
Organic Radicals in Solution

BY B. C. GILBERT

This Report is devoted to recent developments involving the application of e.s.r. spectroscopy in organic chemistry and is primarily concerned with three areas in which research has been particularly active. These are, firstly, the analysis of spectra in terms of the shape, delocalization, and conformation of free radicals; secondly, the derivation of kinetic and thermodynamic parameters for a variety of processes involving radicals; and, thirdly, the diagnosis of their participation in chemical reactions.

The scope of the coverage will be similar to that of an earlier review [1] in that attention will be almost completely restricted to the isotropic spectra of organic radicals in fluid solution. Spectra from radicals in solids will only be discussed where the results are especially relevant to the characterization of species in solution. Triplets and biradicals will be specifically excluded, as will applications of spin-labelling; further, applications of theoretical methods will usually be referred to only where they have been ancillary to essentially experimental investigations. Finally, for aromatic radicals, discussion will be mainly restricted to uncharged radicals, since radical-ions and the dynamic processes associated with them are reviewed elsewhere in this volume (Chapter 5).

Relevant books which have now become available include English translations of works on nitroxides [2] and on the high-resolution spectra of organic radicals,[3] a monograph in German,[4] and, very recently, a comprehensive textbook.[5] A recent Chemical Society Special Publication devoted to free radicals contains several detailed pertinent accounts of various aspects of the application of e.s.r. to their study, notably on the generation of short-lived radicals, both photochemically [6] and with flow systems,[7] and

1 B. C. Gilbert, *Ann. Reports* (*B*), 1970, **67**, 18.
2 E. G. Rozantsev, 'Free Nitroxyl Radicals', Plenum Press, New York, 1970.
3 F. Gerson, 'High Resolution E.S.R. Spectroscopy', John Wiley–Verlag Chemie, Weinheim, 1970.
4 K. Scheffler and H. B. Stegmann, 'Electronenspinresonanz', Springer-Verlag, Berlin–Heidelberg, 1970.
5 J. E. Wertz and J. R. Bolton, 'Electron Spin Resonance', McGraw-Hill, New York, 1972.
6 J. K. Kochi and P. J. Krusic, in 'Essays on Free-radical Chemistry', Chemical Society Special Publication No. 24, 1970, p. 147.
7 R. O. C. Norman, in 'Essays on Free-radical Chemistry', Chemical Society Special Publication No. 24, 1970, p. 117.

on the technique of spin-trapping.[8] Other relevant reviews include two more on spin-trapping,[9, 10] an account of halogen hyperfine interactions,[11] two[12, 13] concerned with the conformational analysis of free radicals (particularly nitroxides), and reviews of the application of e.s.r. to investigations of radicals formed during peroxide decompositions[14] and during polymerization.[15]

1 The Structure of Organic Radicals

Hyperfine Splittings.—The discussion that follows attempts to summarize the information that has been derived from isotropic splittings, and it separates, where possible, the results for central atom and α-substituents (*e.g.* the ^{13}C and ^{1}H hfs for •C—H) from those for β-substituents (*e.g.* •C—CH). The former category can be further sub-divided according to the type of information revealed, namely about hybridization (for aliphatic radicals in particular) and, especially for aromatic radicals, about delocalization of the unpaired electron. It is perhaps appropriate to begin with reference to the increasing importance attached to the first of these aspects.

The Geometry of Simple Radicals. The t-butoxycarbonyl radical, $\cdot CO_2Bu^t$, has been generated and detected along with the t-butyl radical during the photolysis of di-t-butyl peroxide in the presence of t-butyl formate: its low *g*-factor (2.0011) and high ^{13}C splitting (183.5 G) are entirely consistent with its formulation as a σ-radical (*cf.* benzoyl).[16] Phenyl radicals still defy detection in liquid solution, but information about isotropic splittings for these and other σ-radicals often results from the analysis of solid-state spectra. Thus, details for the acetyl and benzoyl radicals,[17] as well as for the phenyl and some pyridyl radicals,[18, 19] are now available (the large ^{13}C splittings are as expected) and some interesting ring-opening reactions, induced by photolysis for the last-mentioned species, have been revealed.[19] INDO and Extended Hückel (EH) calculations have been carried out for some of these aromatic σ-radicals[18, 19] and calculations on vinyl include those employing INDO (results were also reported[20] for

[8] M. J. Perkins, in 'Essays on Free-radical Chemistry', Chemical Society Special Publication No. 24, 1970, p. 97.
[9] E. G. Janzen, *Accounts Chem. Res.*, 1971, **4**, 31.
[10] C. Lagercrantz, *J. Phys. Chem.*, 1971, **75**, 3466.
[11] A. Hudson and K. D. J. Root, *Adv. Magn. Resonance*, 1971, **5**, 1.
[12] A. Rassat, *Pure Appl. Chem.*, 1971, **25**, 623.
[13] E. G. Janzen, *Topics in Stereochemistry*, 1971, **6**, 177.
[14] B. C. Gilbert and A. J. Dobbs, 'Organic Peroxides', ed. D. Swern, Wiley, New York, 1972, Vol. 3, p. 271.
[15] H. Fischer, *Nuclear Magn. Resonance*, 1971, **4**, 301.
[16] D. Griller and B. P. Roberts, *Chem. Comm.*, 1971, 1035.
[17] J. E. Bennett and B. Mile, *Trans. Faraday Soc.*, 1971, **67**, 1587.
[18] J. E. Bennett and B. Mile, *J. Phys. Chem.*, 1971, **75**, 3432.
[19] P. H. Kasai and D. McLeod, *J. Amer. Chem. Soc.*, 1972, **94**, 720.
[20] M. F. Chiu, B. C. Gilbert, and B. T. Sutcliffe, *J. Phys. Chem.*, 1972, **76**, 553.

other σ-radicals including benzoyl and iminoxyl radicals), those accomplished[21] with a perturbational configuration interaction using localized orbitals (where direct and through-bond effects could be contrasted), and those with a modified EH procedure[22] which allows contributions to splittings both from positive spin density and from spin-polarization.

It is interesting that the α-styryl radical (•CPh=CH_2), although isoelectronic with benzoyl, appears to be a π-radical with the bonds linear at the radical centre and with the methylene protons in the anti-nodal plane, as implied by their large coupling (41.5 G).[23] The radical PhC≡C•, detected[24] during the photolysis of phenyliodoacetylene, also has the unpaired electron in a delocalized π-orbital, as evidenced by the *g*-factor (2.0030) and the ring-proton splittings.

The isotropic e.s.r. spectra of the bridgehead radicals 1-adamantyl (1) and bicyclo[2,2,2]octyl (2) have been obtained by employing photolytic decomposition of the appropriate t-butyl peroxycarboxylates in cyclopropane solution at low temperatures.[25] The hyperfine splittings for both species were assigned on the basis of INDO calculations, good agreement being obtained on the assumption of slightly flattened – but still distinctly pyramidal – radical centres. Though no ^{13}C splittings were observed (calculations suggest a value of about 114 G for the α-carbon), interest centres on the low β-proton splittings and on the long-range γ- and δ-proton interactions [see (1) and (2), splittings in gauss]. It was also concluded that

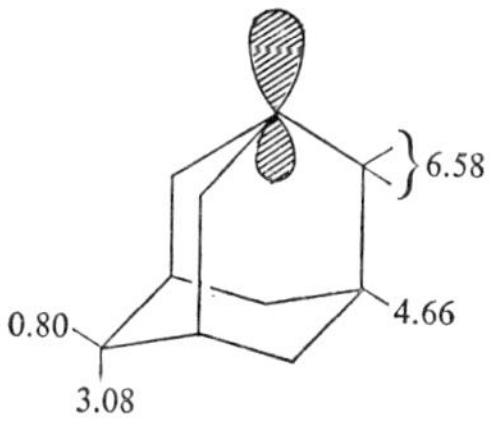

(1)

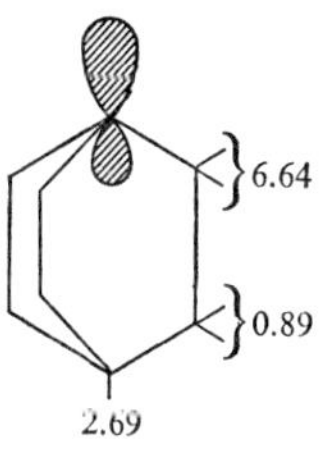

(2)

there is no evidence for a specific 'backlobe' interaction between the unpaired electron and the bridgehead C—H bond in (2), and indeed calculations suggest that closer approach of the orbital concerned leads to smaller splittings. It has now been shown,[26] after some controversy, that *X*-irradiation of solid adamantane at 77 K leads to the 2-adamantyl radical, with $a(\alpha\text{-H}) = 20.5$ G.

[21] J. Langlet, M. Gilbert, and J.-P. Malrieu, *Theor. Chim. Acta*, 1971, **22**, 80.
[22] R. J. Fitzgerald and D. H. Dugre, *Theor. Chim. Acta*, 1971, **22**, 361.
[23] J. E. Bennett and J. A. Howard, *Chem. Phys. Letters*, 1971, **9**, 460; L. Bonazzola, S. Fenistein, and R. Marx, *Mol. Phys.*, 1971, **22**, 689.
[24] J. S. Coleman, A. Hudson, K. D. J. Root, and D. R. M. Walton, *Chem. Phys. Letters*, 1971, **11**, 300.
[25] P. J. Krusic, T. A. Rettig, and P. von R. Schleyer, *J. Amer. Chem. Soc.*, 1972, **94**, 995.
[26] J. R. Ferrell, G. R. Holdren, jun., R. V. Lloyd, and D. E. Wood, *Chem. Phys. Letters*, 1971, **9**, 343.

It now seems clear that many radicals, although not markedly bent, are not planar at the radical centre and therefore not, strictly speaking, of π-type. For example, although $\cdot CH_2Cl$ appears to be essentially planar,[27] as does $\cdot$CHMeCl,[28] $\cdot CCl_3$ is pyramidal as indicated both by a(Cl)[28] and by the splittings for this radical generated[29] by γ-radiolysis of carbon tetrachloride at 77 K [$a(^{35}Cl)$ = 6.25 G, $a(^{13}C)$ = 114 G; another trapped radical, presumably a geometrical modification, has closely similar splittings]. The radical $\cdot CCl_3$ has a structure intermediate between those of $\cdot CH_3$ and $\cdot CF_3$ and it has been estimated that $Cl\widehat{C}Cl$ is 116°.[29] These results are in reasonable agreement with predictions[30] of equilibrium geometries and of hyperfine splittings made on the basis of INDO calculations for this and other radicals containing second-row elements, *e.g.* $\cdot CFCl_2$, $\cdot SiCl_3$, and $\cdot SiH_3$. (Relevant parameters for use with this method were reported: alternative C, H,[20] and N[20,31] parameters have also been proposed for specific radical types by other workers.)

There has been considerable interest in possible effects on the hybridization when other heteroatoms, and oxygen in particular, are substituted at a radical centre.[32–34] For instance, for $\cdot CH_2OH$ it has been argued[32] that the room-temperature ^{13}C-splitting of 4.74 mT (47.4 G) is greater than that for the methyl radical (3.85 mT) because of bending at the radical centre of the former, in accord with INDO predictions based on the consideration both of splittings and of total energy.[32] A detailed study of $\cdot CH_2OH$ (and its ^{2}H-, ^{13}C-, and ^{17}O-substituted analogues), generated photolytically over a range of temperature, has now been reported, and the temperature variations are in accord with increasing out-of-plane vibration as the temperature is raised.[35] Although the detailed study cannot be said either to corroborate the claim for non-planarity or to disprove it, it is interesting that[34] the large $a(^{13}C)$ for $\cdot CH_2OH$ cannot be accounted for using reasonable values of ρ_C and ρ_O with typical Karplus–Fraenkel parameters.

The magnitudes of $a(^{13}C)$ and $a(\alpha$-H) for $\cdot CH_2OH$ become easier to understand if a variety of substituted radicals is considered.[32–34] For example, increased bending at the radical centre with the introduction of methyl groups seems the most reasonable explanation for the increased $a(\alpha$-$^{13}C)$ splittings for $\cdot$CH(OH)Me and $\cdot CMe_2OH$, 5.41 and 6.50 mT respectively, in aqueous solution at room temperature (this is in the opposite direction to that predicted by Karplus–Fraenkel theory for planar radicals).[34]

[27] J. P. Michaut and J. Roncin, *Chem. Phys. Letters*, 1971, **12**, 95.
[28] J. Cooper, A. Hudson, and R. A. Jackson, *Mol. Phys.*, 1972, **23**, 209.
[29] C. Hesse, N. Leray, and J. Roncin, *Mol. Phys.*, 1971, **22**, 137.
[30] H. G. Benson and A. Hudson, *Theor. Chim. Acta*, 1971, **23**, 259.
[31] D. M. Hirst, *Theor. Chim. Acta*, 1971, **20**, 292, 409.
[32] A. J. Dobbs, B. C. Gilbert, and R. O. C. Norman, *J. Chem. Soc.* (*A*), 1971, 124.
[33] A. L. J. Beckwith and P. K. Tindal, *Austral. J. Chem.*, 1971, **24**, 2099.
[34] A. J. Dobbs, B. C. Gilbert, and R. O. C. Norman, *J. C. S. Perkin II*, 1972, 786.
[35] P. J. Krusic, P. Meakin, and J. P. Jesson, *J. Phys. Chem.*, 1971, **74**, 3438.

Deviations from planarity are much more pronounced when more than one oxygen atom is substituted at the radical centre:[32] for example, in (3), $a(\alpha\text{-}^{13}C)$ is (+)9.98 mT and $a(\alpha\text{-H})$ is (±)0.047 mT, the very small value for the latter reflecting not the absence of spin density on the central carbon but rather a degree of bending such that positive contributions to $a(\alpha\text{-H})$ cancel out the (normally) negative splitting. The mesomeric (+ *M*) effect of the oxygen ($H_2\dot{C}\text{—}\ddot{O}H \leftrightarrow H_2\overset{-}{\ddot{C}}\text{—}\overset{+}{\dot{O}}H$) is thought to be

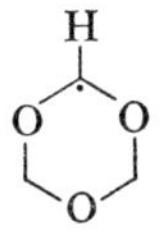
(3)

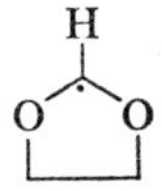
(4)

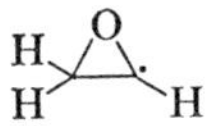
(5)

important and it is significant that in (4), where overlap between the oxygen *p*-orbitals and the radical centre can be more effective, the large $a(\alpha\text{-H})$ value of 2.15 mT (21.5 G) must surely be positive, corresponding to a greater degree of bending.[32–34] Constraining two of the bonds at the radical centre into a small ring also appears to increase the *s*-character of the orbital of the unpaired electron (*cf.* cyclopropyl): in the oxiran radical (5) the α-proton splitting (2.33 mT) is again probably positive, the increased bending also reflecting the effect of oxygen.[32] Other interesting features of this radical, which shows no evidence of ring-opening in aqueous solution at room temperature, are that the *β*-protons exhibit a low splitting (*cf.* 1-adamantyl) and that they appear equivalent; evidently inversion is rapid (*i.e.* $> 10^8\ s^{-1}$; see also ref. 36). For a variety of related radicals (R•), e.s.r. parameters correlate fairly well with n.m.r. parameters for the parent compounds (RH) [*e.g.* $a(\alpha\text{-H})$ and $a(\alpha\text{-}^{13}C)$ with $J(^{13}CH)$, and $a(\alpha\text{-H})$ with $J(HH_{gem})$].[32, 33] It has also been proposed that, for radicals of the type •CHMeX, the ratio $a(\text{Me}) : a(\alpha\text{-H})$ serves as a guide to the degree of bending.[32, 34] Thus, for planar radicals (*e.g.* X = Ph) the ratio is approximately 1.2, but, because $a(\alpha\text{-H})$ decreases markedly on bending, a larger ratio implies deviations from planarity. In this way it is argued that bending occurs for X = NH_2 and NR_2, and also for OH, OR, and O^-, the bending being greatest in the last-named case. The two factors which appear to be important in determining the degree of bending at the tervalent carbon atom are thought to be the presence of one or more conjugatively electron-releasing substituents and the inductive removal of electron density. It is interesting that a trifluoromethyl group evidently has a radical-flattening effect, possibly because of interaction between the singly occupied orbital and an unoccupied π^*-orbital of CF_3.[32] It has been independently proposed,[37] on the basis on INDO calculations, that aliphatic

[36] L. J. Altman and R. C. Baldwin, *Tetrahedron Letters*, 1972, 981.
[37] G. R. Underwood, *Mol. Phys.*, 1971, **22**, 729.

ketyls ($R_2CO^{-}\cdot$)are non-planar at the radical centre, the effect being electronic in origin [though other workers [38] argue that results for $a(\alpha\text{-}^{13}C)$ *can* be accommodated on the assumption of a planar structure]. INDO calculations also support the claim for non-planarity about the radical centres for some 6-oxabicyclo[2,2,1]hept-2-yl radicals.[39]

It is clear that substitution of a carboxylate group together with an oxygen substituent at a radical centre overcomes any tendency to bending, as illustrated by results for radicals of the type $\cdot CR(O^{-})CO_2^{-}$ obtained by electron bombardment of aqueous solutions of carboxylic acids.[40] For $\cdot CMe(O^{-})CO_2^{-}$, $a(\alpha\text{-}^{13}C)$ is 14.09 G, consistent with a planar structure, and the $a(\beta\text{-}^{13}C)$ values are 8.19 and 6.35 G. For $\cdot CH(CO_2^{-})_2$, $a(\alpha\text{-}^{13}C)$ is 31.44 and $a(\beta\text{-}^{13}C)$ (−) 11.92 G.[40] Interestingly, the solid-state data for $\cdot ^{13}CF_2CO_2^{-}$ suggest that this radical is probably non-planar at the radical centre,[41] the bending effect of α-fluorine presumably being dominant.

Other radicals reported to be planar include $\cdot CHMe\cdot N{=}C{=}O$ and $\cdot CHMe\cdot N{=}C{=}S$ (for which INDO calculations were performed) [42] and a variety of sulphur-conjugated radicals.[43] Indeed, it has been concluded [43] that not only is a sulphur substituent less efficient than oxygen at inducing bending [*cf.* the ratio of a(Me) : a(α-H) for $\cdot CHMeSR$ [34]] but also that delocalization on to sulphur is more extensive than on to oxygen. Since the $+M$ effect of sulphur is weaker, perhaps it might be tentatively suggested that the interaction occurs *via d*-orbitals, *i.e.* $\rangle\dot{C}{-}\ddot{S}{-} \leftrightarrow \rangle C{=}\dot{S}{-}$ (see ref. 44 for example). There also appears to be little, if any, bending for $(PhS)_3C\cdot$ [$a(\alpha\text{-}^{13}C)$ = 43.69 G] [45] and $\cdot CPh_2OSiPh_3$ [$a(\alpha\text{-}^{13}C)$ = 25 G].[46]

The view [34, 47] that radicals of the type $\cdot CR_2NR_2$ are non-planar around the radical centre has been challenged:[48] an alternative explanation proposed is that these π^3-radicals are locally planar at carbon but slightly pyramidal at nitrogen, the magnitude of the α-proton splitting for $\cdot CH_2NR_2$ depending on the degree of twist about the C—N bond (see also ref. 49). The π^1-radical $\cdot CH_2BMe_2$ appears to be strictly planar.[48]

Interest in the hybridization of some small nitrogen-centred free radicals has stimulated some interesting and sometimes controversial work. For example, the aziridinyl- (6) and azetidinyl- (7) radicals have been generated [50]

[38] G. A. Russell, D. F. Lawson, H. L. Malkus, and P. R. Whittle, *J. Chem. Phys.*, 1971, **54**, 2164.
[39] T. Kawamura, T. Koyama, and T. Yonezawa, *J. Amer. Chem. Soc.*, 1970, **92**, 7222.
[40] G. P. Laroff and R. W. Fessenden, *J. Chem. Phys.*, 1971, **55**, 5000.
[41] M. T. Rogers and L. D. Kispert, *J. Chem. Phys.*, 1971, **55**, 2360.
[42] D. E. Wood, R. V. Lloyd, and W. A. Lathan, *J. Amer. Chem. Soc.*, 1971, **93**, 4145.
[43] I. Biddles, A. Hudson, and J. T. Wiffen, *Tetrahedron*, 1972, **28**, 867.
[44] A. Ohno, N. Kito, and Y. Ohnishi, *Bull. Chem. Soc. Japan*, 1971, **44**, 463, 467, 470.
[45] D. Seebach, H. B. Stegmann, and A. K. Beck, *Angew. Chem. Internat. Edn.*, 1971, **10**, 500.
[46] A. R. McIntosh and J. K. S. Wan, *Mol. Phys.*, 1971, **22**, 183.
[47] D. E. Wood and R. V. Lloyd, *J. Chem. Phys.*, 1970, **53**, 3932.
[48] A. R. Lyons and M. C. R. Symons, *J. C. S. Faraday II*, 1972, 502.
[49] R. Poupko, A. Loewenstein, and B. L. Silver, *J. Amer. Chem. Soc.*, 1971, **93**, 580.
[50] W. C. Danen and T. T. Kensler, *Tetrahedron Letters*, 1971, 2247.

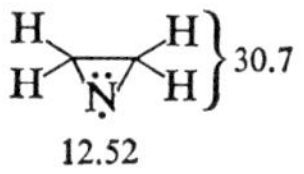

(6)

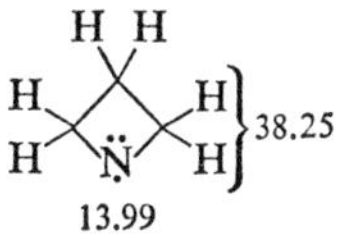

(7)

by reaction of the parent amines with photolytically prepared t-butoxyl radicals in cyclopropane at − 100 °C and the splittings (gauss) suggest, in agreement with INDO calculations, that for both species the unpaired electron resides in a $p(\pi)$-orbital.

Similarly, *N*-alkoxy-*N*-alkylamino-radicals, generated [51] by photolysis of solutions of t-butyl *N*-alkylperoxycarbamates or *NO*-dialkylhydroxylamines, appear to be of π-type, with considerable delocalization of the unpaired electron on to oxygen [for $\cdot NMeOBu^t$ $a(N) = 14.47$ and $a(Me) = 21.51$ G, and $g = 2.0048$], as do the pyrrolidinyl and pyrrolinyl radicals (generated [52] in an adamantane matrix at 77 K). In both studies, INDO calculations are in good agreement with experiment. However, although a spectrum attributed to the 2,2,3,3-tetramethylaziridine nitroxide (8) has been described,[53] other workers [54] have questioned the assignment of the observed spectrum to this radical, and the splittings for this interesting species must be awaited. The corresponding azetidine nitroxide (9) is

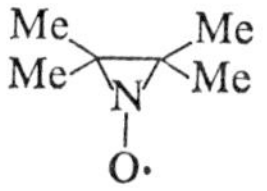

(8)

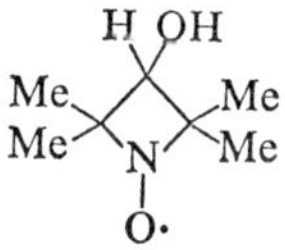

(9)

stable [55] (the small size of this radical lends weight to the suggestion that derivatives in this series could find useful employment as spin-labels) and the fairly high nitrogen splitting (17.75 G in an aqueous solution containing lithium chloride) may indicate a slight deformation from planarity around nitrogen.

Although alkoxy nitroxides, RN(OR)O•, and aliphatic nitro radical-anions, $RNO_2^{-}\cdot$, probably owe their large nitrogen splittings (*ca.* 2.5 mT) to bending [32] (*cf.* the effect of O^- and OR in carbon-centred radicals) there is little evidence to suggest that ordinary nitroxides (especially aromatic ones) are bent. However, a small degree of deformation from planarity

[51] W. C. Danen and C. T. West, *J. Amer. Chem. Soc.*, 1971, **93**, 5582.
[52] D. W. Pratt, J. J. Dillon, R. V. Lloyd, and D. E. Wood, *J. Phys. Chem.*, 1971, **75**, 3486.
[53] G. R. Luckhurst and F. Sundholm, *Tetrahedron Letters*, 1971, 675.
[54] P. Singh, D. G. B. Boocock, and E. F. Ullman, *Tetrahedron Letters*, 1971, 3935; J. F. W. Keana, R. J. Dinerstein, and D. P. Dolata, *ibid.*, 1972, 119.
[55] J.-C. Espie and A. Rassat, *Bull. Soc. chim. France*, 1971, 4385.

may account for variations in $a(\beta\text{-}^{13}C)$ for some aliphatic nitroxides [56] and an electron-diffraction study [57] has established that the bonds about nitrogen are not coplanar in $(CF_3)_2NO\cdot$ (though they are, interestingly, for di-t-butyl nitroxide; see also ref. 12).

Electrochemical oxidation of *N*-(*p*-nitrosophenyl)-*N*-phenylhydroxylamine gives a radical, Ph(*p*-C_6H_4NO)NO•, recognizable [58] as a π-species [a(N) = 9.14, a(N′) = 1.36 G] rather than a σ(iminoxyl) radical (compare the nitrogen splitting of 32.2 G for the moderately stable iminoxyl radical $Bu^t_2C{=}NO\cdot$, obtained by oxidation of the parent ketoxime [59]). Another novel (amido) radical, •N(Me)COMe, is thought to be produced when either 1,4-diacetyl-1,4-dimethyltetrazene-2 or the nitrosamine MeN(NO)-COMe is irradiated in toluene at low temperature;[60] the observed hyperfine splittings [a(N) = 6.9, $a(NCH_3)$ = 8.1 G] certainly support the hypothesis that this, and some related radicals, have a π-ground state, as predicted by INDO calculations: this is in contrast to the conclusions of other workers based on their product studies of the photolytic decomposition of *NO*-diacylhydroxylamines.[61] [However, confirmation of the e.s.r. assignment is awaited since these signals could presumably derive from carbonyl *nitroxides* formed in the reaction; the high *g*-values for the reported radicals, *ca.* 2.0066, would certainly support the alternative formulation for the radicals as acyl alkyl nitroxides, which have closely similar splittings.[62]] Another novel radical, detected during the rearrangement of a benzophenone oxime thionocarbamate [*e.g.* $Ph_2C{=}N{\cdot}OC({=}S)NMe_2$] to a thioxime carbamate, has been assigned [63] the structure $Ph_2C{=}N\cdot$: the nitrogen splitting (10.3 G) is typical of these (σ) radicals and it would be interesting to have a detailed analysis of the ring-proton splittings in the light of earlier results on aliphatic analogues (see ref. 1, for example).

Several (arylazo)oxyl radicals (ArN=NO•) have been detected, along with nitroxides, during the decomposition of some acylarylnitrosamines.[64] Their formulation as σ-radicals (*cf.* iminoxyl radicals) is consistent with their *g*-values (*ca.* 2.0015), the typical nitrogen splittings of *ca.* 31 and 2 G, and the invariance of these parameters for a series of radicals with *para*-substituents of markedly different electronic types. In addition, CNDO/2 and INDO calculations suggest that the unpaired electron should be assigned to an orbital of σ-symmetry, and for PhN=N′O• the INDO

[56] R. Brière, G. Chapelet-Letourneux, H. Lemaire, and A. Rassat, *Mol. Phys.*, 1971, **20**, 211.
[57] C. Glidewell, D. W. H. Rankin, A. G. Robiette, G. M. Sheldrick, and S. M. Williamson, *J. Chem. Soc.* (*A*), 1971, 468.
[58] G. Cauquis, A. Rassat, J.-P. Ravet, and D. Serve, *Tetrahedron Letters*, 1971, 971.
[59] J. L. Brokenshire, G. D. Mendenhall, and K. U. Ingold, *J. Amer. Chem. Soc.*, 1971, **93**, 5278.
[60] P. Tordo, E. Flesia, and J.-M. Surzur, *Tetrahedron Letters*, 1972, 183; P. Tordo, E. Flesia, G. Labrot, and J.-M. Surzur, *ibid.*, p. 1413.
[61] B. Danieli, P. Manitto, and G. Russo, *Chem. and Ind.*, 1971, 203.
[62] A. Mackor, Th. A. J. W. Wajer, and Th. J. de Boer, *Tetrahedron*, 1968, **24**, 1623.
[63] R. F. Hudson, A. J. Lawson, and E. A. C. Lucken, *Chem. Comm.*, 1971, 807.
[64] J. I. G. Cadogan, R. M. Paton, and C. Thomson, *J. Chem. Soc.* (*B*), 1971, 583.

calculation gives an energy minimum for $\widehat{NNO} = 125°$, with $a(N) = 8$ and $a(N') = 23$ G.[64] In contrast, INDO calculations for $PhN_2\cdot$ lead to predicted nitrogen splittings for this species of 4.4 and 5.1 G.[64]

The reaction of chlorine atoms (from Cl_2) or phenyl radicals (from dibenzoyl peroxide) with the chlorinated nitrone $CCl_2{=}NO(CCl_3)$ leads to a radical with splittings from nitrogen (7.4 G) and chlorine (four atoms equivalently, 1.25 G), which has been assigned [65] to a novel π-allyl radical $\cdot CCl_2{-}N(O){=}CCl_2$. (However, perhaps an alternative halogenocarbonyl nitroxide formulation, *e.g.* $Cl\cdot CO\cdot N(CCl_3)O\cdot$ cannot be completely ruled out, especially in view of the unexpected detection [66, 67] of this type of species in the reaction of trihalogenomethyl radicals with Bu^tNO.)

Other results of interest concerning splittings for nitroxides include the small values of $a(N)$, *ca.* 6.5 G, for some bis(organosilyl) nitroxides, for which the delocalization could presumably involve silicon *d*-orbitals.[68] Further factors which affect $a(N)$ include, for aromatic nitroxides, twisting of the plane of the ring out of the CNO plane,[69] and a variety of solvent effects (for acyl alkyl nitroxides) [70, 71] and substituent [71] effects, including the presence of Br or I (for nitronyl nitroxides). These and some related examples are further discussed in the next section. For a [2,2]paracyclophane nitroxide, an intramolecular hydrogen bond raises $a(N)$ [72] (see also ref. 73), and high nitrogen splittings are also observed for acid–base complexes between di-t-butyl nitroxide (and 2,2,6,6-tetramethylpiperidine *N*-oxyl) and Lewis acids such as aluminium chloride and the boron halides.[74]

Novel nitrogen-centred radicals whose e.s.r. spectra have been recently reported include a series of aryl cobalt nitroxides,[75] the nitroxides $PhN(SO_2H)O\cdot$ and $PhN(SO_3H)O\cdot$,[76] and the π-radicals *N*-nitrosophenyl nitroxide [77] and nitrosamine radical-anions.[78]

Splittings in Delocalized Species. This section is restricted to delocalized radicals, usually aromatic, and to the attempts that have been made to calculate hyperfine splitting constants, using various molecular orbital procedures. Only uncharged species will be primarily considered here, since radical-cations and -anions are treated separately elsewhere (Chapter 5).

[65] V. Astley and H. Sutcliffe, *Chem. Comm.*, 1971, 1303.
[66] J. W. Hartgerink, J. B. F. N. Engberts, and Th. J. de Boer, *Tetrahedron Letters*, 1971, 2709.
[67] R. J. Holman and M. J. Perkins, *Chem. Comm.*, 1971, 244.
[68] R. West and P. Boudjouk, *J. Amer. Chem. Soc.*, 1971, **93**, 5901.
[69] A. R. Forrester and S. P. Hepburn, *J. Chem. Soc. (C)*, 1971, 3323.
[70] H. G. Aurich and J. Trösken, *Annalen*, 1971, **745**, 159.
[71] E. F. Ullmann, L. Call, and J. H. Osiecki, *J. Org. Chem.*, 1970, **35**, 3623.
[72] A. R. Forrester and R. Ramasseul, *J. Chem. Soc. (B)*, 1971, 1638.
[73] T. Nishikawa and K. Someno, *Bull. Chem. Soc. Japan*, 1971, **44**, 851.
[74] T. B. Eames and B. M. Hoffmann, *J. Amer. Chem. Soc.*, 1971, **93**, 3141.
[75] M. G. Swanwick and W. A. Waters, *J. Chem. Soc. (B)*, 1971, 1059; J. Basters, *Rec. Trav. chim.*, 1972, **71**, 50.
[76] B. Lakatos, B. Turcsanyi, and F. Tudos, *Acta Chim. Acad. Sci. Hung.*, 1971, **70**, 225.
[77] A. T. Balaban, N. Negoita, and I. Pascaru, *Rev. Roumaine Chim.*, 1971, **16**, 721.
[78] G. R. Stevenson and C. J. Colón, *J. Phys. Chem.*, 1971, **75**, 2704.

Perhaps the most significant advance of general interest is reflected by the reports of INDO calculations on the benzyl radical.[79, 80] Two groups have found that the calculated minimum energy corresponds to a distorted structure, rather than to a regular hexagon, for the aromatic ring, and it is notable that this geometry leads to much better agreement with experiment for the splittings than is achieved by most orthodox methods. In particular, the *para*-proton splitting is predicted to be almost as large as the *ortho*-proton splittings. The energy minimum with respect to skeletal deformations [79] is the structure (10), with the bond lengths in Å; the calculated

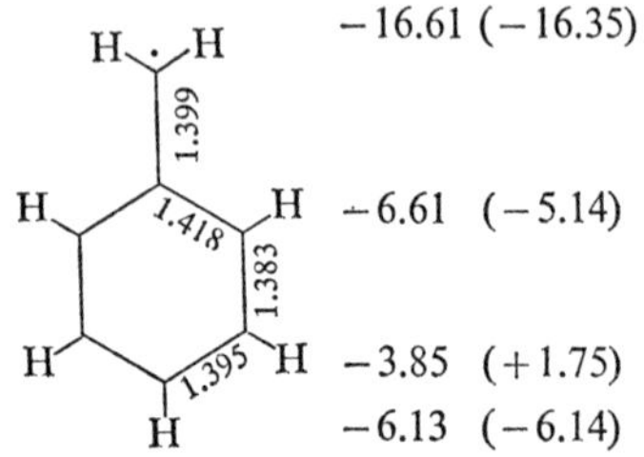

(10)

splittings are given together with, in parentheses, those observed (see also ref. 80).

It has also been pointed out [80] that the tendency towards a *para*-quinonoid structure is probably more pronounced for the phenoxyl radical than for benzyl [for the phenoxyl radical $a(p\text{-H}) = 10.1$ and $a(o\text{-H}) = 6.65$ G [81]].

It had previously been shown [82] that HMO–McLachlan calculations for benzyl can give improved agreement with the experimental splitting constants if the geometry is varied and if relationships are assumed between the bond length and the resonance integral β_{rs} (essentially the same geometry as that reported above gives the best fit). This approach has now been extended [83] to the trityl radical and to a series of phenyl-substituted imidazyl and pyrryl radicals, whose spectra have been recorded and analysed using e.s.r. and ENDOR. Good agreement was obtained if, in addition to slight variations in bond length, twisting of some of the phenyl rings was allowed; results were consistent with the view that these radicals are delocalized π-radicals. It was pointed out that the nitrogen splitting in tetraphenylpyrryl (1.905 G) is low because the pyrrole nitrogen lies at a node in the HMO containing the odd electron, and the appropriate σ–π parameters have been discussed. Comparison of results for some *p*-substituted tetraphenylpyrryl radicals with those for the correspondingly

[79] D. L. Beveridge and E. Guth, *J. Chem. Phys.*, 1971, **55**, 458.
[80] H. G. Benson and A. Hudson, *Mol. Phys.*, 1971, **20**, 185.
[81] T. J. Stone and W. A. Waters, *J. Chem. Soc.*, 1964, 213.
[82] P. V. Schastnev and G. M. Zhidomirov, *Teor. i eksp. Khim.*, 1967, **3**, 541.
[83] R. D. Allendoerfer and A. S. Pollock, *Mol. Phys.*, 1971, **22**, 661.

substituted trityl radicals shows that the substituent effects are different for the two classes;[83] thus, whereas for the trityl radicals all substituents produce bigger proton splittings than for the unsubstituted radical (*i.e.* Walter's[84] class S), the pyrryl radicals appear[84] to belong to class O (*i.e.* the shifts on substitution can be positive or negative and correlate with the sign and magnitude of the substituent's Hammett σ-value).

The authors suggest that this behaviour can be accounted for in terms of orbital symmetry considerations and that this approach can be used to divide radicals into classes S and O. It is interesting that for some *N*-t-butylanilino-radicals, generated[85] by photolysis of 1,4-diaryl-1,4-di-t-butyl-2-tetrazenes (in carefully degassed solutions to prevent nitroxide formation), any aromatic ring substituent lowers *a*(N); *i.e.* this type of radical belongs to Walter's class S.[84]

Another type of molecular distortion, towards a non-planar ten-membered π-perimeter, is thought to account for the low α-proton splittings for the radical-anion of 1,6-methano[10]annulene.[86] A modification of the McConnell equation has been proposed for such species to allow for direct transfer of spin from a non-planar radical centre both to the proton at the centre (*cf.* the bending discussed earlier) and to a proton at an adjacent carbon atom, which will not now be in the nodal plane of the unpaired electron.

The approach to the calculation of hyperfine splittings using the INDO method and optimized geometry has also been applied with some success to some σ-radicals,[20] and to ring-substituted fluorobenzyl radicals and the anilino-radical.[87] In all cases reasonable agreement was obtained, especially after spin annihilation had been carried out. The calculations for fluoro-radicals support the assertion that in these species *a*(F) has the *same* sign as the spin density at the adjacent carbon atom, in contrast to *a*(H) for a ·C—H fragment (this has now also been confirmed[88] for some fluorinated phenoxyl radicals from their ^{1}H and ^{19}F n.m.r. spectra and this long-standing problem appears to be resolved). Values for *a*(F) in the fluorobenzyl radicals were also computed[87] from calculated spin densities and appropriate σ–π parameters ($Q^{F}_{CC} = 4$, $Q^{F}_{CF} = -64$, and $Q^{F}_{FF} = 264$ G; see also ref. 88). For fluorinated phenoxyl radicals,[88] it was found that the ratio of a fluorine splitting to the corresponding proton splitting in the unsubstituted radical was approximately constant (except, perhaps not surprisingly, for the *ortho* position); moreover, for these calculations it was assumed that substitution of fluorine for hydrogen does not disturb the overall spin distribution. This appears not to be the case, however, for some fluorinated benzophenone ketyls and their protonated derivatives.[89]

[84] R. I. Walter, *J. Amer. Chem. Soc.*, 1966, **88**, 1923.
[85] S. F. Nelsen, R. T. Landis, L. H. Kiehle, and T. H. Leung, *J. Amer. Chem. Soc.*, 1972, **94**, 1610.
[86] F. Gerson, K. Müllen, and E. Vogel, *Helv. Chim. Acta*, 1971, **54**, 2731.
[87] H. G. Benson, A. Hudson, and J. W. E. Lewis, *Mol. Phys.*, 1971, **21**, 935.
[88] S. Icli and R. W. Kreilick, *J. Phys. Chem.*, 1971, **75**, 3462.
[89] F. P. Sargeant and M. G. Bailey, *Canad. J. Chem.*, 1971, **49**, 2350.

E.s.r. spectra from some perchlorinated diphenylmethyl and triphenylmethyl radicals have also been reported,[90] and the magnitudes of both chlorine and ^{13}C-splittings have been discussed.

The *g*-values and hyperfine splittings for anilino and the azabenzyl radicals have been calculated (the INDO method being used for the splittings) and the results have been compared with experimental values for radicals trapped in an adamantane matrix.[91] Collections of suitable σ–π parameters for other nitrogen-containing radicals include those for some hydrazine radical-cations [92] and some aromatic nitro radical-anions.[93]

Nitroxide radicals have been prepared from a series of bis(*N*-arylnaphthylamines) by oxidation with *m*-chloroperbenzoic acid in benzene.[94] The correlation of *a*(N) with Hammett σ-values for a variety of substituents in radicals from phenyl-substituted 1,1′-bis(*N*-aryl-2-naphthylamines) appears to reflect the stabilization of the resonance structure $R_2\overset{+\bullet}{N}-O^-$ by electron-donating substituents (they are in Walter's [84] type O class). Molecular orbital calculations were performed for *N*-phenyl-1-naphthyl and *N*-phenyl-2-naphthyl nitroxides and for di-2-naphthyl nitroxide. Other studies of substituent effects in related radicals have been made for some imino and amidino nitroxides [71] (where the nitroxide nitrogen splitting decreases with the electron-withdrawing capacity of the substituent, σ) and for some aryl cobalt nitroxides.[75] For the latter series, *a*(N) appears to correlate quite well with Hammett σ-values for *meta*- and *para*-substituents (except *p*-NO_2 and *p*-CO_2^-) and changes in *a*(Co) and *a*(H) can also be similarly accommodated. As expected, steric decoupling caused by *ortho*-substituents leads to higher values for the nitrogen and cobalt splittings.

It is interesting to compare e.s.r. results for delocalized aromatic nitroxides (*e.g.* $Ph_2NO\cdot$) with the spectrum reported [95] for $Ph_2PO\cdot$ (this radical being generated by *X*-irradiation of a single crystal of diphenylphosphine oxide). The large isotropic phosphorus splitting (1092 MHz, *i.e.* 390 G) indicates that the radical is bent (*cf.* $PO_3^{2-}\cdot$, $HPO_2^-\cdot$) and there is no evidence for delocalization of the unpaired electron into the aromatic rings (*cf.* $PhSO_2\cdot$,[96] but contrast $Ph_2NO\cdot$ and $PhNO_2^-\cdot$). Dialkoxyphosphoranyl radicals, $(RO)_2PO\cdot$, have also been reported,[97] formed in solution by reaction of photolytically generated $\cdot OBu^t$ with dialkylphosphoranes or (*via* S_H2 reaction) with tetra-alkyl pyrophosphates. Large phosphorus splittings (*ca.* 700 G) are typical of these radicals.

90 M. Ballester, J. Riera, J. Cartãner, C. Badía, and J. M. Monšo, *J. Amer. Chem. Soc.*, 1971, **93**, 2215.
91 R. V. Lloyd and D. E. Wood, *Mol. Phys.*, 1971, **20**, 735.
92 T. V. Atkinson and A. J. Bard, *J. Phys. Chem.*, 1971, **75**, 2043; P. Smith, R. D. Stevens, and R. A. Kaba, *ibid.*, p. 2048.
93 D. N. Nanda, J. Subramanian, and P. T. Narasimhan, *Theor. Chim. Acta*, 1971, **22**, 369.
94 R. F. Bridger and E. T. Strom, *J. Org. Chem.*, 1971, **36**, 560.
95 M. Geoffroy and E. A. C. Lucken, *Mol. Phys.*, 1971, **22**, 257.
96 M. McMillan and W. A. Waters, *J. Chem. Soc.*, 1966, 422.
97 A. G. Davies, D. Griller, and B. P. Roberts, *J. Amer. Chem. Soc.*, 1972, **94**, 1782.

The e.s.r. spectra of the methyl- and cyclopropyl-tropenyl radicals have been compared [98] with those of the corresponding dianion-radicals and the results correspond well with the predictions of simple molecular orbital theory; thus, in the uncharged radicals, the unpaired electron occupies an orbital which is predominantly the antisymmetric tropenyl-like ABMO, whereas in the dianion-radicals this orbital is filled and the symmetric orbital contains the unpaired electron.

Hyperfine Splittings for β-Atoms. The general usefulness of β-proton splittings as a measure of the spin density ρ_α at an adjacent carbon atom (especially for freely rotating •C—Me groups) and for the conformational analysis of radicals [12, 13] [*i.e.* on the assumption that a(β-H) also depends on $\cos^2 \theta$] has not been challenged: indeed, these relationships continue to find wide application. However, it is now clear that the simple $B \cos^2 \theta$ rule for protons is not in every case strictly obeyed, and the factors which lead to disagreement are apparently related to important intramolecular interactions. These will be discussed here together with current views about the factors which decide, for β-substituents other than protons, the size of the splitting and the mechanism of spin-transmission.

Conventional application of e.s.r. to conformational analysis of radicals, particularly for stable nitroxides, has been reviewed elsewhere.[12, 13] Other conformational studies include those on the 9,10-dihydrophenanthrene radical-anion (for which the dihedral angle between the rings has been determined as 21°),[99] on the radical-cations of *cis*- and *trans*-1,2,3,6-tetramethylhexahydropyridazine (for which it was concluded [100] that the hydrazinium portion is not planar, in contrast to semidiones) and on the 4-formyl-2,6-di-t-butylphenoxyl radical.[101] An e.s.r. and ENDOR investigation [102] of the temperature dependence of the β-splitting for tris(*p*-isopropylphenyl)methyl (*i.e.* for —CHMe$_2$) leads to the suggestion that a $(B_0 + B \cos^2 \theta)$ relationship is appropriate, B_0 having a small positive value ($B_0/B = 0.119$) which should not be ignored.

Although a β-proton splitting reflects more accurately than an α-proton splitting the spin density at the radical centre, the anomalously small values of a(β-H) for radicals in which the orbital of the unpaired electron has considerable *s*-character (*e.g.* 1-adamantyl,[25] oxiranyl [32]) indicates that the hybridization of the trigonal carbon must be considered before quantitative relationships can be applied with certainty. (It seems possible that through-bond spin polarization becomes more important with increasing deviation from planarity.)

Perhaps the most significant developments in the understanding of the factors affecting β-splittings have involved the increasing application of the

[98] F. Farr, Y. S. Rim, and N. L. Bauld, *J. Amer. Chem. Soc.*, 1971, **93**, 6888.
[99] F. C. Adam, *Canad. J. Chem.*, 1971, **49**, 3524.
[100] S. F. Nelsen and P. J. Hintz, *J. Amer. Chem. Soc.*, 1971, **93**, 7105.
[101] R. D. Allendoerfer and D. J. Eustace, *J. Phys. Chem.*, 1971, **75**, 2765.
[102] N. L. Bauld, C. E. Hudson, and J. S. Hyde, *J. Chem. Phys.*, 1971, **54**, 1834.

photolytic technique [35, 103] for generating short-lived radicals in sufficient concentration for detection (the details of the experimental arrangement employed have now been reported [103]). This has led to the production of well-resolved spectra from a variety of radicals over a considerable temperature range, and a wealth of structural detail has resulted. For example,[103] radicals of the type $\cdot CH_2CH_2X$ (X = SR, SiR_3, GeR_3, or SnR_3), which have been produced by the addition of first-formed radicals (•X) to olefins, have been found to exhibit low β-proton splittings with a marked temperature dependence; for $\cdot CH_2CH_2SMe$, $a(\beta\text{-H})$ is 14.89 G at − 70 °C and 13.50 G at − 124 °C. These results are consistent with there being hindered rotation about $C_\alpha—C_\beta$ in these radicals, with a preferred conformation (11) in which the β-substituent eclipses the orbital of the unpaired electron. In contrast, the size of $a(\beta\text{-H})$ and the occurrence of specific line-broadenings (see p. 236), indicates [103] that the preferred conformations for the β-hydroxyethyl and n-propyl radicals are (12; X = OH) and (12; X = Me), respectively and that the n-butyl radical exists preferentially as (13). For the t-amyl radical, conformation (14) is apparently favoured.[103]

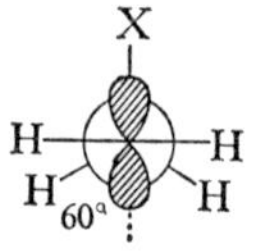

(11)

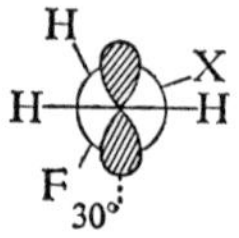

(12)

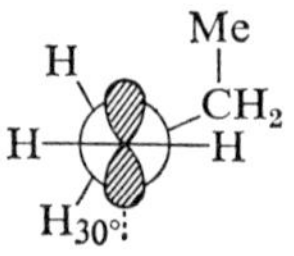

(13)

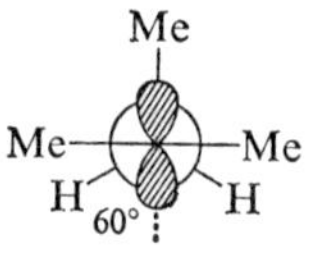

(14)

The temperature dependencies for the β-proton splittings in some alkyl radicals have been calculated [103] by an approach employing classical anisotropic averaging for $a(\beta\text{-H})$ as a function of barrier height; a best fit was obtained with 300 cal mol^{-1} for isobutyl, 420 cal mol^{-1} for n-propyl, and 230 cal mol^{-1} for 3-butenyl. Calculations employing a quantum mechanical approach are in fairly good agreement.[35, 104] Calculated values are considerably higher (1—2 kcal mol^{-1}) for silicon-, germanium-, and tin-substituted radicals,[103] providing another contrast in behaviour with simple alkyl radicals.

Sulphur-substituted radicals have been examined in detail [103] and it has been advocated that the particularly small $a(\beta\text{-H})$ values for $\cdot CH_2CH_2SR$,

[103] P. J. Krusic and J. K. Kochi, *J. Amer. Chem. Soc.*, 1971, **93**, 846.
[104] R. W. Fessenden, *J. Chim. phys.*, 1964, **61**, 1570.

together with their unusual temperature dependence, arise because of an interaction between the unpaired electron and the sulphur atom which not only favours conformation (11) but also distorts the radical at the β-carbon away from being tetrahedral (15). This is supported by INDO calculations on a distorted ethyl radical which do indeed reproduce the low β-proton splittings (below the lower limit assuming $B\cos^2\theta$ behaviour); for a barrier to rotation of *ca.* 1500 cal mol^{-1} and an angle of distortion at the β-carbon of 15°, it was found that the calculated temperature dependence would fit the observed dependence quite closely. It is clear that although there is convincing evidence for 1,3-interaction, these bridged radicals cannot be regarded as symmetrical species with the sulphur equivalently sited with respect to the two carbon atoms.

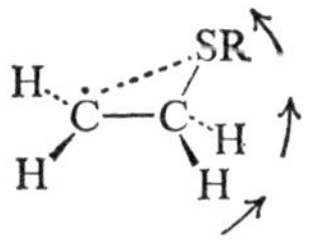

(15)

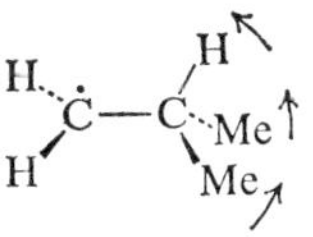

(16)

A distorted model has also been proposed [35] for some simple alkyl radicals in order to give a better account of the temperature dependence of their spectra. For example, a distorted structure (16) for isobutyl, with an angle of distortion of 6°, yields improved agreement between calculated (INDO) and observed splittings over a range of temperature; unusual splittings for cyclopropylmethyl may have a similar origin.[35]

It has been concluded [35] that the equilibrium conformations of alkyl radicals are determined by a balance between steric repulsion and the optimization of C–H hyperconjugation, but there appears to be little evidence that C–C hyperconjugation plays an important part (in contrast to the situation for cations). For the radicals with β-S, -Si, -Ge, or -Sn substituents, it has been suggested [35, 103] that *d*-orbital participation by the heteroatom may be involved to account for the specific 1,3-interaction. Before attempting to clarify the exact nature of this interaction, it should be pointed out that other examples of the preferred eclipsed conformation for suitably substituted radicals have been observed. For instance, this appears to be so for the sulphur atom in a variety of radicals of the type RS·CHX·CHY· [prepared [105] from the titanium(III)–hydrogen peroxide reaction in the presence of added thiols and olefins] and for some radical adducts obtained [106] from ·SH and S^{-}· [from titanium(III) and hydrogen peroxide in the presence of sulphide] with unsaturated aliphatic compounds, including nitro *aci*-anions. In all cases, the interaction was revealed by the occurrence of unusually small β-proton splittings; large *g*-shifts are not

[105] T. Kawamura, M. Ushio, T. Fujimoto, and T. Yonezawa, *J. Amer. Chem. Soc.*, 1971, **93**, 908.

[106] R. O. C. Norman and P. M. Storey, *J. Chem. Soc.* (*B*), 1971, 1009.

produced. A tendency towards the eclipsing conformation is also shown by some radicals formed by the reaction between $\cdot OBu^t$ and some cyclic thioethers;[43] the adduct of $\cdot SnBu^n_3$ and butadiene has been shown [107] to be a *trans* α-substituted allylic radical ($CH_2\cdots CH\cdots CH{-}CH_2SnBu^n_3$) and the low-temperature spectra indicate a preferred conformation with the tin atom optimally located with respect to the π-orbitals (the barrier to rotation was estimated as 5 kcal mol^{-1}). The possibility of hyperconjugative interaction, as an alternative to *d*-orbital overlap, was mooted.[107]

For a variety of β-substituted alkyl radicals trapped in solid matrices, observation of large isotropic splittings from appropriate nuclei confirms the preference for eclipsing conformations,[108, 109] (though again there is no evidence for symmetrical bridged structures). This includes PR_2, PR_3^+, AsR_3^+, SnR_3, and halogen atoms as substituents; for example, for $\cdot CHEt\cdot CH_2\cdot SnBu^n_3$, $a(^{117}Sn/^{119}Sn)$ is 420 G,[108] and for $\cdot CH_2CH_2X$ splittings of 34 ($X = {}^{35}Cl$), 250 ($^{79}Br/^{81}Br$), and 400 G (^{127}I) have been reported.[109] These results, together with the anisotropy of the splittings, lead to the calculated [109] ratio of approximately 1 : 3 for the % *s* : % *p* spin populations on the heteroatoms (with *ca.* 15% *p*-orbital occupancy) as expected for hyperconjugative rather than *d*-orbital interaction. It now appears likely [110] that both effects may contribute to the phenomena involved; thus the large magnitude of the β-atom splitting for $\cdot CH_2CH_2X$ [in contrast to the relatively small values of a(X) for $\cdot CH_2X$ radicals where *d*-orbital effects may operate] favours hyperconjugation [108, 110] rather than polarization of σ- or *d*-electrons. However, trends in the *g*-values for radicals with β-Si, -Ge, or -Sn, where *g decreases* with increasing atomic weight of the substituent (*cf.* trends for $\cdot CH_2MR_3$) suggest *d*-orbital overlap (homoconjugation).[110]

The e.s.r. evidence concerning the marked effects of certain β-heteroatom substituents in radicals also appears to have relevance to various mechanistic studies. For example, although there appears to be no anchimeric assistance from sulphur in the homolysis of $EtSCH_2CMe_2N{=}NCMe_2CH_2SEt$, it appears that, for suitable azo-compounds which decompose to substituted cyclohexyl radicals [*e.g.* azobis(1-cyano-3-thiacyclohexane)], β- or γ-sulphur atoms do stabilize the incipient radical by achieving a favourable conformation for transannular participation in bridging.[111] The enhanced hyperconjugation discussed above could also be analogous to that postulated to account for the *o*/*p* directive effects of CH_2SnR_3, CH_2X (X = Cl, Br, or I), and $CH_2PR_3^+$ (despite the positive charge) in electrophilic aromatic substitution.[108, 112] Similar σ–π interactions of a hyperconjugative nature could also be responsible for trends observed in

[107] T. Kawamura and J. K. Kochi, *J. Organometallic Chem.*, 1971, **30**, C8.
[108] A. R. Lyons and M. C. R. Symons, *Chem. Comm.*, 1971, 1068,
[109] A. R. Lyons and M. C. R. Symons, *J. Amer. Chem. Soc.*, 1971, **93**, 7330.
[110] T. Kawamura and J. K. Kochi, *J. Amer. Chem. Soc.*, 1972, **94**, 648.
[111] A. Ohno and Y. Ohnihsi, *Tetrahedron Letters*, 1972, 339.
[112] M. C. R. Symons, *Tetrahedron Letters*, 1971, 4919.

the polarographic reduction potentials of some *p*-nitrobenzyl halides and related compounds,[113] for the large increase in the rate of protodemetallation reactions of R_3MPh when M is varied from C to Pb along Group IV,[114] and for the lowering of the oxidation potential for *N*-methylaniline on substituting a trimethylsilyl group for an aliphatic hydrogen.[115]

It is interesting that alkyl radicals with first-row elements in the β-position (*e.g.* $\cdot CH_2CH_2OH$ and $\cdot CH_2CH_2F$ [116]) do not show a preference for the eclipsed conformation (and its low associated β-proton splitting). However, the anomalously low β-H splittings [34] for a variety of radicals which have an OH or an OR group bonded to the α-carbon and an electron-withdrawing group in the β-position [*e.g.* $\cdot CH(OH)CH_2OH$, $\cdot CH(OMe)CH_2OAc$, and $\cdot CH(OH)CH_2NH_3^+$] merit some attention; thus, for $\cdot CH(OH)CH_2OH$, $a(\beta\text{-H}) = 0.99$ mT (9.9 G) at room temperature, this being well below the minimum allowed if a $B\cos^2\theta$ relationship is obeyed (*i.e.* $B/4$, *ca.* 1.35 mT). It has been argued that the conformation (17) is strongly favoured; evidently

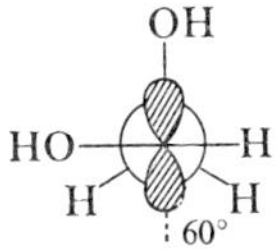

(17)

the stabilizing interaction involves a co-operative effect of suitable α- and β-substituents, with $+M$ and $-I$ effects, respectively, and this can perhaps be represented by contributions from structures (19) and (20). This also leads to 'flattening' of the β-C—H bonds and hence to an anomalously low $a(\beta\text{-H})$ (*cf.* distortion in $\cdot CH_2CH_2SR$ [103] discussed earlier)

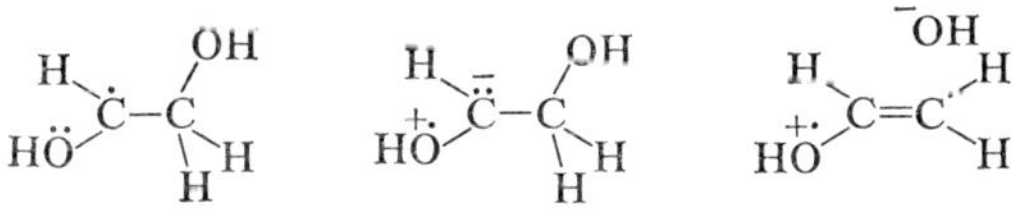

(18) (19) (20)

and, interestingly, to a more nearly planar radical centre than is customary for hydroxyalkyl radicals. Since some of the α,β-disubstituted radicals for which (17) is the preferred conformation undergo [117] ready elimination of the β-substituent (either homolytically or heterolytically) it is envisaged that this structure fulfils a stereoelectronic requirement for reaction (see p. 250). As a matter of ancillary interest it was shown that for radicals of the type $\cdot CH(OH)CH_2NR_3^+$ the nitrogen splittings exhibit a $B\cos^2\theta$-type

[113] M. Mohammad, J. Hajdu, and E. M. Kosower, *J. Amer. Chem. Soc.*, 1971, **93**, 1792.
[114] H. J. Berwin, *J. C. S. Chem. Comm.*, 1972, 237.
[115] B. E. Cooper and W. J. Owen, *J. Organometallic Chem.*, 1971, **29**, 33.
[116] T. Kawamura, D. J. Edge, and J. K. Kochi, *J. Amer. Chem. Soc.*, 1972, **94**, 1752.

dependence, and this also appears to be true for $a(^{31}P)$ in some β-substituted aliphatic phosphorus radicals [118, 119] and some substituted aromatic nitro radical-anions.[120] Hyperconjugation may well be involved in these examples.

The experimental temperature dependencies of the hydroxyl proton splittings in a variety of radicals have been analysed and compared with those calculated using different barrier heights to rotation about the •C—OH bonds.[121] It was concluded that consistent results are obtained if contributions to the splitting from hyperconjugation (from spin on carbon) and from spin-polarization (from spin on oxygen) are balanced as follows, where θ is the angle of twist of the OH bond out of the radical plane:

$$a^{H}_{OH} = -(35 \pm 5)\,\rho^{\pi}_{O}\langle\cos^2\theta\rangle + (50 \pm 10)\,\rho^{\pi}_{C}\langle\sin^2\theta\rangle$$

For $\cdot CH_2OH$, the barrier was calculated [121] as 3.5—5 kcal mol^{-1}, and for the hydroquinone radical-cation the value of 10 ± 2 kcal mol^{-1} which resulted is in agreement with that obtained from studies of linewidth alternation. For the protonated biacetyl semidione, •CMe(OH)•COMe, rotation around the C—OH bond causes all the splittings to be temperature-dependent because of variations in the degree of conjugation of the unpaired electron with the hydroxyl oxygen's lone pairs.[122] A McLachlan MO model was proposed, and a single value of Q^{H}_{CMe} (25 G) appears adequate.[122] The potential energy barrier to rotation was estimated as 2.3—2.8 kcal mol^{-1}, depending on the solvent.[123]

A review of the stereochemistry of nitroxides [13] has presented the conflicting views on the angular dependence of β-halogen splittings in those radicals and on the mechanism of spin-transmission (1,3-*p*,π-overlap and hyperconjugation have both been postulated). It was concluded that the β-fluorine splitting does depend on $\cos^2\theta$ and that for a β-chlorine the splitting also increases with decrease in the dihedral angle. It was also argued that the average conformations for nitroxides with difluoromethylene groups are not the same as those for the corresponding unfluorinated radicals; for various fluorine-containing radicals, conformation (21) appears to be favoured.

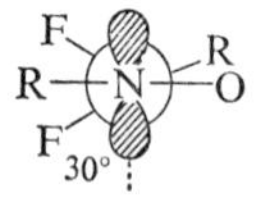

(21)

[117] B. C. Gilbert, J. P. Larkin, and R. O. C. Norman, *J. C. S. Perkin II*, 1972, 794.
[118] W. Damerau, G. Lassmann, and K. Lohs, *Z. Chem.*, 1971, **11**, 182.
[119] W. Damerau, G. Lassmann, and K. Lohs, *J. Magn. Resonance*, 1971, **5**, 408.
[120] W. M. Gulick, jun., *J. Amer. Chem. Soc.*, 1972, **94**, 29.
[121] P. D. Sullivan, *J. Phys. Chem.*, 1971, **75**, 2195.
[122] P. B. Ayscough and R. E. D. McClung, *Mol. Phys.*, 1971, **20**, 35.
[123] P. B. Ayscough, M. C. Brice, and R. E. D. McClung, *Mol. Phys.*, 1971, **20**, 41.

A relationship of the type $a(\mathrm{F}) = (B_0 + B\cos^2\theta)$ has been proposed for the β-fluorine splittings in some aliphatic radicals in single crystals, but an overlap between the half-filled p-orbital and the fluorine $2p$-orbitals can apparently account for this, and such an interaction may well dominate hyperconjugative and spin polarization effects in these examples.[124] This is also the conclusion from a detailed experimental study and INDO calculations on the ketyls of hexafluoroacetone, perfluorodiethyl ketone, and perfluorocyclobutanone.[125] The results indicate that for the acyclic radicals the perfluoroalkyl groups are effectively freely rotating (in contrast to some fluorinated nitroxides – this may reflect non-planarity or be a consequence of the shorter C—N bond length in the latter) and attention has been drawn to the marked difference between $a(\beta\text{-F})$ for $(CF_3)_2CO^{-}\cdot$ (34.94 G in MeCN between 23 and − 30 °C) and that [126] for $(CF_3)_2NO\cdot$ (8.26 G at 24 °C, 8.46 G at − 37 °C, both in $CFCl_3$). INDO calculations [125, 127] suggest that unpaired electron density is delocalized into a vacant anti-bonding orbital of the CF_3 group in both radicals (see also ref. 32) but that the extent of this effect is more marked for the ketyl on account of the relevant orbitals being closer in energy for that species. Thus, although the $B\cos^2\theta$ relationship may be obeyed for β-fluorines, it appears that the constant of proportionality (B) depends critically upon the *type* of radical; further, the degree of delocalization may also depend on the nature of substituents on the β-carbon.[125]

Further examples of the non-equivalence of β-methylene protons due to the presence of an adjacent chiral atom have been reported.[128–131] For example, for a series of nitro radical-anions $XYZC\cdot CH_aH_b\cdot NO_2^{-}\cdot$ the two methylene protons have different splittings when X, Y, and Z are different groups,[128] even when rotation is rapid around the $N—C_\alpha$ and $C_\alpha—C_\beta$ bonds. Line-broadening has also been detected in some cases, indicating the additional occurrence of restricted rotation, but the high-temperature limit still has a doublet of doublets from the β-protons. Previously reported examples of non-equivalence in a range of radicals (which have been customarily attributed to restricted rotation) may well be due to the influence of chirality.[128] The recognition of complete inequivalence in this way [128, 129] has led to some interesting applications and the potential is illustrated by reference to the nitronyl-nitroxide spin-labelled analogue (22) of histidine and to investigations of some related species.[129] The differences between the β-proton splittings in these radicals have been used as a diagnosis of

124 M. Iwasaki, *Mol. Phys.*, 1971, **20**, 503.
125 W. R. Knolle and J. R. Bolton, *J. Amer. Chem. Soc.*, 1971, **93**, 3337.
126 W. R. Knolle and J. R. Bolton, *J. Amer. Chem. Soc.*, 1969, **91**, 5411; P. J. Scheidler and J. R. Bolton, *ibid.*, 1966, **88**, 371.
127 K. Morokuma, *J. Amer. Chem. Soc.*, 1969, **91**, 5412.
128 B. C. Gilbert, J. P. Larkin, and R. O. C. Norman, *J. C. S. Perkin II*, 1972, 1272.
129 R. J. Weinkam and E. C. Jorgensen, *J. Amer. Chem. Soc.*, 1971, **93**, 7028.
130 R. J. Weinkam and E. C. Jorgensen, *J. Amer. Chem. Soc.*, 1971, **93**, 7033.
131 E. Flesia, A. Croatto, P. Tordo, and J.-M. Surzur, *Tetrahedron Letters*, 1972, 535.

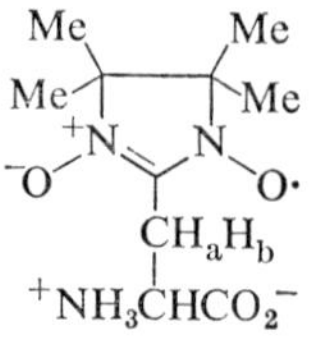

(22)

conformational equilibria; in particular, Δa depends on the nature of the groups on the chiral carbon, and in some cases a reduced rate of conformational interconversion leads to broadening. Ion–dipole interaction between a carboxylate ion and the ring for suitable amino-acids accounts for the pH dependence of their spectra. Further, a(N) reflects the environment around the nitroxide fragment, and it is also pointed out that effects analogous to those observed with simple spin labels result when the radical has a low tumbling rate. The technique has been extended [130] to some labelled peptides in the angiotensin series, both for the investigation of ion–dipole interactions and for the measurement of activation energies for conformational interconversion. Nitroxides with chiral groups have also been reported.[131]

Long-range Splittings. As improved methods of radical production lead to a variety of fairly stable radicals (suitable for study with e.s.r., n.m.r., or ENDOR) and to larger concentrations of specific short-lived species – and hence allow low-power, low-modulation recording – resolution improves and additional small splittings can often be observed. These are not only of intrinsic interest from the point of view of the mechanism of spin-transmission to remote parts of the radical, but they are often of crucial importance in the correct assignment of an observed signal [*e.g.* γ- and δ-splittings of 0.055 mT (0.55 G) allowed the recognition [117] of the unexpected intermediate $\cdot CH(OH)\cdot CH_2\cdot CH_2\cdot CHO$ during the oxidation of glycol with $\cdot OH$]. This may be particularly important in the unambiguous application of spin-trapping methods.

Observation [128] of appreciable γ- and δ-splittings for the acyclic radical $OCH\cdot CH_2\cdot CH_2\cdot NO_2^{-}\cdot$ and for the above-mentioned example suggests that the mechanism of homoconjugation proposed [103] to account for the apparently unusual long-range splitting in the allylmethyl radical (23), for

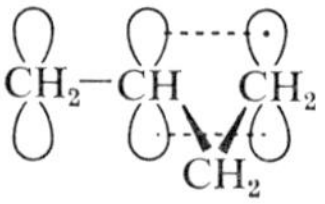

(23)

which $a(\gamma$-H) is 0.61 G and $a(\delta$-H), for one proton, is 0.35 G, is not necessarily of paramount importance. However, the stereochemical demands for spin transmission in some bicyclic radicals and related species have been considerably unravelled recently and results for different radical types indicate that a more or less unified model can be proposed. For instance, the INDO method (with reparametrization) has been applied to the investigation of long-range effects in rigid bicyclic semidiones [132] and the results have been rationalized in terms of different spin-delocalization mechanisms, namely homohyperconjugation, (24) ↔ (25), and, as illustrated by (26), spin polarization (*N.B.* unmodified EH theory is incapable

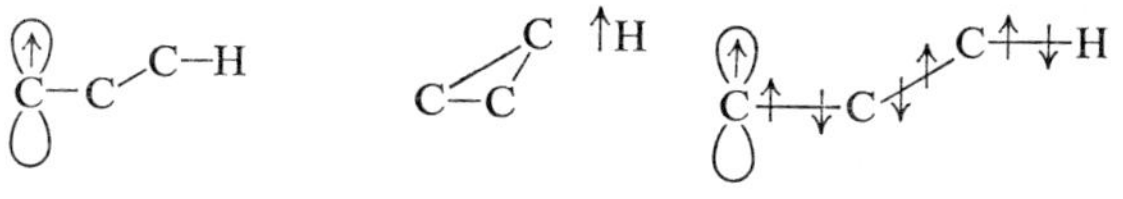

(24) (25) (26)

of predicting spin densities from the latter). Quite good agreement was obtained for γ-protons in general; these appear to be normally negative (*i.e.* they arise *via* spin-polarization) unless the geometry is such that the W-plan arrangement [*e.g.* (24)] ensures a large positive splitting from homohyperconjugation (25) (this will be strongly orientation-dependent). The splittings also depend on the symmetry properties of the singly-occupied orbital: contrast the large bridgehead splitting for the semidione (27) with the small value for the corresponding *p*-semiquinones and semifuraquinones where the bridgehead carbon and protons are in the nodal

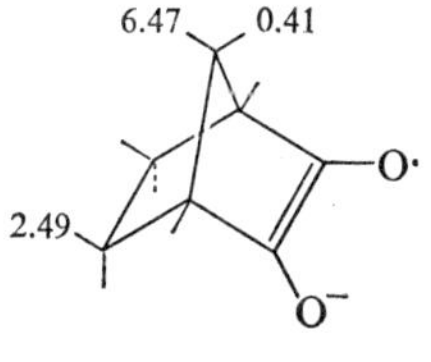

(27)

plane of the orbital containing the odd electron (*i.e.* when the electron occupies an antisymmetric orbital).

The homohyperconjugation delocalization mechanism, rather than spin-polarization, is also apparently dominant for long-range effects in some bicyclo[3,1,0]hexan-2,3-semidiones [*e.g.* (28)], where EH calculations give good agreement;[133] in these radicals the zig-zag coplanar arrangement

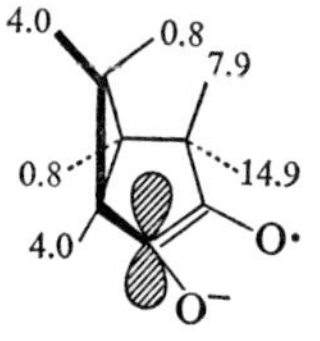

(28)

[132] G. R. Underwood and V. L. Vogel, *J. Amer. Chem. Soc.*, 1971, **93**, 1058.
[133] G. A. Russell, J. J. McDonnell, P. R. Whittle, R. S. Givens, and R. G. Keske, *J. Amer. Chem. Soc.*, 1971, **93**, 1452.

apparently favours this type of long-range interaction. (This detailed paper also reports a highly stereoselective H–D exchange for the *exo*-hydrogen in the α-methylene group in (28) in basic DMSO; and for *syn*-6-alkylbicyclo[3,1,0]hexan-2,3-semidiones, rearrangement to the *anti*-isomers probably proceeds *via* electrocyclic ring opening followed by ring closure and aromatization.) Long-range interactions are smaller for the bicyclo-[4,1,0]heptane, -[5,1,0]octane, and -[6,1,0]nonane systems.[133] E.s.r. spectra with long-range splittings have also been reported for some semidiones from bicyclo[*n*,1,1]alkanes [134] and for some polycyclic derivatives of cyclobutane-semidione [135] (in both series, good agreement has been achieved using EH calculations with appropriate variations in the polycyclic geometries). The reader is referred to the original literature for the wealth of structural detail involved and for descriptions of insertion reactions involving $^{-}CH_2\cdot SO\cdot Me$ [134] and the conversion of 7-norbornenone into the corresponding semidione.[136]

Long-range splittings for furasemiquinones [137] have been discussed in terms of spin-delocalization and spin-polarization, and SCF–LCAO calculations for bicyclo[1,1,1]pentyl yield [138] a large positive value for a(γ-H, *anti*), as expected for W-plan interaction. For the 6-nitrobenzonorbornene radical-anion (29) the large long-range splitting from the

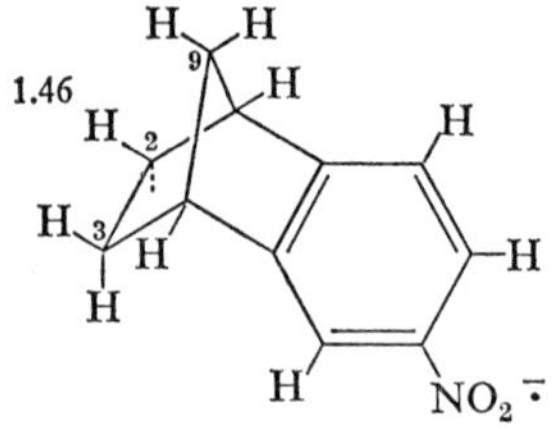

(29)

exo-C-2 proton (homo-*para-exo*), in contrast to that for the *exo*-C-3 proton (homo-*meta-exo*), which appears to be either 0.37 or 0.24 G, reflects the spin density in the benzene ring; the *anti*-C-9 proton splitting (0.37/0.24 G) is probably small because of the opposite signs of the HMO coefficients for the unpaired electron at the *meta* and *para* ring positions (as discussed above and as found for the analogous norbornyl *p*-semiquinone but in contrast to norbornyl *o*-semiquinone).[139]

[134] G. A. Russell, P. R. Whittle, and R. G. Keske, *J. Amer. Chem. Soc.*, 1971, **93**, 1467.
[135] G. A. Russell, P. R. Whittle, R. G. Keske, G. Holland, and C. Aubuchon, *J. Amer. Chem. Soc.*, 1972, **94**, 1693.
[136] J. P. Dirlam and S. Winstein, *J. Org. Chem.*, 1971, **36**, 1559.
[137] S. F. Nelsen, E. F. Travecedo, and E. D. Seppanen, *J. Amer. Chem. Soc.*, 1971, **93**, 2913.
[138] Y. Ellinger, A. Rassat, and R. Subra, *J. Chim. phys.*, 1971, **68**, 730.
[139] S. Terabe and R. Konaka, *J. Amer. Chem. Soc.*, 1971, **93**, 6323.

The mode of transmission of electron spin density through σ-orbitals in bicyclic systems has also been investigated [140] for 6-aminobenzonorbornene with a paramagnetic shift reagent [bis(acetylacetonato)nickel(II)]. Relative contact shifts correspond reasonably well with the e.s.r. splittings for (29) in magnitude and allow signs to be proposed for the latter; small negative spin densities (from polarization) and larger positive spin densities (for appropriate W-plan homoconjugation) can be characterized.

Long-range splittings have been reported for cyclohexyl t-butyl nitroxide (30) [141] and, with the help of n.m.r., for 1-adamantyl t-butyl nitroxide (31);[142] splittings for both are in gauss. For (30) the equatorial δ-proton

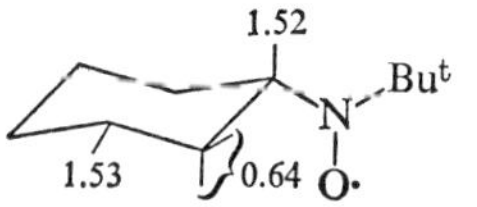

(30)

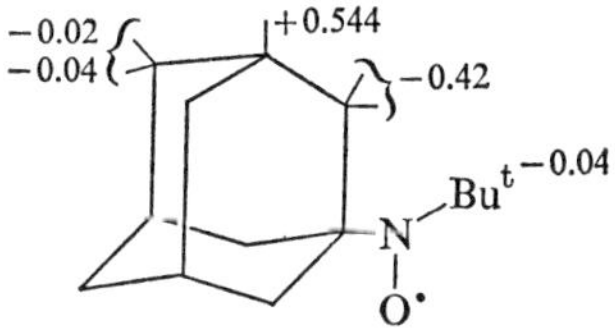

(31)

satisfies the zig-zag arrangement for appreciable interaction and there is good agreement with Barfield's calculations (though signs were not measured).[143] Long-range splittings in a bicyclic [3,2,1] nitroxide have also been described [144] and compared with the calculated spin distributions given by Barfield's approach,[143] by INDO, and by the method of Luz,[145] which incorporates through-bond spin-polarization. The last-named procedure gives a good account of the (usually) negative γ- and positive δ-splittings, but not the positive stereoselective γ-proton splitting associated with the W-plan arrangement (and therefore homohyperconjugation). INDO calculations give good agreement. Long-range coupling has also been detected for an isoquinuclidine nitroxide [146] where the radical centre is situated in a six-membered, boat-shaped ring.

The employment of n.m.r. in conjunction with e.s.r. has also led to the analysis of the hfs for some stable hindered aromatic nitroxides.[147, 148] This has allowed extension of previous work on aromatic t-butyl nitroxides to some fluorinated analogues;[147] the splitting from a ring-substituted fluorine has been shown to have the same sign of spin density as the adjacent carbon (*cf.* fluoro-benzyl and -phenoxyl radicals), and it appears that the dominant interaction for ^{19}F in a CF_3 substituent and for ^{1}H in an OMe

[140] K. Tori, Y. Yoshimura, and R. Muneyuki, *J. Amer. Chem. Soc.*, 1971, **93**, 6324.
[141] Th. A. J. W. Wajer, A. Mackor, and Th. J. de Boer, *Rec. Trav. chim.*, 1971, **90**, 568.
[142] C. Morat and A. Rassat, *Bull. Soc. chim. France*, 1971, 891, 893.
[143] M. Barfield, *J. Phys. Chem.*, 1970, **74**, 621.
[144] A. Rassat and J. Ronzaud, *J. Amer. Chem. Soc.*, 1971, **93**, 5041.
[145] Z. Luz, *J. Chem. Phys.*, 1968, **48**, 4186.
[146] A. Rassat and P. Rey, *Tetrahedron*, 1972, **28**, 741.
[147] H. J. Jakobsen, T. E. Petersen, and K. Torssell, *Tetrahedron Letters*, 1971, 2913.
[148] J. A. Pedersen and K. Torssell, *Acta Chem. Scand.*, 1971, **25**, 3151.

group is hyperconjugation (but see ref. 125, for example). Twisting of the plane of the aromatic ring out of the C′CNO plane has been investigated for some *ortho*-substituted aryl t-butyl nitroxides.[148] Since the spectrum from *o*-hydroxyphenyl t-butyl nitroxide differed from the spectra of all other *ortho*-substituted derivatives and resembled the spectra from unhindered derivatives, it appears that the nitroxide group is forced into the plane of the aromatic ring by internal hydrogen bonding.[148]

Long-range hfs had also been reported for some bicyclic iminoxyl radicals,[149] where the 'W-effect' appears to be operative, and for some fairly stable keto-iminoxyl radicals generated from acyl radicals and nitrile *N*-oxides.[150]

g-Values.—The usefulness of a careful measurement of the *g*-value of a radical in aiding both the confirmation of a proposed structure and the diagnosis of electronic interactions still appears not to have been universally recognized. However, the potential usefulness of this parameter is clearly indicated by the results for some radicals of the type $\cdot CH_2CH_2X$ (X = Si, Ge, Sn, or Cl).[110, 116] It is argued [110] that the decrease in *g* with atomic weight for the Group IV-substituted radicals (being the reverse of the trend for metal-centred radicals $\cdot MR_3$, where M = C, Si, Sn, or Pb) supplies crucial evidence for *d*-orbital participation (*cf.* the same trend for $\cdot CH_2MR_3$). However, molecular orbital calculations suggest that the very low *g*-value for $\cdot CH_2CH_2Cl$ (2.00214) [116] reflects a *p–p* (homoconjugative) overlap rather than *p–d* overlap. It is pointed out that delocalization involving the C—Cl σ-bond should give a positive shift.

INDO and CNDO/2 calculations have been performed for the *g*-tensors of the radical $H_2NO\cdot$ and for a variety of σ-radicals including vinyl,[151, 152] phenyl, acetyl, and nitrogen dioxide, with good agreement in each case. The observed isotropic values are, of course, well below the free-spin value, and this appears to be a useful criterion for classification of radicals into the σ-category.

For some radical-cations, obtained from bis(4-alkoxyphenyl)sulphides, ethers, and the corresponding bis-hydroxy-compounds with $AlCl_3$–$MeNO_2$, the *g*-values have been calculated [153] using HMO–McLachlan theory and Stone's model; good agreement for *g*'s and splittings leads to consistent parameters for O and S. For some oxygen-containing radical-anions (*e.g.* from fluorenone and anthraquinone), the changes in *g* and splittings produced by ion-pairing have been studied;[154] it has been shown that $a(^{13}C)$ for the carbonyl group and the *g*-value are correlated, and also that

[149] H. Căldăraru and M. Moraru, *Tetrahedron Letters*, 1971, 3183.
[150] B. C. Gilbert, V. Malatesta, and R. O. C. Norman, *J. Amer. Chem. Soc.*, 1971, **93**, 3290.
[151] T. Morikawa, O. Kikuchi, and K. Someno, *Bull. Chem. Soc. Japan*, 1971, **44**, 2256.
[152] T. Morikawa, O. Kikuchi, and K. Someno, *Theor. Chim. Acta*, 1971, **22**, 224.
[153] P. D. Sullivan and H. J. Shine, *J. Phys. Chem.*, 1971, **75**, 411.
[154] T. Takeshita and N. Hirota, *J. Amer. Chem. Soc.*, 1971, **93**, 6421.

the use of the *g*-value can provide additional information concerning the spin density on oxygen.

Theories relating CIDNP effects to the isotropic e.s.r. parameters of the intermediate radicals involved have been tested with known radicals and also used to predict e.s.r. parameters for elusive species. Thus, for $\cdot CHCl_2$ it was predicted[155] that $g = 2.0080 \pm 0.0003$ and that $a(H) = -17 \pm 1.0$ G; the subsequently reported values (footnote 28 of ref. 155) are in excellent agreement. Confirmation of another *g*-value obtained similarly, 2.0058 for acetoxyl (see ref. 8 of ref. 156), should prove rather more difficult.

For those interested in the precise determination of absolute *g*-values, attention is drawn to the note concerning a small error in a standard procedure.[157]

2 Thermodynamic and Kinetic Data from E.S.R. Spectra

This section is concerned essentially with those variables which contribute to the height of an e.s.r. signal, and in particular with the measurement of concentration and the recognition of processes which cause overall or selective line-broadening.

Conventional Methods.—Although it is not possible here to review details of instrument design, attention is drawn to the report[158] of a simple integrator which has a useful application in the quantitative aspects of spin trapping, to advances in the application of ancillary electrochemical apparatus (for studying the kinetics of electrode processes[159] and for determining absolute concentrations[160]), and to details of another method for obtaining reliable spin concentrations.[161] Other methods are reported in detail in some of the kinetic investigations discussed below.

E.s.r. has been successfully employed in conjunction with different approaches to the determination of the kinetics of radical reactions (*e.g.* direct measurement of decay, use of a rotating sector, competitive reactions, *etc.*), depending on the life-time of the species concerned. For instance, a steady-state method has been used[162, 163] to determine the rate constants for reactions of hydrogen atoms (produced by continuous *in situ* electron-bombardment of aqueous solutions) with a variety of added solvents; since the spectrum of $\cdot H$ was initially observed with abnormal intensities (because of non-equilibrium population of spin-levels), with a relaxation time to

155 H. Fischer and M. Lehnig, *J. Phys. Chem.*, 1971, **75**, 3410.
156 J. Bargon, *J. Polymer Sci., Part B, Polymer Letters*, 1971, **9**, 681.
157 R. D. Allendoerfer, *J. Chem. Phys.*, 1971, **55**, 3615.
158 R. Barr, R. J. Holman, and M. J. Perkins, *Chem. and Ind.*, 1972, 43.
159 I. B. Goldberg and A. J. Bard, *J. Phys. Chem.*, 1971, **75**, 3281.
160 J. K. Dohrmann and F. Gallusser, *Ber. Bunsengesellschaft phys. Chem.*, 1971, **75**, 432.
161 E. Ohmes, G. Kothe, A. Naujok, and H. Zimmerman, *Ber. Bunsengesellschaft phys. Chem.*, 1971, **75**, 895.
162 P. Neta, R. W. Fessenden, and R. H. Schuler, *J. Phys. Chem.*, 1971, **75**, 1654.
163 P. Neta and R. H. Schuler, *J. Amer. Chem. Soc.*, 1972, **94**, 1056.

equilibrium distribution of *ca.* 50 μs under the conditions employed, chemical reactions could be studied which effectively competed with this.[162] Observations for various substrates gave second-order rate constants in the range 10^5—10^9 l mol^{-1} s^{-1};[162] as expected, for aliphatic compounds, where abstraction is presumably involved, rate constants increase with molecular dimensions from 10^5 l mol^{-1} s^{-1}, reaching a plateau at *ca.* 10^8 l mol^{-1} s^{-1} for C_6 and above. Higher values, near the diffusion-controlled limit, for ethylenic compounds and aromatics, and for halides (iodides in particular), probably indicate that the reactions involved are addition and halogen-abstraction reactions, respectively. The reactivity of •H reflects that of •OH (a hydroxy-substituent activates and a carbonyl group deactivates an adjacent C—H bond to attack) and its electrophilic character is confirmed by a study of substituent effects for a range of aromatic and heterocyclic compounds (electron-donating substituents enhance the rate).[163] A pulse radiolysis study has also been carried out.[164]

The continuous photolysis of a mixture of a phenol and an α-diketone (*i.e.* an H-atom acceptor) and the measurement[165] of the steady-state concentrations of the radicals produced [reaction (1)] have been used to

$$\text{MeCOCOMe} + \text{ArOH} \xrightarrow{h\nu} \cdot\text{C(OH)MeCOMe} + \text{ArO}\cdot \qquad (1)$$

derive the termination rate constants for dimerization of phenoxyl radicals (the rate constant for dimerization of the semidione being previously obtained with a rotating sector). The particular production of the unsubstituted phenoxyl radical and the protonated biacetyl semidione leads to values of 6.3×10^8 l mol^{-1} s^{-1} (in benzene) and 0.55×10^8 l mol^{-1} s^{-1} (in chlorobenzene) for the dimerization of the former. For a variety of radicals there appears to be no correlation between their rates of reaction and spin-density distributions. However, the method has been extended[166] to the determination of activation energies for a variety of phenoxyl radicals, and two classes of behaviour can be distinguished: in the first, radicals which yield keto-dimers [*e.g.* (32)] whose heats of formation from the radicals exceed 20 kcal mol^{-1}, exhibit values of $\log k_{30\,°C}$ and ΔE consistent with diffusion theory for a rapid irreversible reaction (*e.g.* 9.5 and 4.2 kcal mol^{-1}, respectively, for 3,5-dimethylphenoxyl in benzene). In the second category are radicals which yield dimers whose heats of formation are smaller, and the anomalous values of $\log k_{30\,°C}$ and ΔE (*e.g.* 7.2 and 1.6 kcal mol^{-1}, respectively, for 4-methoxyphenoxyl) suggest a reversible formation of keto-dimer followed by rate-determining enolization [(32) $\rightarrow$ (33)]. These suggestions are supported by the observed isotope effects.[167]

[164] B. Smaller, E. C. Avery, and J. R. Remko, *J. Chem. Phys.*, 1971, **55**, 2414.
[165] S. A. Weiner, *J. Amer. Chem. Soc.*, 1972, **94**, 58.
[166] L. R. Mahoney and S. A. Weiner, *J. Amer. Chem. Soc.*, 1972, **94**, 585.
[167] L. R. Mahoney and S. A. Weiner, *J. Amer. Chem. Soc.*, 1972, **94**, 1412.

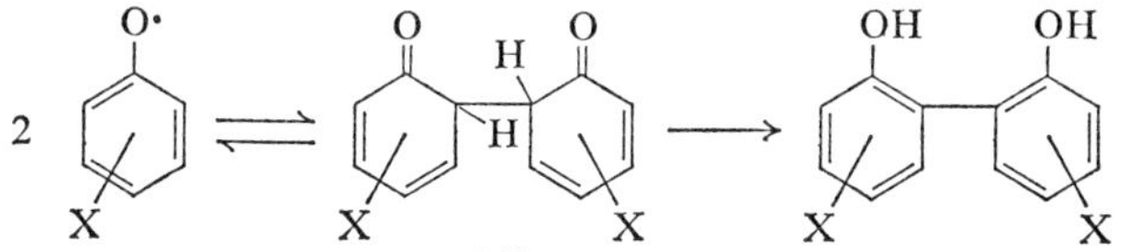

(32) (33)

E.s.r. spectroscopy has also been employed,[168, 169] in conjunction with absorption spectroscopy in one case,[168] for determining rate constants for termination, disproportionation, and proton transfer for some ketyl radicals, and for obtaining equilibrium constants for the ketyl–pinacol dimerization equilibrium.[168] For example,[168] for benzophenone ketyls produced by thermolysis of the corresponding pinacols between 80 and 141 °C, a *para*-bromine substituent shifts the equilibrium in favour of the radical, relative to the unsubstituted example (a *p*-methoxy-group has the opposite effect); *p*-Br also causes more rapid formation, a lower rate constant for termination, and a higher ratio of recombination to disproportionation than for the related radicals. Hydrogen-bonding with the protic solvent accounts for the substantial differences between the thermodynamic and kinetic parameters for solutions in propan-2-ol and benzene.

Another group has been actively involved in kinetic studies of a variety of nitrogen-centred radicals. For example, for diethyl nitroxide [prepared [170] both from the hydroxylamine with cerium(IV) and by photolysis, with or without di-t-butyl peroxide] the decay rate has been studied using stop-flow and intermittent radiation methods (*e.g.* with a rotating sector and CAT). A reaction mechanism is proposed which involves reversible formation of a diamagnetic dimer [probably as indicated by reaction (2)] followed by

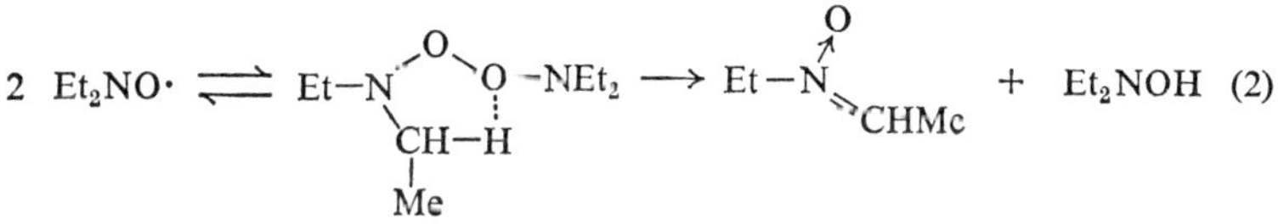

its slow decomposition to *NN*-diethylhydroxylamine and a nitrone (*cf.* tetroxide formation from peroxyl radicals). Between − 100 °C and − 145 °C in CF_2Cl_2 a reversible dimerization has been demonstrated ($\Delta H = 6 \pm 3$ kcal mol^{-1}, $\Delta S = 18 \pm 20$ cal K^{-1} mol^{-1}); measurements at higher temperatures lead to an expression for k (in the same solvent) as $10^9 \exp(-7000/RT)$ s^{-1}. For nortropane *N*-oxyl, which is stable in solution at room temperature, values for the dissociation of the corresponding dimer in isopentane are $\Delta H = 7.2$ kcal mol^{-1}, $\Delta S = +14.8$ cal K^{-1} mol^{-1}; the lower values obtained for both parameters for solutions in CF_2Cl_2

[168] S. A. Weiner, *J. Amer. Chem. Soc.*, 1971, **93**, 6978.
[169] H. Yoshida and T. Warashina, *Bull. Chem. Soc. Japan*, 1971, **44**, 2950.
[170] K. Adamic, D. F. Bowman, T. Gillan, and K. U. Ingold, *J. Amer. Chem. Soc.*, 1971, **93**, 902.

probably reflect the heat of solvation of the nitroxide and the ordering produced by the radicals in this solvent.

Further studies have confirmed the general applicability of reaction mechanism (2) for other dialkyl nitroxides [171] and have demonstrated that the reversible self-reaction (3) occurs for a number of mono-substituted

$$2RNHO \underset{k_{-1}}{\overset{k_1}{\rightleftharpoons}} RNHOH + RNO \qquad (3)$$

nitroxides.[172] For t-butyl nitroxide in benzene, the appropriate parameters for the dissociation of the dimer are $\Delta H = -10.4$ kcal mol^{-1} and $\Delta S = 8.2$ cal K^{-1} mol^{-1}, with $k_1 = 6.7 \times 10^6$ l mol^{-1} s^{-1} at 25 °C [$k_1 = 10^{8.7} \exp(-2500/RT)$ l mol^{-1} s^{-1}]. Reversible dimer formation has also been demonstrated [173] for some dialkylamino-radicals generated photolytically from tetrazenes; for the reversible formation of $Pr^i_2NNPr^i_2$ from di-isopropylamino, $\Delta H = 25 \pm 2$ kcal mol^{-1} (less than the N—N bond strength for hydrazine because of steric effects, presumably). The kinetics and mechanism of irreversible radical decay were also discussed. Lastly, bimolecular rate constants for the self-reactions of some radicals containing Group IV elements ($Me_3M\cdot$, $Me_3MCH_2\cdot$, $Me_3MMMe_2\cdot$, and $Me_3MMMe_2CH_2\cdot$, where M = C, Si, Ge, or Sn) have been found to be in the range 1.4—11 $\times 10^9$ l mol^{-1} s^{-1};[174] an anomalously low value for $\cdot CCl_3$ has also been recorded.[174]

Other workers have concentrated upon kinetic investigations of homolytic substitution (S_H2) reactions of a variety of heteroatom centres, brought about by t-butoxyl and other photolytically generated radicals.[175] For instance, competition by $\cdot OBu^t$ between S_H2 t-butoxydealkylation of tri-n-butylborane [reaction (4)] and abstraction from cyclopentane (the

$$\cdot OBu^t + BBu^n_3 \longrightarrow Bu^tOBBu^n_2 + \cdot Bu^n \qquad (4)$$

rate of which had been previously determined) leads, *via* the measurement of the relative concentrations of $\cdot Bu^n$ and cyclopentyl, to the rate constant for (4). For this example, k at 30 °C is 3×10^7 l mol^{-1} s^{-1} with $A = 3.5 \times 10^7$ l mol^{-1} s^{-1} and $E \sim 0$. This type of reaction has been demonstrated for a range of heteroatoms (Zn, Cd, Al, P, As, Sb, and Bi). It is not clear whether the reactions are stepwise or concerted, but it is concluded that there is no evidence for 'hot' radicals in this system. A similar method [176] has been used to obtain kinetic data for homolytic alkylthiyldealkylation of boron compounds with $\cdot SBu^t$; it was found that, as with $\cdot OBu^t$, reactivity decreases in the direction $BBu^n_3 > BBu^i_3 > BBu^s_3$ (this series

[171] D. F. Bowman, T. Gillan, and K. U. Ingold, *J. Amer. Chem. Soc.*, 1971, **93**, 6555.
[172] D. F. Bowman, J. L. Brokenshire, T. Gillan, and K. U. Ingold, *J. Amer. Chem. Soc.*, 1971, **93**, 6551.
[173] J. R. Roberts and K. U. Ingold, *J. Amer. Chem. Soc.*, 1971, **93**, 6686; 1972, **94**, 2159.
[174] G. B. Watts and K. U. Ingold, *J. Amer. Chem. Soc.*, 1972, **94**, 491.
[175] A. G. Davies, D. Griller, and B. P. Roberts, *J. Chem. Soc.*, (*B*), 1971, 1823.
[176] A. G. Davies and B. P. Roberts, *J. Chem. Soc.* (*B*), 1971, 1830.

probably reflects the increased crowding in the transition states compared with the planar boron compounds); oxygen and sulphur substituents on boron also decrease the reactivity. Similar trends have been found [177] for the S_H2 reactions of the triplet states of ketones at boron. The rate constants are similar to those reported for homolytic alkoxy- and alkylthiyl-dealkylation of boron; for instance, acetone and BBu^n_3 on photolysis give e.s.r. signals of $\cdot Bu^n$ and $\cdot CMe_2OBBu^n_2$, with $k = 7 \times 10^6\ l\ mol^{-1}\ s^{-1}$ at 20 °C.

In the reaction between the t-butoxyl radical and triethyl phosphite, both the intermediate phosphoranyl radical (34), with $a(^{31}P) = 891$ G, and the t-butyl radical were detected [178] [reaction (5)]. By monitoring concentra-

$$\cdot OBu^t + P(OEt)_3 \longrightarrow \underset{(34)}{Bu^tO\dot{P}(OEt)_3} \longrightarrow \cdot Bu^t + OP(OEt)_3 \quad (5)$$

tions of these and of the cyclopentyl radical (produced in a competitive reaction) it was possible to derive a rate constant for the overall reaction of $1.6 \times 10^8\ l\ mol^{-1}\ s^{-1}$ at 30 °C with a lower limit for decomposition of the intermediate as $4.4 \times 10^6\ s^{-1}$ at the same temperature. Kinetic results on the decarbonylation of the t-butoxycarbonyl radical have also been presented.[16]

The photoreduction of biacetyl and pyruvic acid [179, 180] and of aromatic nitroso- and nitro-compounds [180] in organic solvents has been studied kinetically. Bimolecular termination rate constants for $\cdot CMe(OH)COMe$ in different solvents at 25 °C are close to those predicted for diffusion-controlled reactions if one-quarter of the encounters lead to product formation *via* a singlet transition state.[180] With nitrosobenzene, dissociation to Ph· and NO (not *via* the triplet) leads to the trapping of phenyl radicals (and of radicals derived from the solvent) by the parent compound; with nitrobenzene the picture looks particularly complex.[180]

Kinetic studies have also been carried out on the decay of some aryl alkoxy nitroxide radicals, detected during the u.v. photolysis of a variety of substituted nitrobenzene derivatives in ethers,[181] which are evidently formed *via* hydrogen atom abstraction by the nitro-compound followed by addition of the derived aliphatic radical to an oxygen atom in another molecule of the nitro-aromatic. Values of a(N) for these radicals correlate quite well with Hammett σ constants for *m*- and *p*-substituents; for *o*-substituents, steric effects can cause an increase in a(N). The mechanism for the decomposition of the sterically unhindered alkoxy nitroxides is not clear, and it is thought unlikely that a homolytic fission to an alkoxyl

177 A. G. Davies, D. Griller, B. P. Roberts, and J. C. Scaiano, *Chem. Comm.*, 1971, 196; A. G. Davies, B. P. Roberts, and J. C. Scaiano, *J. Chem. Soc.* (*B*), 1971, 2171.
178 A. G. Davies, D. Griller, and B. P. Roberts, *Angew. Chem. Internat. Edn.*, 1971, **10**, 738.
179 P. B. Ayscough and M. C. Brice, *J. Chem. Soc.* (*B*), 1971, 491.
180 P. B. Ayscough, R. C. Sealy, and D. E. Woods, *J. Phys. Chem.*, 1971, **75**, 3454.
181 R. B. Sleight and L. H. Sutcliffe, *Trans. Faraday Soc.*, 1971, **67**, 2195.

radical and a nitroso-compound, followed by trapping of an alkyl fragment from the former by the latter, provides the main reaction pathway (in contrast to results for some sterically hindered nitrobenzene derivatives). Studies have also been made for alkyl alkoxy nitroxides prepared [182] by the addition of other radicals to aliphatic nitro-compounds. Steric interactions between alkyl and alkoxy-groups apparently lead to restricted rotation about the carbon–nitrogen single bonds.

Reactions between aromatic nitro-compounds and hydroxylamines in basic solution have been followed using e.s.r.[183] The mechanism of reaction appears to involve electron donation from $RNHO^-$ to $ArNO_2$ to give detectable signals from RNHO• and $ArNO_2^{\overline{\bullet}}$; the decay of the latter was monitored. Kinetic and e.s.r. studies also suggest [184] that the oxidation of hydroxamic acids (RCONHOH) to *NO*-diacylhydroxylamines (RCO-NHOCOR) with alkaline ferricyanide proceeds *via* the dimerization of the first-formed radical-anion $[RCO{-}N{=}O]^{\overline{\bullet}}$ followed by intramolecular rearrangement of the intermediate so produced.

Kinetic studies, using e.s.r., of the transition-metal-induced decomposition of peroxides have promoted considerable controversy (although there now seems agreement on the assignment of the singlets detected in the absence of substrate to [metal ion—HO_2•] complexes; see ref. 1, for example). It has been claimed [185] that the rates of reaction of H_2O_2 with •CMe_2OH and •$CH_2CHMeOH$ [formed from propan-2-ol with titanium(III) –hydrogen peroxide in a flow system] can be determined as 4.0×10^2 $l\,mol^{-1}\,s^{-1}$ and $2.5 \times 10^2\,l\,mol^{-1}\,s^{-1}$ respectively, from the variation of their measured absolute concentrations with $[H_2O_2]$ and with time after mixing. The rate constant for dimerization of the former was reported as $1.6 \times 10^7\,l\,mol^{-1}\,s^{-1}$. It was also suggested that radical–H_2O_2 reactions do not perpetuate chain reactions since they produce an 'unreactive' form of •OH. However, this kinetic analysis has been challenged [186] (it is suggested that such a scheme would lead to much higher radical concentrations than were actually observed) and an alternative description has been proposed. Firstly, it was pointed out that the original analysis assumes that the generating reaction between Ti^{III} and H_2O_2 [reaction (6)] is over

$$Ti^{III} + H_2O_2 \xrightarrow{k} Ti^{IV} + OH^- + \cdot OH \qquad (6)$$

before the solutions reach the cavity, which is probably not the case [a value for (6) of $k = 1000\,l\,mol^{-1}\,s^{-1}$ gives a half-life of 14 ms with initial concentrations of 0.005M Ti^{III} and 0.05M H_2O_2 (*cf.* time between mixing and observation, 20—70 ms)]. Secondly it is shown that the observed

[182] R. B. Sleight and L. H. Sutcliffe, *Spectrochim. Acta*, 1972, **28A**, 95.
[183] D. J. Cowley, M. H. Millen, and W. A. Waters, *J. Chem. Soc.* (*B*), 1971, 2393.
[184] T. R. Oliver and W. A. Waters, *J. Chem. Soc.* (*B*), 1971, 677.
[185] R. E. James and F. Sicilio, *J. Phys. Chem.*, 1970, **74**, 1166.
[186] C. E. Burchill, *J. Phys. Chem.*, 1971, **75**, 167.

kinetic behaviour is consistent with a scheme whereby reaction (6) is rate-determining, with more usual bimolecular radical termination rates of *ca.* $4 \times 10^9\, l\, mol^{-1}\, s^{-1}$ and with $\cdot CMe_2OH$ being more effective than the primary radical in the decomposition of H_2O_2 to give hydroxyl radicals (as previously suggested on the basis of three-way mixing experiments).[187] In this way a chain mechanism [186, 187] is supported, and these suggestions have the advantage that anomalous rate constants [185] need not be postulated. The original authors [188] disagree with this reinterpretation and prefer to attribute the anomalies to the reaction between Ti^{IV} and $\cdot OH$, but Burchill's suggestion [186] finds support from other workers' detailed treatment of the kinetics involved in the Ti^{III}–H_2O_2 and other metal–ion reactions.[189, 190] They suggest, on the basis of a kinetic analysis, that radicals detected by e.s.r. in a flow system are formed *within* the cavity and therefore that results obtained with multiple-stage mixing chambers have no significance. However, there seems to be agreement [187, 189, 190] on the ease of oxidation of oxygen-conjugated radicals (*e.g.* $\cdot CMe_2OH$) with H_2O_2 and with high-valent metal ions [Ti^{IV}, Fe^{III}, Cr^{III}] and on the importance of chain mechanisms in these reactions. There also appears to be no valid evidence for metal-complexed radicals other than complexes of $HO_2\cdot$. Confirmation of some of these aspects also derives from a photolytic investigation [191] of the effects of adding H_2O_2 to solutions containing oxygen-conjugated free radicals (from propan-2-ol, methanol, and some ethers) obtained by abstraction from the parents using the photosensitization of acetone. For the oxidation of $\cdot CMe_2OH$ by H_2O_2 at 30 °C, $k = 5 \times 10^5\, l\, mol^{-1}\, s^{-1}$. This report also presents an analysis of the Ti^{III}–H_2O_2–propan-2-ol reaction consistent both with the suggestions discussed above [186, 189, 190] and with photolytic results.

A kinetic analysis [192] of the VO^{2+}–H_2O_2 reaction has also been presented (it was possible to monitor the concentrations of VO_2^+ and of vanadium(v)–peroxyl radical complexes) and an e.s.r. study has been made of the reaction between Co^{III} acetylacetonate and t-butyl hydroperoxide.[193] In the latter example, a broad singlet ($\Delta H = 20$ G, $g = 2.0147$) was detected which showed an unusual (reversible) temperature dependence. Although the authors prefer to regard the signal as that from a radical–metal complex, the parameters would also fit fairly well with those expected for uncomplexed $\cdot O_2Bu^t$, with spin rotation and reversible dimerization to tetroxide [1] accounting for the anomalous behaviour. However, observation [194] of signals from other radicals, in the presence of added substrates, which

[187] R. O. C. Norman and P. R. West, *J. Chem. Soc.* (*B*), 1969, 389.
[188] R. E. James and F. Sicilio, *J. Phys. Chem.*, 1971, **75**, 1326.
[189] G. Czapski, *J. Phys. Chem.*, 1971, **75**, 2957.
[190] G. Czapski, A. Samuni, and D. Meisel, *J. Phys. Chem.*, 1971, **75**, 3271.
[191] C. E. Burchill and P. W. Jones, *Canad. J. Chem.*, 1971, **49**, 4005.
[192] H. B. Brook and F. Sicilio, *Inorg. Chem.*, 1971, **10**, 2530.
[193] A. Tkáč, K. Veselý, and L. Omelka, *J. Phys. Chem.*, 1971, **75**, 2575.
[194] A. Tkáč, K. Veselý, and L. Omelka, *J. Phys. Chem.*, 1971, **75**, 2580.

showed definite interaction with Co ($I = \frac{7}{2}$), provides clear evidence that cobalt-containing radical-complexes can be formed.

Chemical Information from Line Broadening.—Line broadening due to exchange involving $\cdot CH(NH_2)CO_2H$ and $\cdot CH(NH_2)CO_2^-$ in the presence of phosphoric acid has been used [195] to obtain the rate constants for the processes involved, and the pH dependence of the apparent splittings leads to the pK_a of the radical (6.6). Similarly, broadening is observed [34] because of exchange between $\cdot CH_2OH$ and $\cdot CH_2O^-$, the separation between the lines of the averaged spectrum being pH dependent and consistent with the pK_a of the species (10.7). Another source of broadening for $\cdot CH_2O^-$ is probably the same as that discussed [196] for some other radicals with a linear heavy-atom framework (*e.g.* $\cdot CH_2CN$, $\cdot CH_2C{\equiv}CH$, and $\cdot CH_2{-}C{\equiv}CMe$) and attributed [196] to spin-rotational interaction involving relatively free rotation about the long molecular axis. INDO calculations on these radicals were also reported.[196] Spin-rotation probably also accounts for the line broadening, and its anomalous temperature dependence, for $\cdot SO_3F$.[197] Other relaxation effects are treated elsewhere in this volume (Chapter 3).

The observation of a specific broadening of the central lines of the α-proton 1 : 2 : 1 splitting pattern (*i.e.* $M_I = 0$) for $\cdot CH_2CH_2Me$ and for $\cdot CH_2CH_2OH$ in contrast [103] to the broadening of the central line of the β-splitting pattern for $\cdot CH_2CH_2CH_2Me$ confirms information derived from the size and temperature dependence of β-splittings; the first two have a favoured conformation of the type (35) with the β-protons equivalent [this interconverts with (36) and thus modulates a(α-H)] and the last-mentioned radical prefers the conformation (37) [and interconverts with (38), thereby modulating the β-splittings].

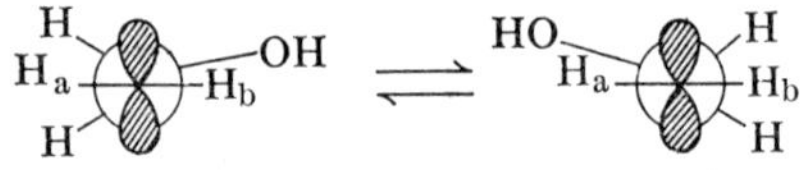

(35) (36)

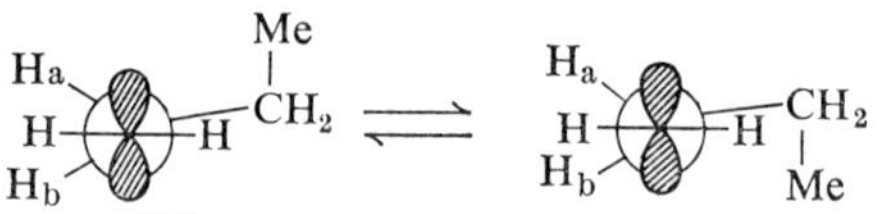

(37) (38)

Detailed calculations for the extent of the broadening for the n-propyl radical using a density matrix method lead to an energy barrier to rotation of 3.1 kcal mol^{-1},[35] much larger than that estimated from the temperature

[195] H. Paul and H. Fischer, *Helv. Chim. Acta*, 1971, **54**, 485.
[196] H. G. Benson, A. J. Bowles, A. Hudson, and R. A. Jackson, *Mol. Phys.*, 1971, **20**, 713.
[197] P. M. Nutkowitz and G. Vincow, *J. Phys. Chem.*, 1971, **75**, 712.

dependence of $a(\beta\text{-H})$ for this radical; the reason for this discrepancy is not clear. The activation energy for the interconversion (39) ⇌ (40) for the hydroxymethyl radical, and that for the same process for $\cdot CH_2OMe$, have also been obtained from a density matrix treatment of observed line

(39) (40)

broadening: for $\cdot CH_2OH$, $E_0 = 4.6$ kcal mol^{-1}, $\nu_0 = 10^{12.6}$ s^{-1}; for $\cdot CH_2OMe$, $E_0 = 5.3$ kcal mol^{-1}, $\nu_0 = 10^{12.5}$ s^{-1} (*cf.* $\cdot CH_2COMe$, for which $E_0 = 9.4$ kcal mol^{-1}, $\nu_0 = 10^{12.9}$ s^{-1}). The potential energy barrier for $\cdot CH_2OH$ is considerably higher than that previously reported [198] but it appears that the latter result was in error because of neglect of second-order effects on the central lines.[43] For $\cdot CH_2OH$ there is reasonable agreement [35] between calculations made on the basis of the linewidths and those based on the temperature dependence of a(OH), a quantum mechanical approach (with INDO calculations) to the analysis of the latter giving better results than a classical approach.[103] Sullivan [121] obtained an energy barrier for $\cdot CH_2OH$ of 3.5 kcal mol^{-1} from calculations based on the temperature dependence of a(OH), INDO results, and a classical approach.

Linewidth variation of similar type has been reported for $\cdot CH_2SBu^t$, $\cdot CH_2OBu^t$, and some related radicals.[43] Calculations for the barrier to internal rotation utilized the Bloch equation for a two-site model, and experimental spectra were matched with predicted ones using an on-line computer. For $\cdot CH_2SBu^t$, $E_0 = 25.4$ kJ mol^{-1}, with $\nu_0 = 10^{12.5}$ s^{-1}, and the lower limit of E_0 for $\cdot CH_2SMe$ appears to be *ca.* 30 kJ mol^{-1}, the higher barriers for the sulphur radicals than for their oxygen counterparts (by about 4—8 kJ mol^{-1}) being understandable in terms of the greater delocalization for sulphur-substituted radicals.

Restricted rotation also leads to line broadening for (41),[107] (42),[177] and (43):[199] for the 2-furanylmethyl radical, the energy barrier to rotation about $C—CH_2\cdot$ has been calculated as 25.16 kcal mol^{-1}, and the minimum-energy geometry (as given by INDO calculations) has also been discussed.[199] Conformational changes in some polyphenoxyl radicals also cause broadening.[200]

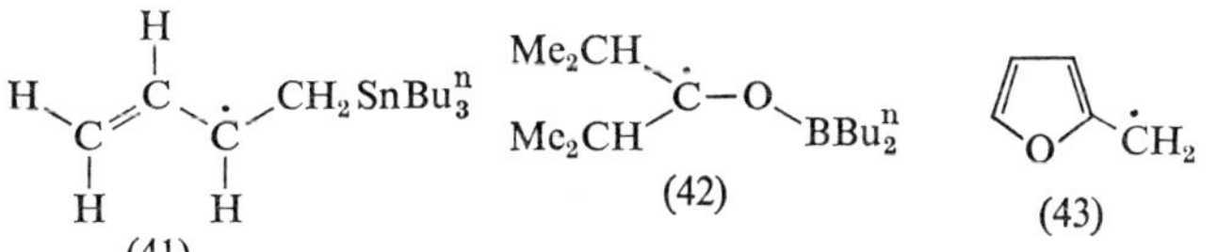

(41) (42) (43)

198 A. Hudson, *J. Chem. Soc.* (*A*), 1969, 2513.
199 L. D. Kispert, R. C. Quijano, and C. U. Pittmann, jun., *J. Org. Chem.*, 1971, **36**, 3837.
200 W. J. Van den Hoek, J. F. Th. de Winter, and J. Smidt, *J. Magn. Resonance*, 1972, **6**, 15.

Evidence for intramolecular hydrogen transfer results[201] from the alternating line-widths observed for ·ON(Bu^t)·CO·N(OH)Bu^t, the two nitrogen splittings being equivalent at high temperature (E_a was estimated as 2.7 ± 0.3 kcal mol^{-1}). Geometrical isomerism, tautomeric interconversions, and hydroxyl-proton exchange have been described[202] for some semidione species obtained from the reduction of dicarbonyl compounds (*e.g.* biacetyl, glyoxal, and pyruvic, maleic, and fumaric acids and their esters) with the radicals ·CO_2H, ·CO_2^-, and ·CMe_2NH_2. Particular effects specific to maleic acid include the intramolecular proton transfer involving (44) (exchange being rapid at pH 1 but giving broadening at around pH 4), the recognition of ·C(OH)$_2$CH=CHCO$_2^-$ at pH 4—5, and the detection of rapid intramolecular transfer in (45) at pH 7.

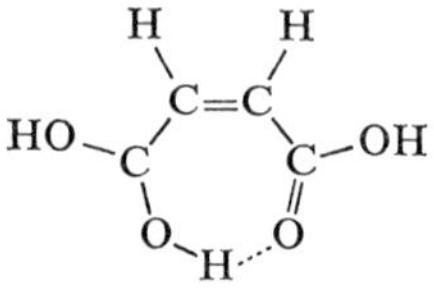

(44)

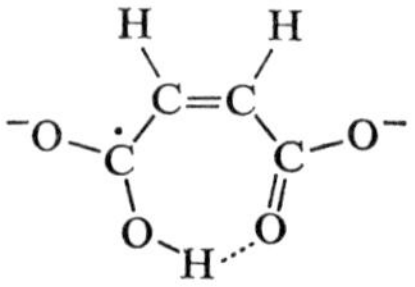

(45)

3 Free-radical Reactions

This final section is concerned mainly with information derived from e.s.r. about the development of methods for the preparation and trapping of free radicals and about our understanding of their mechanisms of reaction.

Generation of Radicals.—*Radiolysis and Photolysis.* Experimental details have now been described for the continuous *in situ* bombardment of aqueous solutions with a 2.8 MeV electron beam.[203] The particular advantages offered by this combination of radiolysis and e.s.r. over other methods for studying transient species in aqueous solution include not only the inherent sensitivity and unambiguous assignment of radicals with e.s.r. but also the possibility of pH-variation over an extensive range (without modification of conditions). Further, the scope is extended by the possibility of producing, selectively, the primary reactive species e$^-$, ·H, and ·OH by the use of selective scavengers (*e.g.* N_2O for e$^-$, which gives ·OH, H$^+$ for e$^-$ to give H·, and t-butyl alcohol for ·OH). For example, the reactions of ·OH have been investigated at high pH (see also ref. 34); ·CH_2O^- and ·CH(O$^-$)CO$_2^-$ have been obtained from methanol and glycollic acid, respectively. A variety of adducts from ·H and ·OH with benzenoid derivatives have been detected, and reactions of the hydrated electron demonstrated include its addition to acetone to give the corresponding

[201] H. G. Aurich and K. Stork, *Tetrahedron Letters*, 1972, 555.
[202] N. H. Anderson, A. J. Dobbs, D. J. Edge, R. O. C. Norman, and P. R. West, *J. Chem. Soc.* (*B*), 1971, 1004.
[203] K. Eiben and R. W. Fessenden, *J. Phys. Chem.*, 1971, **75**, 1186.

ketyl and the reductive elimination of chloride ion from chloroacetate to yield $\cdot CH_2CO_2^-$. Some spectra show anomalous intensity distributions, this being particularly marked for cyclohexadienyl radicals, presumably on account of their smaller anisotropic interactions (and hence longer T_1) than for aliphatic species with larger α-splittings. Other applications will be described later.

In addition to the continued application of photolysis for the decomposition of di-t-butyl peroxide and peresters (*cf.* the many examples already discussed), and also for the generation of the thiyl radical $\cdot SBu^t$ from di-t-butyl disulphide,[176] some other novel applications have been reported. For example, photolysis (230—340 nm) of solutions of aliphatic ketones ($R_2C{=}O$) in methylcyclopentane, with a 1 kW high-pressure mercury lamp, gave [204] e.s.r. signals not only of the radicals expected from abstraction by the triplet ketone ($\cdot CR_2OH$ and solvent-derived species) but also to intense signals from alkyl radicals ($R\cdot$) formed by Norrish Type I cleavage. Irradiation of ketones in CCl_3F led to mixtures of radicals including $\cdot CCl_2F$ and, for acetone, $\cdot CH_2COMe$. Since photolysis of CCl_3F itself gave no detectable radical, it is argued that its decomposition proceeds *via* exciplex formation involving the excited ketone and that the abstraction radical results from the reaction of chlorine atoms:

$$CCl_3F \xrightarrow{Me_2C=O^*} \cdot Cl + \cdot CCl_2F$$

$$\cdot Cl + CH_3COMe \longrightarrow HCl + \cdot CH_2COMe$$

Photolysis of citric acid in aqueous solution leads, *via* loss of $\cdot CO_2H$ from the excited molecule, to $\cdot C(OH)(CH_2CO_2H)_2$ and to $\cdot CH_2\cdot C(OH)(CO_2H)\cdot CH_2CO_2H$, and, by abstraction of hydrogen by these radicals, to $\cdot CH(CO_2H)\cdot C(OH)(CO_2H)\cdot CH_2CO_2H$.[205] In the presence of hydrogen peroxide, stronger signals of the last-named radical were obtained; the use of deuterium oxide facilitated spectral assignment. Photolysis of sodium citrate gave signals from $\cdot CH(CO_2^-)\cdot C(OH)(CO_2^-)\cdot CH_2CO_2^-$ exclusively, this radical having clearly observed inequivalent γ-proton splittings. This behaviour may be the result of restricted rotation [205] but, alternatively, it appears that it could be because the γ-carbon atom is attached to a chiral (β) carbon atom.

Photolysis of alkaline aqueous solutions of thiosulphate leads,[206] from excitation of the $n \rightarrow \pi^*$ band, to the formation of $S^-\cdot$ and $SO_3^-\cdot$, which have been recognized by the adducts they form with the *aci*-anion of nitromethane; further applications of this type of technique for flow systems are described later. The production of the hydrated electron by excitation of a charge-transfer complex was also demonstrated. Another technique successfully employed for the detection of radicals in solution involves [207]

204 H. Paul and H. Fischer, *Chem. Comm.*, 1971, 1038.
205 H. Zeldes and R. Livingston, *J. Amer. Chem. Soc.*, 1971, **93**, 1082.
206 D. Behar and R. W. Fessenden, *J. Phys. Chem.*, 1971, **75**, 2752.
207 T. J. Kemp, H. D. Burrows, and D. Greatorex, *J. Amer. Chem. Soc.*, 1971, **93**, 2539.

photo-oxidation of organic molecules in the presence of transition metals. For example, photolysis ($\lambda > 300$ nm) of flowing ethanolic solutions of cerium(IV) ammonium nitrate (5×10^{-3} M) at 185 K leads to e.s.r. signals from $\cdot CH_3$, whereas use of uranyl perchlorate as the oxidant leads to $\cdot$CHMeOH (this is also formed almost exclusively from Ce^{IV} at 77 K). E.s.r. results also reveal the oxidative decarboxylation of benzilic acid to $\cdot CPh_2OH$ by complexed oxidant (*e.g.* Fe^{III} at 185 K).

Flow Methods. Developments in the employment of low-valent transition-metal ions for homolytic decompositions amenable to e.s.r. study include the reaction between titanium(III) and *N*-alkylhydroxylamines to give alkylamino-radicals.[208] These intermediates were not directly detectable, but their participation was inferred from the spectra of adducts given with the *aci*-anion of nitromethane (and other spin traps) and of the α-aminoalkyl radicals produced by rearrangement:

$$R^1R^2CHNHOH \xrightarrow{Ti^{III}} R^1R^2CHNH\cdot \xrightarrow{^-CH_2NO_2} R^1R^2CH\cdot NH\cdot CH_2NO_2^{\overline{\cdot}}$$

$$R^1R^2CHNH\cdot \longrightarrow \cdot CR^1R^2\cdot NH_2$$

A spectrum assigned to $SO_3^-\cdot$ (or $\cdot SO_3H$) was observed when solutions of hydroperoxides and sulphur dioxide flowed together through the cavity of a spectrometer.[209] Radicals were evidently formed *via* reaction (7) or (8)

$$RO_2H + SO_2 \longrightarrow RO\cdot + HOSO_2\cdot \qquad (7)$$

$$RO_2H + SO_2 \longrightarrow HO\cdot + ROSO_2\cdot \qquad (8)$$

$$HO\cdot \xrightarrow{SO_2} HOSO_2$$

Further, for t-butyl hydroperoxide in methanol, a signal due to $Bu^tOSO_2\cdot$ [$g = 2.0034$, a(H) $= 0.28 \pm 0.02$ G] was observed, whereas in water $MeSO_2\cdot$ was detected [with $g = 2.0055$ and a(H) $= 0.9 \pm 0.1$ G]. These probably arise *via* $\cdot OBu^t$ and $CH_3\cdot$ respectively; adducts of $HOSO_2\cdot$, $MeSO_2\cdot$, $Bu^tOSO_2\cdot$, and $\cdot OBu^t$ with a variety of acrylic derivatives were detected and corroborative evidence for their participation in polymerization was forthcoming.

It has also been shown [106] that sulphite ion is oxidized to $SO_3^-\cdot$ by $\cdot$OH, $\cdot NH_2$, $SO_4^-\cdot$, and Ce^{IV}, that $S_2O_3^-\cdot$ results from the reaction of thiosulphate with hydroxyl, and that sulphide ion is oxidized to $S^-\cdot$ or $\cdot$SH (depending on the pH) by both the hydroxyl and the amino-radicals. Further reactions of

[208] N. H. Anderson and R. O. C. Norman, *J. Chem. Soc.* (*B*), 1971, 993.
[209] B. D. Flockhart, K. J. Ivin, R. C. Pink, and B. D. Sharma, *Chem. Comm.*, 1971, 339.

these radicals with a variety of unsaturated aliphatic compounds have been reported.[106]

Spin Trapping. Two trapping systems with special application to aqueous systems have been proposed. In the first, employed successfully to trap radicals formed by the reactions of e^-, •H, and •OH (from *in situ* electron bombardment) with a variety of substrates, fumarate ion was used to scavenge the organic radicals produced.[210] Its use as a trap was justified, at least in part, by the high radical-adduct concentrations that could be achieved, the second-order disappearance of these radicals being relatively slow on account of their double charge. In a few cases, hyperfine splittings from the addend could be obtained, (*e.g.* γ-hfs for the adducts of •CH_3, •CF_3, and •CH_2Cl), an advantage shared by the second trap, the nitromethane *aci*-anion, $^-CH_2NO_2$.[128] The usefulness of this trap has been reviewed and it has been extended to some novel reactions of radicals in flow systems (this method has also been applied to photolytic systems [206]). Long-range γ- and, sometimes, δ-proton splittings facilitate the recognition of radical adducts and, in addition, a radical with three different γ-substituents is immediately recognizable because of the non-equivalence of the β-proton splittings in the adduct (due to chirality). An illustration is provided by •CHMeOH (from ethanol with •OH) which gives [128] strong signals from $HOCHMe{\cdot}CH_AH_B{\cdot}NO_2^-{\cdot}$ with $a(N) = 2.51$, $a(\beta\text{-}H_A) = 1.12$, $a(\beta\text{-}H_B) = 0.97$, and $a(\gamma\text{-}H) = 0.06$ mT.

Some improvements in more conventional types of spin-trap (see also refs. 9 and 10) have been described; thus, the increased resolution possible by employing perdeuterionitrosobutane ($[^2H_9]Bu^tNO$) rather than the corresponding protio-compound has been exemplified,[211] as in the detection of long-range splitting in the adduct from •COOMe (from methyl formate and •OBu^t). Another bifunctional trap (*cf.* the phenolic nitrone described [212] previously) is 2,4,6-tri-t-butylnitrosobenzene (46) which combines the advantages of increased thermal and photochemical stability compared with many nitroso-compounds with the ability [213] to trap radicals at nitrogen (to give a nitroxide) or oxygen (to give a substituted amino-radical) depending apparently on the steric requirements of the radical, thus, secondary and tertiary radicals exhibit different behaviour [reactions (9) and (10) for example]. Whereas (47) has a g-value of 2.0060, a nitrogen splitting of 13.29 G, and proton splittings of 0.76 (m-H) and 22.19 G ($CHMe_2$; this unusually large splitting presumably reflects the conformation of the radical), the anilino-radical (48) is clearly distinguished by its g-value (between 2.0036 and 2.0040 for radicals in this series) and splittings [$a(N) = 11.01$ and $a(3H) = 1.82$ G]. The observation of splittings, in the resulting adducts, from the attacking species is also feasible for the iminoxyl

[210] P. Neta, *J. Phys. Chem.*, 1971, **75**, 2570.
[211] R. J. Holman and M. J. Perkins, *J. Chem. Soc.* (*C*), 1971, 2324.
[212] J. G. Pacifici and H. L. Browning, jun., *J. Amer. Chem. Soc.*, 1970, **92**, 5231.
[213] S. Terabe and R. Konaka, *J. Amer. Chem. Soc.*, 1971, **93**, 4306.

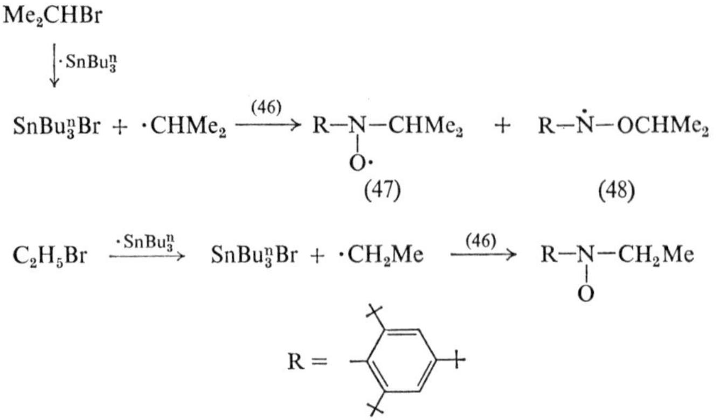

(9)

(10)

radicals, whose detection [150] during the trapping of a variety of acyl and alkyl radicals with nitrile *N*-oxides confirms their proposed [214] intermediacy in this reaction.

Despite the obvious usefulness of the fashionable spin-trapping technique in its many guises, several warning notes have been sounded about the care which must be taken over its application if an unequivocal reaction mechanism is to be demonstrated. Thus, with nitrones and with nitroso-compounds, an alternative (and in some cases more likely) route to nitroxides lies in nucleophilic attack followed by oxidation of the resulting anion to a nitroxide.[215] For instance, acetate ion reacts with *N*-benzylidene-t-butylamine *N*-oxide [PhCH=N(Bu^t)O] *via* addition, which is followed by adventitious oxidation, to give the radical PhCH(OAc)N(Bu^t)O·; an essentially similar mechanism, rather than one involving acetoxyl radicals, is envisaged to account for the formation of the same adduct in the presence of lead tetra-acetate. Nucleophilic addition perhaps also explains the effect of base on the formation of nitroxides from the same nitrone in the presence of peroxides.[216]

Other interesting reactions which complicate trapping experiments have been described:[66, 67] for example mono-, di-, and tri-halogenomethyl radicals, generated photolytically from halogenomethanes in the presence of Bu^{t}NO, lead to the detection of carbonyl nitroxide radicals, namely Cl·CO·N(Bu^t)O· and Br·CO·N(Bu^t)O· from ·CCl_3 and ·CBr_3, respectively, and H·CO·N(Bu^t)O· from ·$CHBr_2$, ·CHI_2, and ·CH_2I.[66] The bromocarbonyl nitroxide has a(N) = 6.7 G, a(^{79}Br/^{81}Br) = 2.0 G, and g = 2.0071.[66] It seems agreed that the reaction proceeds *via* the initial adduct of the type CX_3·N(Bu^t)O· with subsequent ready elimination of a halogen atom to give a nitrone X_2C=N(Bu^t)O·. Suggestions for the ensuing mechanism [Schemes 1 and 2, respectively, for refs. 66 and 67] diverge at this point.

[214] T. Caronna, A. Quilico, and F. Minisci, *Tetrahedron Letters*, 1970, 3633.
[215] A. R. Forrester and S. P. Hepburn, *J. Chem. Soc.* (*C*), 1971, 701.
[216] B. I. Shapiro, S. D. Staurova, and I. P. Chikhacheva, *Doklady Akad. Nauk S.S.S.R.*, 1970, **195**, 104.

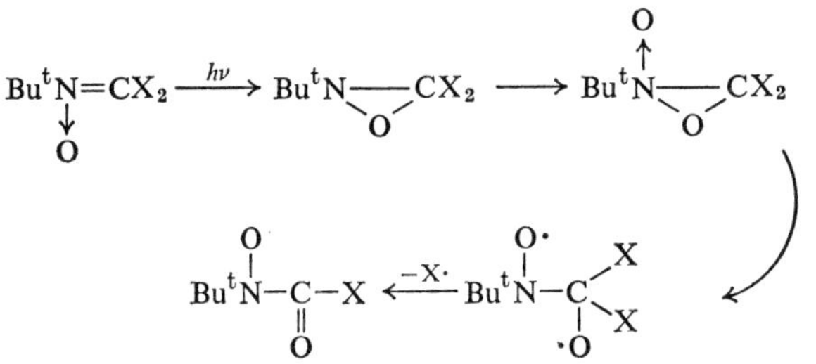

Scheme 1

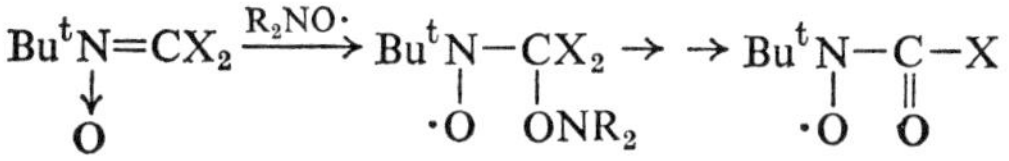

Scheme 2

The halogenocarbonyl nitroxides also react with nucleophiles;[67] for instance, $Me_2N \cdot CO \cdot N(Bu^t)O\cdot$ results from the treatment of either chloro- or bromo-carbonyl t-butyl nitroxide with dimethylamine.

Other complications may be that traps can react with starting materials by non-radical pathways (*e.g.* the reaction [217] of 2-methyl-2-nitrosopropane with organoboranes). In another example, it was demonstrated [208] that reaction between a nucleophile (*e.g.* NH_3) and a radical derived from a trap (*e.g.* $\cdot CH_2NO_2$ from $^-CH_2NO_2$) can lead to observed signals from $^+NH_3CH_2NO_2^-\cdot$, as a possible alternative to the expected reaction between $NH_3^+\cdot$ and $^-CH_2NO_2$. Careful experiments are needed to distinguish these cases to obtain correct mechanistic interpretations.

Radicals in other Reactions. Other reactions reported which lead to the production of radicals include the photolytic synthesis of trimethyliso-quinuclidine *N*-oxyl (50), *via* the photochemical dissociation of 1,8-dinitroso-*p*-menthane (49) in solution in cyclohexane to a nitrosoalkyl radical and its subsequent (internal) trapping (in 72% yield):[218]

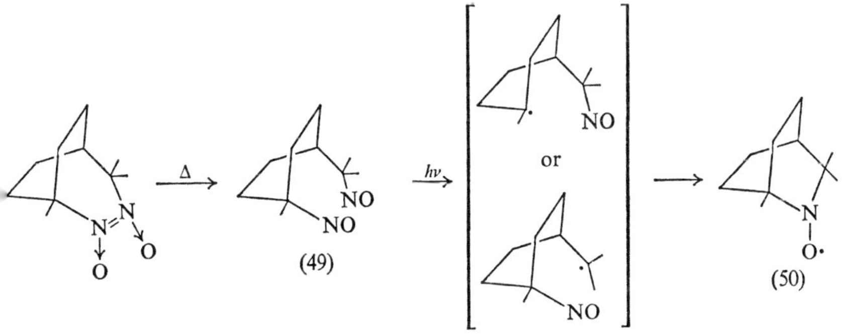

[217] K. G. Foot and B. P. Roberts, *J. Chem. Soc.* (*C*), 1971, 3475.

Other convenient nitroxide syntheses have been applied to give stable pyrrolidinyl radicals in reasonable yields:[219]

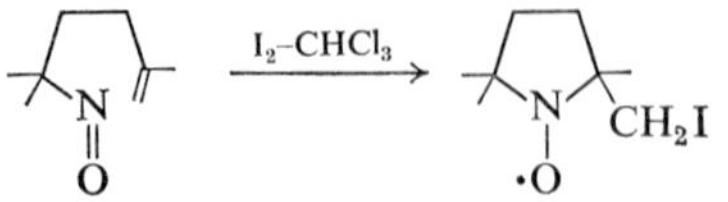

(11)

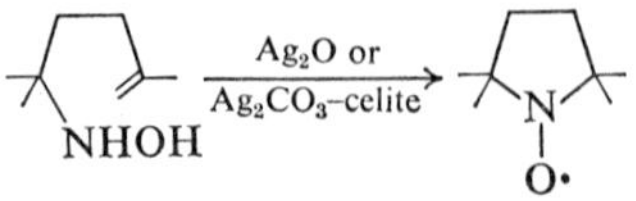

(12)

An e.s.r. signal from diphenyl nitroxide was detected during the thermal decomposition of dibenzoyl in peroxide in aromatic solvents in the presence of nitrobenzene;[220] this is thought to arise by the scavenging of phenyl radicals by nitrosobenzene formed in the reaction. The same nitroxide apparently can oxidize the phenylcyclohexadienyl radical (formed in the phenylation) to biphenyl, hence accounting for the prevention of the usual side-reactions by the addition of nitrobenzene. Diphenyl nitroxide also results from the molecule-induced homolysis of dibenzoyl peroxide by diphenylhydroxylamine, which probably proceeds *via* initial electron transfer from the hydroxylamine to the peroxide.[221]

E.s.r. results relevant to the mechanism of decomposition of acylarylnitrosamines in various solvents have now been reported in detail [64] and reviewed.[222] For a range of substrates, signals from (arylazo)oxyl radicals (aryl diazotate, ArN=NO•) and nitroxide radicals •ON(Ar^1)•N(Ar^2)COR have been observed; the latter arise by the trapping of intermediate aryl radicals by the parent nitrosamine. In certain solvents, radicals derived by reaction with aryl radicals (*e.g.* •CHMeOCH_2Me from diethyl ether) are also trapped by the nitroso-compound. It has been suggested [64] that the (σ) radical ArN=NO• is probably the more significant chain-carrying radical in certain solvents and that a further key step may involve electron transfer to ArN_2^+ (to give phenyl radicals and continue the chain) from the intermediate arylcyclohexadienyl radical or from the ether-derived radical. Aryl radicals have also been trapped during the reaction of *N*-aryl-*N'*-aroylhydrazines (*e.g.* PhNHNHCOPh) with either mercuric oxide or sodium hydride in the presence of oxygen;[223] in both cases, homolysis of the intermediate azo-compound PhN=NCOPh is probably involved.

Nitroxides have been detected during the oxidation of some bicyclic oximes with lead tetra-acetate.[224] Although it is clear that these probably

[218] A. Rassat and P. Rey, *Chem. Comm.*, 1971, 1161.
[219] W. B. Motherwell and J. S. Roberts, *J. C. S. Chem. Comm.*, 1972, 328.
[220] G. R. Chalfont, D. H. Hey, K. S. Y. Liang, and M. J. Perkins, *J. Chem. Soc.* (*B*), 1971, 233.
[221] G. R. Chalfont and M. J. Perkins, *J. Chem. Soc.* (*B*), 1971, 245.
[222] J. I. G. Cadogan, *Accounts Chem. Res.*, 1971, **4**, 186.
[223] R. A. W. Johnstone, A. F. Neville, and P. J. Russell, *J. Chem. Soc.* (*B*), 1971, 1183.
[224] A. Caragheorgheopol, H. Căldăraru, T. Constantinescu, and V. E. Sahini, *J. Amer. Chem. Soc.*, 1971, **93**, 6766.

arise from further reaction of intermediate nitroso-compounds [*gem*-nitrosoacetates, $R^1R^2C(NO)OAc$, result from the reaction of some ketoximes] it appears questionable whether the nitroso-compounds result from radical reactions of first-formed iminoxyl radicals ($R^1R^2C{=}NO\cdot$). Thus, not only are the latter σ-radicals – $R^1R^2\dot{C}{-}N{=}O$ cannot surely be a contributing structure, as required – but, in addition, it seems most unlikely that acetoxyl radicals, if formed, could remain intact to participate. Heterolytic oxidation to the nitroso-compound followed by its thermolysis (Scheme 3) could perhaps present a likely alternative.

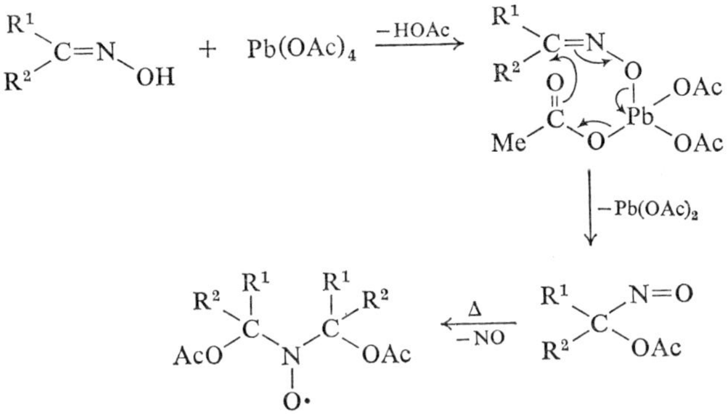

Scheme 3

Spin-trapping methods using the nitrone $PhCH{=}N(Bu^t)O$ have been used to demonstrate the oxidation of methanol to methoxyl radicals, $\cdot OCH_3$ (probably *via* the radical cation $MeOH^{+\cdot}$) by SO_4^{-}, produced thermally or photochemically from potassium persulphate.[225] The hydroxymethyl radical, obtained by reaction of excess of methanol with $\cdot OCH_3$, was also trapped. Electron-transfer oxidation of methanol was also brought about by photochemical reaction with paraquat (1,1′-dimethyl-4,4′-bipyridylium dichloride) and with UO_2^{2+}. Methods employed for the preparation of stable radical-cations from aromatic compounds, including amines,[226] include the use of lead tetra-acetate [226] and anodic oxidation,[227] both in trifluoroacetic acid.

Mechanisms of Radical Reactions.—Of the variety of possible radical reaction types, dimerization and substitution (S_H2) have been discussed above (see pp. 231 and 232). Discussion here will centre around abstraction, rearrangement, fragmentation, and electron-transfer reactions.

Abstraction and Addition. Interest here has been mainly with reactions of organic compounds with the reactive species $\cdot H$, $\cdot OH$, $\cdot NH_2$, $SO_4^{-\cdot}$, $\cdot SR$,

225 A. Ledwith, P. J. Russell, and L. H. Sutcliffe, *Chem. Comm.*, 1971, 964.
226 F. A. Neugebauer and S. Bamberger, *Angew. Chem. Internat. Edn.*, 1971, **10**, 71.
227 O. Hammerich, N. S. Moe, and V. D. Parker, *J. C. S. Chem. Comm.*, 1972, 156; U. Svanholm and V. D. Parker, *Tetrahedron Letters*, 1972, 471.

and the hydrated electron, and with comparisons of their behaviour (see also pp. 238 and 247).

For example, the reactions of the hydrated electron and the hydroxyl radical with some thiols have been studied;[228] alkaline solutions of mercaptoacetate gave $\cdot CH_2CO_2^-$ (from e^-) and $\cdot CH(S^-)CO_2^-$ (from $\cdot OH$), the latter having $g = 2.0036$ and $a(\alpha\text{-H}) = 13.4$ G, and a large linewidth ($\Delta H \sim 0.25$ G). This probably results from the expected large anisotropy in g, and it allowed high power to be employed. No thiyl radicals were detected in this study, again possibly owing to their marked anisotropy (though photolysis [176] of di-n-butyl sulphide at -60 °C leads to $\cdot CHPr^nSBu^n$ plus a species with a broad signal, ΔH *ca.* 11.5 G, $g = 2.0262$, which may be of RS$\cdot$ type; see also ref. 229). Cysteine reacts with hydroxyl to give $\cdot C(NH_2)(CO_2^-)CH_2S^-$, and dithioglycollic acid, between pH 9 and 11, gives $\cdot CH(CO_2^-)SSCH_2CO_2^-$; different results were obtained when the solutions were irradiated after they had been standing for some time: $^-O_2CCH_2SO_2\cdot$ resulted (evidently *via* radical reaction of a hydrolysis product) with $g = 2.00840$ and $a(\text{H}) = 1.82$ G. This suggests [228] that the radical previously obtained by reaction of $^-O_2CCH_2S\cdot$ with O_2,[229] which has $g = 2.018$, actually possesses the structure $^-O_2CCH_2SOO\cdot$.

The same spectra result [230] from reaction of methionine with Ti^{III}–H_2O_2 as from methionine sulphoxide with this system (the former is probably converted into the latter on standing with H_2O_2); signals from $\cdot CH_3$ and $\cdot CH_2CH_2CH(NH_2)CO_2H$ were reported. (For the latter it appears that non-equivalence of the β-proton splittings due to chirality of the adjacent carbon atom could account for the unusual linewidth effects.) The mode of the reaction between $\cdot OH$ and dialkyl sulphoxides has been the subject of discussion; it was tentatively proposed that addition to sulphur is followed by (or concerted with) radical fragmentation [reaction (13)], but an alternative proposal [10] is that the overall reaction proceeds *via* a hydroperoxide [reaction (14)]. Since alkyl radicals were not trapped when

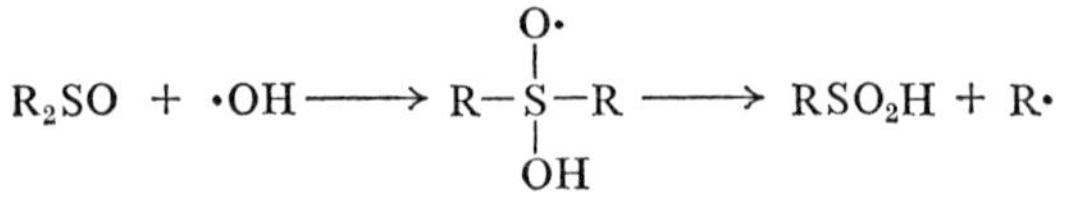
(13)

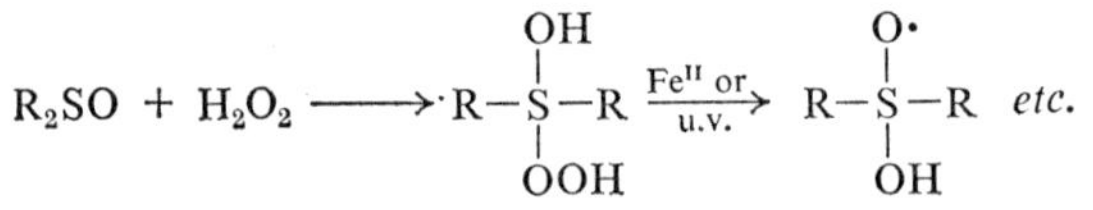
(14)

di-t-butyl peroxide was employed,[13] it seems that the second mechanism is the more likely.[231] It is also interesting that mixtures of hydrogen peroxide

228 P. Neta and R. W. Fessenden, *J. Phys. Chem.*, 1971, **75**, 2277.
229 W. W. Wolf, J. C. Kertesz, and W. C. Landgraf, *J. Magn. Resonance*, 1969, **1**, 618.
230 H. Taniguchi, H. Takagi, and H. Hatano, *J. Phys. Chem.*, 1972, **76**, 135.
231 B.-M. Bertilsson, B. Gustafsson, I. Kühn, and K. Torssell, *Acta Chem. Scand.*, 1970, **24**, 3590.

and sulphoxides in the presence of Fe^{II}, or under photolytic conditions, constitute effective alkylating agents.[231]

The reactions of the hydroxyl radical, generated radiolytically, with a variety of amines in aqueous solution have been investigated.[232] Between pH 7 and 13, $\cdot CH_2NH_2$, $\cdot CH_2NHMe$, and $\cdot CH_2NMe_2$ result from reactions with mono-, di-, and tri-methylamine, respectively. In acid solution protonated analogues were not observed but trimethylamine reacted to give $Me_2\overset{+}{N}H\cdot$ and, even more surprisingly, dimethylamine gave a signal attributed to the ethyleneimine radical-cation; possible secondary reaction mechanisms have been discussed.* Glycine gave signals from $\cdot CH(NH_2)$-CO_2H in acid (this assignment being based, in part, on the results of an oxidation in D_2O) and $\cdot CH(NH_2)CO_2^-$ in alkaline solution; the splittings and assignments for these radicals are in good agreement with those obtained in a previous investigation[195] of the reaction of glycine with Ti^{III}–H_2O_2, except that an extra proton splitting (0.9 G, COOH) was detected[232] for the former radical.

In contrast, reaction of a variety of amines with Ti^{III}–H_2O_2 in a flow system between pH 9 and 12 is reported[49] to give nitroxides of various types, depending on the flow rate; thus dimethylamine gave $Me_2NO\cdot$ and, probably, $MeN(CH_2OH)O\cdot$, and methylamine gave signals attributed to $MeN(CH_2NH_2)O\cdot$. Glycine gave $\cdot CH(NH_2)CO_2^-$ at high pH, and at pH 4 the observed spectrum was attributed to $\cdot CH(NH_3^+)CO_2^-$, extra lines being rationalized in terms of a radical complex with titanium. [However, it seems much more likely, in the light of other results,[195, 232] that the species is actually $\cdot CH(NH_2)CO_2H$, with different carboxy- and amino-proton splittings and with unusual linewidths which probably reflect exchange processes.] Results of structural interest include the observation of inequivalent amino-proton splittings, with an appreciable temperature dependence, for $\cdot CH(NH_2)CO_2^-$;[49] possible bending at nitrogen in this and other radicals containing $\cdot CH{-}NH_2$ has been discussed and some appropriate σ–π parameters have been suggested.[233] Inequivalent amido-proton splittings have also been reported for the hydroxy-adducts of acrylamide and methacrylamide.[234]

Spin-trapping experiments with the *aci*-anion of nitromethane have been employed[128] to detect differences in the behaviour of $\cdot OH$ and $SO_4^-\cdot$ in their reactions with organic compounds. Thus the oxidations of acetate ion to $\cdot CH_2CO_2^-$ by hydroxyl and to $\cdot CH_3$ by the sulphate radical-anion illustrate the increased tendency towards one-electron abstraction by $SO_4^-\cdot$.

Cyclohexadienyl-type adducts of uracil, orotic acid, and some related compounds with $\cdot OH$ (these were produced photolytically) or with $H\cdot$

232 P. Neta and R. W. Fessenden, *J. Phys. Chem.*, 1971, **75**, 738.
233 R. Poupko and B. L. Silver, *J. Amer. Chem. Soc.*, 1971, **93**, 575.
234 A. T. Bullock, G. M. Burdett, and C. M. L. Kerr, *European Polymer J.*, 1971, **7**, 791.

* The spectra have now been reassigned: see R. W. Fessenden and P. Neta, *J. Phys. Chem.*, 1972, **76**, 2857.

(formed in their photochemical reaction in the presence of propan-2-ol) have been described.[235] Hydroxylation of some substituted phenols with the Ti^{III}–H_2O_2 system in a flow system in the pH range 7—9 leads to the spectra from 1,4- or 1,2-benzosemiquinones,[236] and oxy- and dioxy-benzosemiquinones have been detected during the photolysis of alkaline aqueous solutions of phenol.[237] These reactions all apparently proceed *via* hydroxycyclohexadienyl adducts.

Reactions of other unsaturated compounds investigated include the addition of ·OH to some unsaturated bicyclic compounds [39] and the addition of ·OH and $\cdot NH_2$ to some β-dicarbonyl compounds (they react as their enol tautomers) and to some enol ethers.[238] The differences in behaviour of ·OH and $\cdot NH_2$ in some examples have been attributed to the possibility of interesting rearrangements which are discussed in the next section. Thiyl radicals, generated [105] from mercaptans with Ti^{III}–H_2O_2, have been added to a variety of olefins (see also ref. 106), and for propargyl alcohol the following reactions have been inferred:[105]

$$RSH + HC{\equiv}CCH_2OH \xrightarrow{\cdot OH} RSCH{=}CHCH_2OH$$

$$RS\cdot + RSCH{=}CHCH_2OH \longrightarrow \cdot CH(SR){-}CH(SR)CH_2OH$$

An intramolecular addition of an amido-radical to an olefinic double bond leads (Scheme 4) eventually to a nitroxide (51) with inequivalent β-splittings from protons attached to a chiral centre.[131]

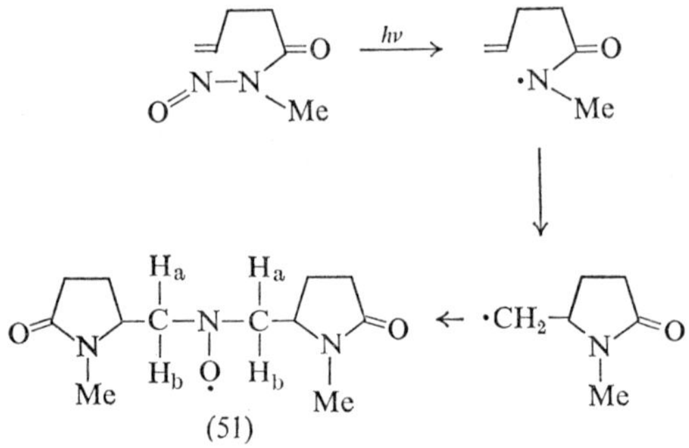

Scheme 4

Fragmentation and Rearrangement. Although, by and large, unambiguous examples of radical rearrangements are scarce, it has been demonstrated that, given suitable structural requirements and, in some cases, acid or

[235] J. K. Dohrmann and R. Livingston, *J. Amer. Chem. Soc.*, 1971, **93**, 5363.
[236] K. Günther, W. G. Filby, and K. Eiben, *Tetrahedron Letters*, 1971, 251.
[237] M. Tomkiewicz, A. Groen, and M. Cocivera, *J. Amer. Chem. Soc.*, 1971, **93**, 7102.
[238] D. J. Edge, B. C. Gilbert, R. O. C. Norman, and P. R. West, *J. Chem. Soc.* (*B*), 1971, 189.

base catalysis, both heterolytic and homolytic transformations can be demonstrated.

For example, the investigation of the rearrangement of the radical $\cdot CH(OH)CH_2OH$ (from glycol with $\cdot OH$) to $\cdot CH_2CHO$ at low pH [reaction (15)] has been extended to other 1,2-disubstituted radicals formed by addition or abstraction reactions.[117, 238] It has also been shown [117] that at pH *ca.* 2, the radical $\cdot CH(OH)CH_2CH_2CHO$ is detected from glycol,

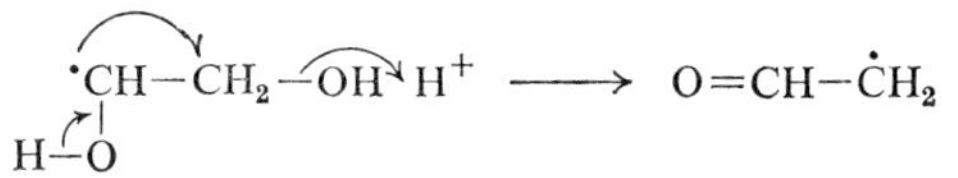
(15)

this presumably arising *via* hydrogen abstraction by $\cdot CH_2CHO$ ($\leftrightarrow CH_2{=}CH{-}O\cdot$) to give the enol form of acetaldehyde which then adds a further rearranged radical.

For instance, a rearrangement which in some ways resembles (15) has been proposed to account for the pH dependence of the two radicals (52) and (53) formed by the addition of $\cdot OH$ to vinyl ethers. At low pH ($\sim$ 1)

$$CH_2{=}CHOMe \xrightarrow{\cdot OH} \underset{(52)}{\cdot CH_2CH(OH)OMe} + \underset{(53)}{\cdot CH(OMe)CH_2OH}$$

radical (52) predominates, whereas at higher pH ($\sim$ 2.5 and above) the reverse is true. This behaviour can be explained if the formation of the delocalized radical (53) is kinetically favoured, if the isomer (52) is thermodynamically favoured (thermochemical calculations support this assertion), and if an acid-catalysed mechanism for their interconversion exists; this probably proceeds *via* an intermediate radical-cation (54) [reaction (16)].

$$RO{-}\dot{C}H{-}CH_2OH + H^+ \underset{}{\overset{-H_2O}{\rightleftharpoons}} \underset{(54)}{\left[R\overset{+}{O}{=}CH{-}\dot{C}H_2\right]} \overset{H_2O}{\rightleftharpoons} \begin{matrix}RO\\HO\end{matrix}{>}CH{-}\dot{C}H_2 + H^+ \quad (16)$$

Support for this mechanism comes also from the observation that reaction of $HO\cdot$ with 2-methoxyethanol leads to $\cdot CH_2OCH_2CH_2OH$, $\cdot CH(OMe)$-CH_2OH, and, at low pH and evidently *via* a mechanism similar to reaction (16), to $\cdot CH_2CH(OH)OMe$. This reaction is not exhibited by the hydroxyl-radical adduct of vinyl acetate [$\cdot CH(OAc)CH_2OH$] nor by the adducts of $\overset{+}{N}H_3\cdot$ (*i.e.* $\cdot NH_2$ at low pH) with vinyl ethers; evidently the structural requirements (a $+M$-substituent at the α-position and a good leaving group at the β-position) are not fulfilled in those cases.[238] The scope of the rearrangement has been extended [117] to some halogen-containing radicals prepared by hydrogen-atom abstraction with $\cdot OH$; for example, the radical $\cdot CH(OH)CH_2Cl$ from 2-chloroethanol loses chloride ion readily and gives, again probably *via* an intermediate radical-cation, the radicals $\cdot CH_2CHO$ (by loss of a proton from the transient cation), $\cdot CH_2CH(OH)_2$, and $\cdot CH(OH)CH_2OH$.

In some cases, for essentially similar radicals, homolytic fragmentation appears to be a favourable alternative.[117] For instance, choline chloride [$CH_2(OH)\cdot CH_2NMe_3^+Cl^-$] reacts with the hydroxyl radical to give e.s.r. signals from $\cdot CH(OH)CH_2NMe_3^+$ and also $\overset{+}{NMe_3}\cdot$ (see also ref. 239 for the detection of this species in a sample of γ-irradiated choline chloride).

It is perhaps relevant that loss of β-halogen atoms in substituted nitroxides appears to be a facile process,[66, 67] and it has also been pointed out [117] that the 1,2-disubstituted radicals described above have anomalously small β-proton splittings, indicating a preferred eclipsing conformation (see p. 222) and, possibly, a stereoelectronic requirement of this sort for fragmentation. (Perhaps it is the essentially eclipsed arrangement of the orbital of the unpaired electron and the incipient leaving group which accounts for the rather surprising ring-opening reactions of pyridyl radicals.[19])

Alicyclic oxygen-conjugated radicals, generated from the reactions of the hydroxyl radical with a series of mono-, di, and tri-oxygen substituted cyclic ethers (*e.g.* oxiranyl, tetrahydrofuran-2-yl, or 1,3-dioxan-2-yl), show no evidence of giving radicals by ring opening.[32, 33] The radical (55), obtained from the parent 1,3-dioxolan, is not, apparently, converted into either (56) or (57), nor is it formed from them. However, the feasibility

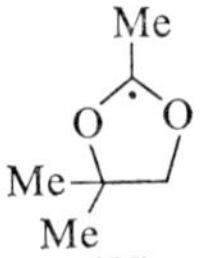

(55)

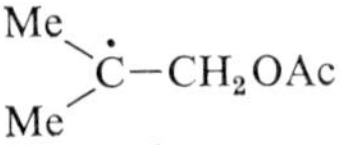

(56)

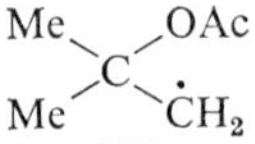

(57)

of acetoxy migration is indicated by the observation of signals from (56), presumably formed *via* (57), in the reaction of $\cdot OH$ with t-butyl acetate.[33] The radicals $\cdot CH_2CMe(OAc)CO_2^-$ and $\cdot CH_2CH_2OAc$ show no tendency towards rearrangement or bridging. It is concluded that the acetoxy migration involved in the conversion of (56) into (57) proceeds *via* fragmentation to isobutylene and the acetoxyl radical, followed by re-arrangement within a solvent cage, rather than by intramolecular addition of the radical to the carbonyl group.[33]

Other aspects of ring-opening reactions of radicals have been investigated with the application of spin-trapping techniques.[240, 241] For example, the decomposition of an alkyl nitroso-compound (RNO) with red light, which leads eventually to the dialkyl nitroxide ($R_2NO\cdot$), has been studied [240] with a view to distinguishing the following possible reaction mechanisms:

$$\mathrm{RNO} \xrightarrow{h\nu} \mathrm{RNO^*} \xrightarrow{\mathrm{RNO}} \mathrm{R_2NO\cdot} + \mathrm{NO} \qquad (17)$$

$$\mathrm{RNO} \xrightarrow{h\nu} \mathrm{RNO^*} \longrightarrow \mathrm{NO} + \mathrm{R\cdot} \xrightarrow{\mathrm{RNO}} \mathrm{R_2NO\cdot} \qquad (18)$$

239 M. C. R. Symons, *J. C. S. Faraday I*, 1972, 216.
240 J. A. Maasen, H. Hittenhausen, and Th. J. de Boer, *Tetrahedron Letters*, 1971, 3213.
241 J. W. Hartgerink, L. C. J. van der Laan, J. B. F. N. Engberts, and Th. J. de Boer, *Tetrahedron*, 1971, **27**, 4323.

With the nitroso-monomer (58), the nitroxide (59) was obtained, in accord with the suggestion that the dissociation pathway [reaction (18)], which gives a radical which can then ring-open, is favoured.

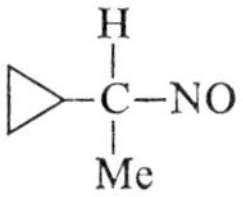

(58)

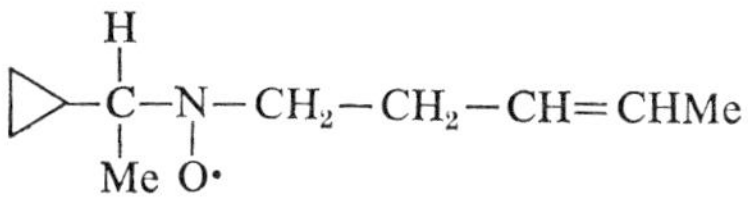

(59)

In another investigation,[241] photochemical excitation of benzophenone was employed to abstract hydrogen atoms from 2-alkyl-1,3-dioxolans and 2-alkyl-1,3-oxathiolans in benzene, and the resulting radicals were trapped with 2-nitroso-2-methylpropane [*e.g.* reaction (19)]. It is significant that

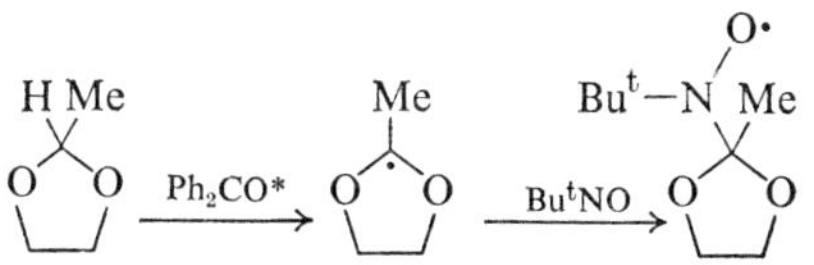

(19)

no ring-opened radicals were detected, and even 2-cyclopropyl-1,3-dioxolan-2-yl remained intact; thus again there is no e.s.r. evidence for the β-scission of cyclic acetal radicals. It appears that the reactions in the presence of $CFCl_3$, which yield exclusively *open* 2-chloroethyl carboxylic esters and *S*-2-chloroethyl thiocarboxylic esters, probably proceed *via* a cyclic chloro-acetal (or -thioacetal) which reacts further according to a *heterolytic* mechanism (Scheme 5).

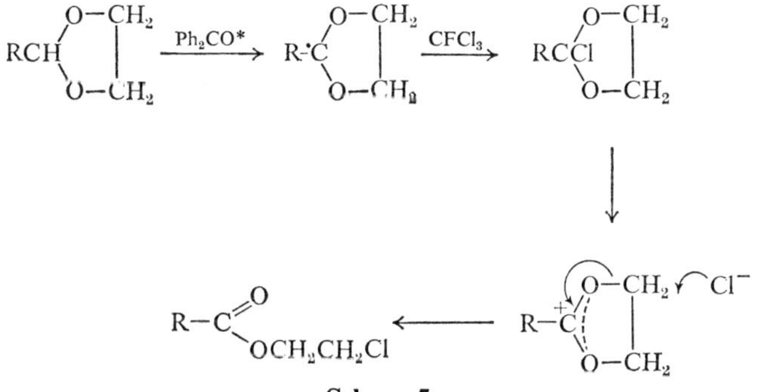

Scheme 5

This mechanism is substantiated by the results for some chiral 1,3-dioxolans, whose reaction involves inversion of configuration at C-4.

Further Radical Reactions. Other interesting radical reactions described include the production[51] of *N*-alkoxy-*N*-alkylamino-radicals from *N*-alkylperoxycarbamic esters [*e.g.* $\cdot N(Me)OBu^t$ from $MeNH\cdot CO\cdot OOBu^t$]

probably *via* alkylamino-radicals. In another study, alkylamino-radicals ($R^1R^2CHNH\cdot$) were obtained by the one-electron reduction (Ti^{III}) of alkylhydroxylamines and were shown [208] to be converted into α-alkylamino-radicals, evidently *via* proton transfer (*cf.* the well-known reaction of aminium radical cations):

$$R^1R^2CHNH\cdot \xrightarrow{H^+} R^1R^2CHNH_2^{+\cdot} \underset{}{\overset{-H^+}{\rightleftharpoons}} R^1R^2\dot{C}NH_2 \underset{}{\overset{H^+}{\rightleftharpoons}} R^1R^2\dot{C}NH_3^+$$

These two types of radical show some interesting contrasts in behaviour; thus whereas alkylamino-radicals are oxidizing agents [*e.g.* with titanium-(III), by reaction with more hydroxylamine to give a nitroxide, or with nitro *aci*-anions, reaction (20)], α-aminoalkyl radicals are effective one-

$$R^1R^2CHNH\cdot + {}^-CHMeNO_2 \longrightarrow R^1R^2CHNH^- + \cdot CHMeNO_2 \qquad (20)$$

$$R^1R^2CHNH^- \longrightarrow R^1R^2CHNH_2 \qquad \cdot CHMeNO_2 \xrightarrow{^-CHMeNO_2} O_2NCHMeCHMeNO_2^{-\cdot}$$

electron reducing agents (*cf.* the analogous behaviour of radicals containing an α-hydroxy- or α-alkoxy-substituent). Thus, radicals were obtained from the reduction of maleic acid and its anions, of t-nitrosobutane, and of some aliphatic nitro-compounds, as well as those derived by the trapping of α-aminoalkyl radicals with maleic acid and the nitroso-compound. It also appears that the radicals are readily oxidized by hydrogen peroxide (*cf.* $\cdot C{-}OR$ radicals):

$$R^1R^2\dot{C}NH_2 + H_2O_2 \longrightarrow R^1R^2\overset{+}{C}{-}\ddot{N}H_2 + \cdot OH + OH^- \longrightarrow\longrightarrow R^1R^2C{=}O$$

For a series of α-oxygenated carbon radicals (*e.g.*, tetrahydrofuran-2-yl, oxetan-2-yl, and oxiranyl) the effectiveness for one-electron donation has been measured by the extent of their reaction with $BrCH_2CO_2H$ to give $\cdot CH_2CO_2H$ and Br^-.[33]

10
Organometallic Radicals

BY A. HUDSON

1 Introduction

The study of organometallic radicals has developed considerably over the past two or three years. The earlier work in this field was mostly confined to long-lived species, particularly radical ions. The current emphasis is on the transient radicals which are involved in many organometallic reactions. The development of steady-state photolysis techniques has been of particular significance. Initial experiments were mainly concerned with Group IV elements, particularly silicon,[1] but there is increasing interest in the e.s.r. spectra of boron- and phosphorus-containing intermediates. This chapter is mainly concerned with radicals containing elements from Groups III, IV, and V; transition-metal systems involving relatively stable species such as ferrocenium salts,[2] $V(CH_2Ph)_4$,[3] $V(CH_2SiMe_3)_4$,[4] bisarene chromium(I) complexes,[5] and bicyclo-octatetraene-vanadium(IV) [6] will not be discussed in detail.

An article [7] on the characterization of organometallic compounds by e.s.r. is mostly concerned with linewidth effects and contains little on organometallic compounds.

2 Group III Radicals

Homolytic substitution reactions of organometallic compounds have been extensively investigated by e.s.r. Alkoxyl radicals react with trialkylboron

$$R_3B + Bu^tO\cdot \longrightarrow R_2BOBu^t + R\cdot \qquad (1)$$

[1] S. W. Bennett, C. Eaborn, A. Hudson, H. A. Hussain, and R. A. Jackson, *J. Organometallic Chem.*, 1969, **16**, P36; S. W. Bennett, C. Eaborn, A. Hudson, R. A. Jackson, and K. D. J. Root, *J. Chem. Soc.* (*A*), 1970, 348; P. J. Krusic and J. K. Kochi, *J. Amer. Chem. Soc.*, 1969, **91**, 3938.

[2] R. Prins and A. T. G. Kortbeek, *J. Organometallic Chem.*, 1971, **33**, C33; A. Horsfield and A. Wassermann, *J. C. S. Dalton*, 1972, 187; D. O. Cowan, G. A. Candela, and F. Kaufman, *J. Amer. Chem. Soc.*, 1971, **93**, 3889; J. J. McDonnell and D. J. Pochopien, *J. Org. Chem.*, 1971, **36**, 2092.

[3] S. D. Ibekwe and J. Myatt, *J. Organometallic Chem.*, 1971, **31**, C65.

[4] W. Mowat, A. Shortland, G. Yagupsky, N. J. Hill, M. Yagupsky, and G. Wilkinson, *J. C. S. Dalton*, 1972, 533.

[5] W. Karthe and W. Kleinwachter, *Z. Phys. Chem.* (*Leipzig*), 1971, **247**, 241.

[6] J. L. Thomas and R. G. Hayes, *Inorg. Chem.*, 1972, **11**, 348.

[7] F. J. Smentowski, *Character. Organometallic Compounds*, 1971, **26**, 481.

compounds to give[8,9] alkyl radicals [equation (1)] and similar reactions have been reported for aluminium and gallium compounds.[8] Zinc, cadmium, antimony, and bismuth alkyls also give detectable concentrations of alkyl radicals.[10] No metal-centred intermediates have been detected in these systems although they have been observed in analogous reactions involving phosphorus (see below). Dimethylamino-radicals[11] and the triplet states of ketones[12] also cause homolytic displacement at boron. More detailed accounts of some of this work are now available. Rate constants for the reaction of t-butoxyl radicals at boron have been obtained[13] by competition experiments in cyclopentane when equation (2)

$$Bu^tO\cdot + C_5H_{10} \longrightarrow Bu^tOH + C_5H_9\cdot \qquad (2)$$

competes with equation (1). The relative concentrations of R· and $C_5H_9\cdot$ are measured by e.s.r. The S_H2 reactions are very fast with low activation energies. Butylthiyl radicals from the photolysis of di-n-butyl disulphide also displace alkyl radicals from trialkyl borons.[14]

It is well known that the $n\pi^*$ triplet states of ketones will abstract hydrogen atoms bonded to carbon. They also react rapidly by an S_H2 process at boron [equation (3)] and competition experiments have again

$$R^1R^2CO + BR^3_3 \longrightarrow R^1R^2\dot{C}OBR^3_2 + R^3\cdot \qquad (3)$$

provided kinetic parameters.[12] The coupling constants of some of the boron radicals are given in Table 1.

Table 1 *Coupling constants* (G) *for radicals* $R^1R^2\dot{C}OBR^3_2$

R^1	R^2	R^3	$a(^{11}B)$	a^H	*Temperature*/°C
Me	Me	Pr^n	8.3	20.6	− 95
Me	Pr^i	Pr^n	8.3	12.3 (CH) 20.5 (Me)	− 110
Pr^i	Pr^i	Bu^n	9.2	12.5	− 70

The methyl proton splitting in $Me_2\dot{C}OBR_2$ is similar to those found in $Me_2\dot{C}OB(OPr^i)_2$ and $Me_2\dot{C}OSiEt_3$. However, the ^{11}B splittings (8.3—9.4 G) for radicals of the type $R^1R^2\dot{C}OBR^3_2$ are much greater than the temperature-dependent splittings of 0.35—0.66 G reported[8,15] for radicals of the type $R^1R^2\dot{C}OB(OR^3)_2$. Moreover, the former splittings are temperature-independent.[14] It has been suggested that π-delocalization on to boron is less

[8] P. J. Krusic and J. K. Kochi, *J. Amer. Chem. Soc.*, 1969, **91**, 3942.
[9] A. G. Davies and B. P. Roberts, *Chem. Comm.*, 1969, 699.
[10] A. G. Davies and B. P. Roberts, *J. Organometallic Chem.*, 1969, **19**, P17.
[11] A. G. Davies, S. C. W. Hook, and B. P. Roberts, *J. Organometallic Chem.*, 1970, **22**, C37.
[12] A. G. Davies, D. Griller, B. P. Roberts, and J. C. Scaiano, *Chem. Comm.*, 1971, 196; A. G. Davies, B. P. Roberts, and J. C. Scaiano, *J. Chem. Soc.* (*B*), 1971, 2171.
[13] A. G. Davies, D. Griller, and B. P. Roberts, *J. Chem. Soc.* (*B*), 1971, 1823.
[14] A. G. Davies and B. P. Roberts, *J. Chem. Soc.* (*B*), 1971, 1830.
[15] A. Hudson and H. A. Hussain, *J. Chem. Soc.* (*B*), 1969, 793.

important in a radical of the type $R^1R^2\dot{C}OB(OR^3)_2$ because the $2p_z$ orbital on boron is already involved in $(p\text{–}p)\pi$-bonding with the oxygen atoms of the alkoxy-groups. The radical $Pr^i_2\dot{C}OBR_2$ (R = Bu^n or Pr^n) exhibits linewidth alternation attributed to restricted rotation about the C—O bond.

Gamma irradiation of BMe_3 at 77 K gives the radical $\dot{C}H_2BMe_2$ with $a^{\alpha}_{H} = -18$ G, $a(^{11}B) = -4$ G, and $a(Me) = 4$ G, corresponding to about 15% spin density on boron and 85% on carbon.[16] Irradiation of aluminium alkyls similarly provides e.s.r. spectra attributed [17] to a $H_2\dot{C}$—Al$\lessdot$ species with $a^{\alpha}_{H} = 19$ G and $a(^{27}Al) = 18.9$ G. These have been compared with other radicals of the type H_2C—X (X = PR_3^+, POR_2, *etc.*). It was concluded that spin is transmitted to X by spin polarization of the C—X σ-bond and a_X is negative. The negative sign for $a(^{11}B)$ in $\dot{C}H_2BMe_2$ fits into this general scheme. The irradiation of aluminium alkyls also gives [18] spectra attributed to AlR_3^-. $AlMe_3^-$ has $g_{\perp} = 2.020$, $g_{\parallel} = 1.979$, $a(^{27}Al) = 324$ G, $A_{\parallel} - 357$ G, and $A_{\perp} = 307$. The $p:s$ ratio of 2.39 corresponds to a pyramidal structure with an out-of-plane angle of 21°. Comparison with SiR_3 and PR_3^+ radicals shows that there is a well-defined trend towards flatter radicals across the periodic table as expected from electronegativity considerations.

3 Group IV Radicals

Trends in the coupling constants [19] of the series of radicals $SiMe_n(SiMe_3)_{3-n}$ indicate (Table 2) that as methyl groups are replaced by less electronegative trimethylsilyl groups there is a trend towards more planar structures. It seems likely that tris(trimethylsilyl)silyl is essentially planar.

Table 2 *Hyperfine splittings* (G) *for the radicals* $SiMe_n(SiMe_3)_{3-n}$

	a(Me)	$a(SiMe_3)$	$a(^{29}Si)$	g
$SiMe_3$	6.34	—	183	2.0031
$SiMe_3SiMe_2$	8.21	0.47	137	2.0037
$(SiMe_3)_2SiMe$	9.36	0.47	71	2.0045
$(SiMe_3)_3Si$	—	0.44	65	2.0050

Similar changes in geometry are exhibited by the fluorosilyl radicals [20] (Table 3) which have been prepared by electron irradiation of fluorosilanes in an SF_6 matrix. As expected SiF_3 is more pyramidal than SiH_3. The large and presumably positive proton coupling in $SiHF_2$ is particularly noteworthy. The radical Me_2SiF has $a_H = 4.53$ and $a_F = 65.21$ G.

The trimethyltin radical is conveniently prepared [21] by an S_H2 reaction (4) of t-butoxyl radicals with hexamethylditin. The β-H splitting is 2.75 G and

[16] A. R. Lyons and M. C. R. Symons, *J. C. S. Faraday II*, 1972, **68**, 502.
[17] A. Begum, A. R. Lyons, and M. C. R. Symons, *J. Chem. Soc.* (*A*), 1971, 2388.
[18] A. Begum, A. R. Lyons, and M. C. R. Symons, *J. Chem. Soc.* (*A*), 1971, 2290.
[19] J. Cooper, A. Hudson, and R. A. Jackson, *Mol. Phys.*, 1972, **23**, 209.
[20] M. V. Merritt and R. W. Fessenden, *J. Chem. Phys.*, 1972, **56**, 2353.
[21] G. B. Watts and K. U. Ingold, *J. Amer. Chem. Soc.*, 1972, **94**, 491.

Table 3 *Coupling constants for the fluorosilyl radicals*

	a_F	a_H	$a(^{29}Si)$	g
SiH_3	—	7.84	(−)190	2.0032
SiH_2F	55.2	34.6	—	2.0024
$SiHF_2$	77.8	89.9	—	2.0013
SiF_3	136.6	—	(−)498	2.0003

$g = 2.017$. The $\cdot SiMe_3$ and $\cdot GeMe_3$ radicals have been reported previously. Using a rotating sector and CAT, Watts and Ingold[21] have determined bimolecular rate constants for the self-reactions of radicals with the structures $Me_3\dot{M}$, $Me_3M\dot{C}H_2$, $Me_2M\dot{M}Me_2$, and $Me_3MMMe_2\dot{C}H_2$, where M is C, Si, Ge, or Sn. The $2k_t$ values lie in the range 1.4×10^9—11×10^9 mol^{-1} s^{-1} and correlate with the molecular weights of the radicals.

$$Bu^tO\cdot + Me_3SnSnMe_3 \longrightarrow Me_3SnOBu^t + \cdot SnMe_3 \qquad (4)$$

The reaction of triethylsilyl radicals with alkyl bromides (or chlorides) which forms a convenient method for preparing specific alkyl radicals[22] has been further investigated[23] [reactions (5)—(7)]. Competition experiments suggest that the ease of removal of halogen atoms increases in the

$$Bu^tO\cdot OBu^t \xrightarrow{h\nu} 2Bu^tO\cdot \qquad (5)$$

$$Bu^tO\cdot + Et_3SiH \longrightarrow Bu^tOH + Et_3Si\cdot \qquad (6)$$

$$Et_3Si\cdot + RBr \longrightarrow Et_3SiBr + R\cdot \qquad (7)$$

order: primary alkyl < secondary < tertiary < acetonyl $\approx$ $\dot{C}H_2CO_2Et$ < allyl $\approx$ propargyl $\approx$ benzyl < cyanomethyl. The presence of electron-withdrawing groups favours radical formation, as expected for attack by a radical centred on an electropositive element. When bromobenzene was used as a substrate, the radical $Et_2SiH\dot{C}HCH_3$ was observed with $a^{\alpha}_{H} = 20.3$, $a^{\beta}_{H} = 25.3$ and $a_{SiH} = 11.8$ G. Phenyl radicals also reacted with trimethylsilane to give $Me_2SiH\dot{C}H_2$ ($a^{\alpha}_{H} = 20.9$ and $a_{SiH} = 14.8$ G). Cyclopropyl bromide gave similar results, but cyclopropyl radicals were observable in trimethylsilane.

Silyl radicals add[22,23] to ketones [reaction (8)], aldehydes, and esters; trifluoroacetone gives $CF_3\dot{C}(OSiEt_3)CH_3$ ($a^{\beta}_{H} = 21.5$ and $a_F = 28.8$ G) and methyl formate $CH_3O\dot{C}(OSiEt_3)H$ [$a^{\alpha}_{H} = 12.5$ and $a(OCH_3) = 0.9$ G].

$$R_2CO + \cdot SiEt_3 \longrightarrow R_2\dot{C}OSiEt_3 \qquad (8)$$

The thermal reaction of triphenylsilane and benzophenone at 400 K gives[24] the radical $Ph_2\dot{C}OSiPh_3$. The spin distribution is similar to that

22 A. Hudson and R. A. Jackson, *Chem. Comm.*, 1969, 1323.
23 A. J. Bowles, A. Hudson, and R. A. Jackson, *J. Chem. Soc. (B)*, 1971, 1947.
24 A. R. McIntosh and J. K. S. Wan, *Canad. J. Chem.*, 1971, **49**, 817; *Mol. Phys.*, 1971, **22**, 183.

found in the benzophenone ketyl with $a_H^o = 3.16$, $a_H^m = 1.23$, $a_H^p = 3.48$, $a^{\alpha}(^{13}C) = 25$, and $a(^{29}Si) = 7.1$ G. The ^{29}Si coupling is much larger than that reported [25] for $\dot{C}H_2OSiMe_3$ (1.5 G), in spite of delocalization on to the phenyl rings. Presumably, steric overcrowding in $Ph_2\dot{C}OSiPh_3$ leads to enhanced hyperconjugation.

There has been considerable interest [25, 26] in β-substituted alkyl radicals (1) with R = H, Me, Et, *etc.*, and X = Si, Ge, or Sn. These are conveniently prepared by the addition of metal-centred radicals to olefins. The coupling constants for the Si and Ge analogues of n-propyl are given in Table 4. Trialkyltin radical adducts have also been observed. Thus

Table 4 *Group IV analogues of the n-propyl radical*

	a_H^{α}	a_H^{β}	a_H^{γ}	*Temperature*/°C
$H_3CCH_2\dot{C}H_2$	22.14	30.33	0.27	− 110
$H_3SiCH_2\dot{C}H_2$	21.39	17.68	2.77	− 70
$H_3GeCH_2\dot{C}H_2$	20.96	15.84	4.32	− 80

photolysis of tri-n-butyl stannane and di-t-butyl peroxide in ethylene at − 148 °C gives $Bu_3^nSnCH_2\dot{C}H_2$ with $a_H^{\alpha} = 19.64$ and $a_H^{\beta} = 15.67$ G. The low values found for a_H^{β}, and the fact that they decrease on lowering the temperature, suggest that these molecules have a preferred conformation (2)

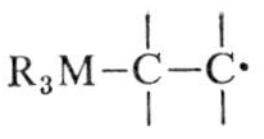

(1)

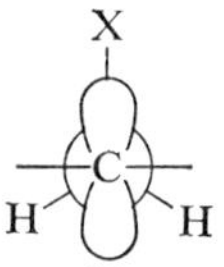

(2)

in which the metal atom eclipses the odd electron orbital. The temperature dependence of the β-proton coupling constants has been analysed [26] in terms of both classical and quantum mechanical models [27] to give barriers to internal rotation of 5.2, 6.7, and 8.4 kJ mol^{-1} for $Et_3SiCH_2\dot{C}H_2$, $Et_3GeCH_2\dot{C}H_2$, and $Bu_3^nSnCH_2\dot{C}H_2$ respectively.

The *g*-factors of these radicals are also of some interest as can be seen by considering the data [28] in Table 5. Whereas the *g*-factors of the metal-centred radicals $Me_3M\cdot$ increase as the atomic number of M increases,[1, 21] the *g*-factors of the β-substituted radicals decrease. A similar trend has been reported [29a] for the radicals $Me_3M\dot{C}H_2$ and attributed to delocalization

25 J. K. Kochi and P. J. Krusic, in 'Essays on Free-radical Chemistry', Chemical Society Special Publication No. 24, London, 1970, Chap. 7.

26 P. J. Krusic and J. K. Kochi, *J. Amer. Chem. Soc.*, 1971, **93**, 846.

27 P. J. Krusic, P. Meakin, and J. P. Jesson, *J. Phys. Chem.*, 1971, **75**, 3438.

28 T. Kawamura and J. K. Kochi, *J. Amer. Chem. Soc.*, 1972, **94**, 648.

29 (*a*) J. H. Mackey and D. E. Wood, *Mol. Phys.*, 1970, **18**, 783; (*b*) A. R. Lyons, G. W. Neilson, and M. C. R. Symons, *J. C. S. Faraday II*, 1972, **68**, 807.

Table 5 *E.s.r. parameters for* $Me_3MCH_2\dot{C}H_2$

M	*Temperature*/°C	a_H^{α}	a_H^{β}	$a(CH_3)$	a(M)	g
C	− 105	22.06	24.71	0.1	—	2.002 68
Si	− 112	21.05	17.68	0.1	37.4 (^{29}Si)	2.002 65
Ge	− 100	20.70	16.57	0.14	—	2.002 55
Sn	− 101	19.69	15.84	0.15	488.9 (^{119}Sn) 467.7 (^{117}Sn)	2.002 05

of the odd electron into *d*-orbitals on the metal. However, this interpretation has recently been criticized.[29b]

It has been estimated that the *g*-shift in $Bu^n_3CH_2\dot{C}H_2$ corresponds to homoconjugative delocalization of an odd electron density of 0.12 on to the metal *d*-orbitals. The large ^{29}Si and ^{119}Sn splittings are also of interest, and suggest that the odd electron is hyperconjugatively delocalized on to the sp^3 hybrid metal orbital. For the tin radical p_σ is estimated to be 0.13. From a valence-bond point of view we can consider that structures such as (3) contribute significantly. Similar conclusions have been arrived at from an investigation of β-substituted alkyl radicals in γ-irradiated polycrystalline solids.[30] Radicals with β-atoms such as Sn, P, or As all have the preferred conformation (2) with large metal splittings corresponding to up to 6% valence *s*-electron spin density on the heavy atom. Examples of β-phosphorus-substituted alkyl radicals studied in solution are mentioned below.

$CH_2{=}CH_2$ $\cdot MEt_3$

(3)

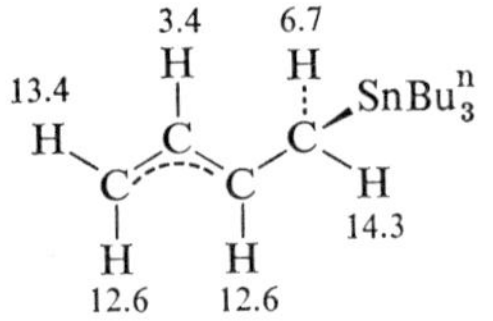

(4)

The adduct of trialkylstannyl radicals and butadiene has been assigned[31] the structure (4). At low temperatures the two hydrogens on C-4 were inequivalent, but at + 16 °C a 1 : 2 : 1 triplet was observed with an averaged splitting of 11.04 G. The barrier to rotation was estimated to be 21 kJ mol^{-1} with a frequency factor of 10^{13} s^{-1}. The most stable conformation is one in which the tin atom is optimally located with respect to the π-orbitals of the allylic radical.

The radical-anion of 1-phenyl-2-trimethylsilylacetylene has been studied[32] in THF with a variety of gegenions, and the kinetics of its dimerization have been determined by following the time dependence of the e.s.r. signal intensity. The coupling constants are $a_H^o = 3.16$, $a_H^m = 0.16$,

[30] A. R. Lyons and M. C. R. Symons, *Chem. Comm.*, 1971, 1068, *J. C. S. Faraday II*, 1972, **68**, 622.
[31] T. Kawamura and J. K. Kochi, *J. Organometallic Chem.*, 1971, **30**, C8.
[32] A. G. Evans, J. C. Evans, P. J. Emes, and T. J. Phelan, *J. Chem. Soc.* (*B*), 1971, 315.

a^{p}_{H} = 7.49, $a_H(SiMe_3)$ = 0.28, and $a(^{29}Si)$ = 4.4 G. The radical-anion of the organosilicon cyclophane (5) in DME at − 80 °C has [33a] a 2.5 G splitting from methyl groups attached to carbon and a 1.25 G splitting from those attached to silicon. The *g*-factor is 2.001 69. The odd electron in $PhSiMe_2Ph^-$ is [33b] delocalized over only one ring below − 70 °C but is delocalized over both rings above − 40 °C.

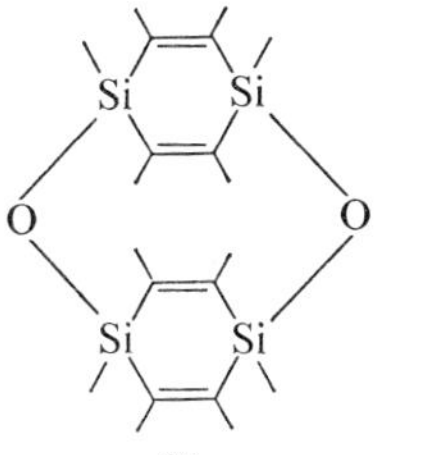

(5)

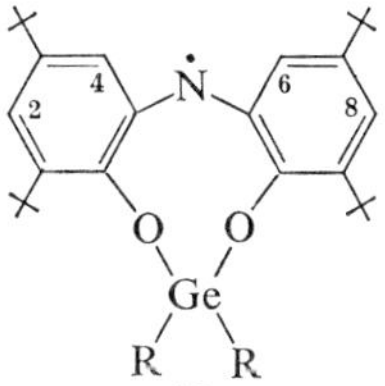

(6)

The 11,11-disubstituted 1,3,7,9-tetra-t-butyldibenzo[*dg*][1,3,6,2]dioxagermanocin-5-yls (6) give well-resolved spectra in ethanol or benzene [34] with (R = Ph) $a^{H}_{4,6}$ = 1.41, $a^{H}_{2,8}$ = 3.02. a_N = 6.13, $a(^{73}Ge)$ = 3.02, and g = 2.002 906. This is not as claimed the first report of a ^{73}Ge coupling in solution. The *p*-triethylgermylnitrobenzene radical-anion has $a(^{73}Ge)$ = 1.25 G.[35]

The bis(organosilyl) nitroxides,[36] $(RMe_2Si)_2NO$, where R = Me, Bu^t, or Ph, have surprisingly low nitrogen splittings near 6.5 G and *g*-values of 2.0092. The ^{29}Si splitting in the bis(trimethylsilyl) nitroxide is 5.9 G. It has been suggested that the low value of a_N (for Bu^t_2NO a_N = 15.2 G) is due to extensive delocalization of the nitrogen lone-pair into *d*-orbitals on silicon and localization of the odd electron on oxygen. The spectrum previously assigned [37] to $(SiMe_3)_2NO$ with a_N = 15 G appears on bubbling oxygen through the solution and is presumably that of a decomposition product. It would be interesting to determine the spin density on oxygen by measuring the ^{17}O hyperfine coupling constant for $(SiMe_3)_2NO$.

4 Group V Radicals

There have been several reports of phosphorus-centred radicals characterized by large isotropic hyperfine interactions with the ^{31}P nucleus. Photolysis of di-t-butyl peroxide in the presence of tertiary phosphines leads to homolytic replacement reactions [S_H2; reaction (9)] and the displaced

33 (*a*) F. Wudl, R. D. Allendoerfer, J. Demirgian, and J. M. Robbins, *Chem. Comm.*, 1971, 50; (*b*) E. S. Kear, K. Fisher, and R. West, *J. Amer. Chem. Soc.*, 1972, **94**, 3246.
34 H. B. Stegmann, K. Scheffler, and F. Stocker, *Angew. Chem. Internat. Edn.*, 1971, **10**, 499.
35 A. Hudson, J. W. E. Lewis, and D. R. M. Walton, *J. Organometallic Chem.*, 1969, **20**, 75.
36 R. West and P. Boudjouk, *J. Amer. Chem. Soc.*, 1971, **92**, 5901.
37 G. Chapelet-Letourneaux, H. Lemaire, and A. Rassat, *Bull. Soc. chim. France*, 1965, 3283.

radicals R· are observable by e.s.r.[38] In some cases the intermediate quadrivalent phosphoranyl radicals are also detectable.[39] It seems likely

$$\mathrm{Bu^tO\cdot + PR_3 \longrightarrow Bu^tO\dot{P}R_3 \longrightarrow Bu^tOPR_2 + R\cdot} \quad (9)$$

these radicals have trigonal-bipyramidal structures (possibly distorted) with the odd electron orbital in an equatorial position. The radical (7) from PH_3 has $a(^{31}P) = 626.7$, $a_H = 10.8$ (triplet) and $a_H = 134.6$ G (doublet). On prolonged photolysis and at higher temperatures a second species was also observed. This has been assigned the structure (8) with $a(^{31}P) = 672.5$ G and a very small triplet splitting of 0.6 G from the two protons.

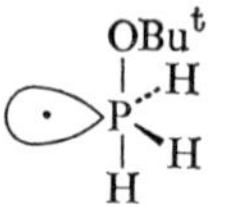
(7)

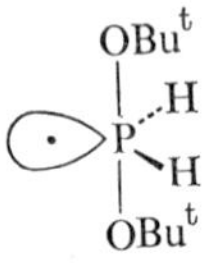
(8)

Under similar conditions, trialkyl phosphites $(RO)_3P$ give t-butyl radicals,[15, 38] again *via* a phosphoranyl radical.[40] When R = Et, $a(^{31}P) = 891$ G, and $g = 2.008$. The overall rate constant k_2 for reaction (10) has been determined [40] from competition experiments in cyclopentane by measuring the relative concentrations of t-butyl and cyclopentyl radicals as a function of temperature. The derived value of k_2 [$6.7 \times 10^9 \exp(-2400/RT)\ \mathrm{l\ mol^{-1}\ s^{-1}}$] indicates that reaction (10) is one of the fastest known bimolecular radical–molecule reactions.

$$\mathrm{Bu^tO\cdot + P(OR)_3 \longrightarrow Bu^tO\dot{P}(OR)_3 \longrightarrow Bu^t\cdot + OP(OR)_3} \quad (10)$$

The reaction of t-butoxyl radicals with dialkyl phosphites $(RO)_2P(O)H$ produces phosphonyl radicals;[41] $(MeO)_2\dot{P}O$ has a phosphorus splitting of 700 G and $g = 2.004$. More intense spectra were obtained from tetraethylpyrophosphite by reaction (11). In the presence of an olefin, addition

$$\mathrm{Bu^tO\cdot + (EtO)_2POP(OEt)_2 \longrightarrow Bu^tOP(OEt)_2 + O\dot{P}(OEt)_2} \quad (11)$$

occurs and the e.s.r. spectra of β-phosphonylalkyl radicals are observed. Thus $H_2C{=}CMe_2$ gives $(EtO)_2P(O)CH_2\dot{C}Me_2$. At − 60 °C the coupling constants are $a(^{31}P) = 112.5$, $a_H^{\beta} = 13.5$, and $a(CH_3) = 22.9$ G. The radical therefore has a preferred conformation in which the C(β)—P bond eclipses the odd-electron orbital, and is similar to those found on adding Group IV organometallic radicals to olefins as discussed earlier. The abstraction of bromine by phosphonyl radicals has also been demonstrated

38 J. K. Kochi and P. J. Krusic, *J. Amer. Chem. Soc.*, 1969, **91**, 3944.
39 P. J. Krusic, W. Mahler, and J. K. Kochi, *J. Amer. Chem. Soc.*, 1972, **94**, 6033.
40 A. G. Davies, D. Griller, and B. P. Roberts, *Angew. Chem. Internat. Edn.*, 1971, **10**, 738.
41 A. G. Davies, D. Griller, and B. P. Roberts, *J. Amer. Chem. Soc.*, 1972, **94**, 1782.

[reaction (12)]. The analogous reaction involving triethylsilyl radicals is mentioned above.

$$(EtO)_2\dot{P}O + RBr \longrightarrow (EtO)_2P(O)Br + R\cdot \qquad (12)$$

In the presence of oxygen, phosphoranyl radicals give [42] observable concentrations of peroxyl radicals with small ^{31}P splittings. $(Bu^tO)_4POO\cdot$ has $a(^{31}P) = 9$ G and $g = 2.02$.

$$X_4\dot{P} + O_2 \longrightarrow X_4POO\cdot \qquad (13)$$

The phosphorus analogue of diphenylnitroxide has been studied [43] in *X*-irradiated single crystals of diphenyl phosphine oxide. Ph_2PO is non-planar ($s : p$ ratio 0.18) with an isotropic ^{31}P splitting of 1092 MHz and anisotropic components 347, − 163, and − 184 MHz.

Radicals of general formula $(R^1O)_2P(O)CHOR^2\dot{C}Cl_2$ derived from phosphinic acid esters have been observed in aqueous solutions.[44] The ^{31}P splittings (*ca.* 60 G) are dependent upon the nature of R^1 and R^2 and also decrease as the temperature is raised, whereas a_H^{β} increases. This behaviour is similar to that discussed above for other β-phosphonylalkyl radicals.

Positive radical ions of triphenyl phosphonium methylides, $Ph_2\dot{C}\overset{+}{P}Ph_3$, have been prepared;[45] the ^{31}P splitting is 26.6 G. There have also been several reports [46–49] of phosphorus-containing radical-anions. PEt_3^+, which has been identified [18] in γ-irradiated triethylphosphine, has $g_{\parallel} = 1.991$, $g_{\perp} = 2.015$, $a(^{31}P) = 384$ G, $A_{\parallel} = 588$ G, and $A_{\perp} = 283$ G. The $p : s$ ratio of 9.29 corresponds to an out-of-plane angle of about 12°.

The u.v. photolysis of a frozen solution of diarsine (9) in ethanol at 96 K gave [50] overlapping spectra attributed to $(diars)^+$ and $AsMe_2$. The cation exhibits a coupling to two equivalent ^{75}As nuclei with $A_{\parallel} = 9.5$ G and $A_{\perp} = 11.5$ G, $g_{\parallel} = 2.070$ and $g_{\perp} = 2.075$. The metal-centred radical has $g_{\parallel} = 2.000$, $g_{\perp} = 2.002$, $A_{\parallel} = 220$ G, and $A_{\perp} = 5$ G. The unpaired electron orbital in $Me_2As\cdot$ is primarily $4p$ in character with 2.2% $4s$ contribution. Arsenic-containing anion-radicals have been detected in the cleavage of 1-arylarsoles by alkali metals.[51]

[42] G. B. Watts and K. U. Ingold, *J. Amer. Chem. Soc.*, 1972, **94**, 2528.
[43] M. Geoffroy and E. A. C. Lucken, *Mol. Phys.*, 1971, **22**, 257.
[44] W. Damerau, G. Lassmann, and K. Lohs, *Z. Chem.*, 1971, **11**, 182; *J. Magn. Resonance*, 1971, **5**, 408.
[45] H. M. Buck, A. H. Huizer, S. J. Oldenburg, and P. Schipper, *Phosphorus*, 1971, **1**, 97.
[46] D. Kilcast and C. Thomson, *Tetrahedron*, 1971, **27**, 5705; *Chem. Comm.*, 1971, 782.
[47] A. V. Il'yasov, Y. A. Levin, I. D. Morozova, A. A. Vafina, I. P. Gozman, and E. I. Zoroatskaya, *Doklady Akad. Nauk S.S.S.R.*, 1971, **201**, 108.
[48] A. V. Il'yasov, Yu. M. Kargin, Y. A. Levin, I. D. Morozova, B. V. Mel'nikov, A. A. Vafina, N. N. Sotnikova, and V. S. Galeev, *Izvest. Akad. Nauk S.S.S.R., Ser. khim.*, 1971, 770.
[49] W. M. Gulick, jun., *J. Amer. Chem. Soc.*, 1972, **94**, 29.
[50] J. R. Preer, F. D. Tsay, and H. B. Gray, *J. Amer. Chem. Soc.*, 1972, **94**, 1875.
[51] G. Markl and H. Hauptmann, *Angew. Chem. Internat. Edn.*, 1972, **11**, 439.

Arsenanyl radicals $\cdot AsX_4$ have been observed in solution during the reactions of thiyl or alkoxyl radicals with trimethylarsine.[52] $Bu^tS\dot{A}sMe_3$ has $a(As) = 648.4$ G and $g = 2.018$. A brief account has been given [53] of some phosphorus spiro-radicals (10).

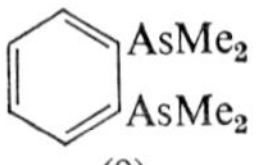

(9)

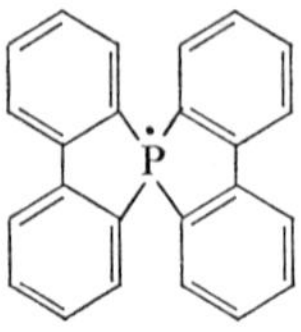

(10)

[52] A. G. Davies, D. Griller, and B. P. Roberts, *J. Organometallic Chem.*, 1972, **38**, C8.
[53] R. Rothuis, T. K. J. Luderer, and H. M. Buck, *Rec. Trav. chim.*, 1972, **91**, 836.

Author Index